# PRINCIPLES OF MICROBIOLOGY

# PRINCIPLES OF MICROBIOLOGY

C. MANOHARACHARY Ph.D., D.Sc, FNASc, FNAAS

and

K.V.B.R. TILAK Ph.D., FNASc, FNAAS
Department of Botany, Osmania University,
Hyderabad, A.P., India

TECHSAR™
NEW DELHI

*Published and Printed by*

TechSar Pvt. Ltd.
4435-36/7, Ansari Road, Daryaganj,
New Delhi–110 002 (India)
E-mail: tech_info@techsarworld.com
Website: www.techsarworld.com

ISBN: 978-93-80578-19-4



Reprint 2025

# Preface

Microbiology has become a front line science subject in this 21st century. It has undergone a sea change in the last three decades. Microorganisms do not follow borders, hence are ubiquitous and cosmopolitan in distribution. They colonize, establish, grow, multiply and survive in/on diversified habitats in nature. The exploration of microbes in extreme environments has changed the total scenario of modern biology. Microbes form an integral part of natural ecosystems and thrive as saprophytes, symbionts and also as parasites. The discovery of deep hot biosphere having unique microbial assemblage and natural dynamics has astounded the scientific fraternity. Microorganisms play an important role in nutrient cycling, organic matter decomposition, soil and crop health management, pesticide degradation, in the production of antibiotics, growth hormones, drugs, therapeutic proteins, vaccines, disease diagnostic tools, in food processing, biofertilizers, in bio-remediation etc. India is rich in microbial diversity and one-third of global microbial diversity exists in India. Unfortunately, 90–95% of microbial forms are yet to be cultured. Microbes pose the greatest challenge to the scientists. It is considered that it is a hidden treasure with greater potential of biotechnological application in search for not only bioactive molecules but also where one expects to find the likely progenitor of all life forms on the globe. Thus the microbial diversity and bio-prospecting has become a new thrust area in this century.

Microbiology being an emerging science, the authors felt the necessisity of bringing out a book entitled "Principles of Microbiology" with their own humble and sincere efforts so as to benefit students, teachers and other users. The book has got 19 chapters dealing with basic concepts, historical aspects, microscopy, diversity, cultivation and control of microorganisms, bacteria and viruses at length, nutrition and physiology of microbes, immunology, taxonomy, microbial genetics, and microbes in human welfare and other related aspects. We are optimistic that readers will find it more useful. Suggestions, positive criticism and lapses are accepted with gratefulness and we assure you that we will rectify them in the next edition. Our thanks are to I.K. International Publ. Pvt. Ltd., New Delhi for bringing out this volume meticulously and in time. We also thank Mr. Y.S.N. Murthy, O.U., for his help in the manuscript preparation and Prof. Gopal Reddy, Dept. of Microbiology, O.U. Hyderabad for his inputs in two chapters.

C. MANOHARACHARY
K.V.B.R. TILAK

# Foreword

Microorganisms are microscopic and ultra-microscopic and live in most diversified habitats and in inhospitable sites across the various ecosystems besides being ubiquitous. The 21st century clearly indicates that study of microorganisms is central to the applied biology. Microorganisms live as mutualists, symbionts, saprobes, pathogens, hyperparasites, etc. They play a major role in the maintenance of biogeochemical cycles which help the soil, plant and human health. Though microorganisms are deeply associated with gut and intestine of the animals, some forms are pathogenic and cause genetic change leading to serious health problems. However solutions are also available within new and emerging microbial diverse gene pool. There are more than 90% of microorganisms which are so far not culturable. It is anticipated that tools of post-genomic era such as the DNA microarrays and proteomics would permit better understanding of the metabolic pathways which will allow the design of improved media for cultivation.

Microorganisms form the backbone of industrial technology, drug and therapeutic protein products, enzyme elaboration, food processing, genetic engineering, recombinant DNA technology, bioremediation, biofertilizer technology, natural product search, functional genomics and many other biotechnological aspects. Present biotechnological searches are however no longer hitting in the dark, but working on specific targets; molecular designs and large data base researches. Most of the microorganisms are beneficial and serve the human beings, but handfuls of microbial forms are causal agents of diseases. The eruption of a new disease, viz., *E. coli* strain causing SARS and its sudden spread is of much concern to the humanity.

The chapters contained in this volume beautifully highlight the concepts, basic information, methodologies, current trends such as microbial genetics, immunology and microbes in relation to human welfare. The authors are well established in the field of Botany/Microbiology and leaders in their own right with academic excellence and rich experience in teaching and research. The authors deserve to be congratulated for this academic venture.

Certainly this excellent volume will serve and solve the needs of teaching fraternity and the students. And this book will gain a respectable place not only in the libraries but in the hearts of undergraduate and post-graduate students of science, agriculture as well as chemical engineering, microbial technology, pharmacy and others.

Prof. Dr. S.M. Paul Khurana

# Contents

# History and Scope of Microbiology

## 1.1 INTRODUCTION

Microorganisms are too small and cannot be seen with naked eyes. This group includes bacteria, viruses, fungi (yeasts and moulds), protozoa, microscopic algae, actinomycetes etc. Viruses are regarded as being at the borderline of life and non-life. Microbes are present in varied habitats such as soil, litter, water, air, animal excreta, extreme environments and others. The microbes colonize, survive and multiply in/on the above substrates in the presence of favourable environmental conditions. Marine and freshwater microorganisms form the basis for the food chain in oceans, lakes and rivers. Microbes help in the breakdown of complex chemical substances including hydrocarbons, pesticides, wastes etc. and convert them into simpler substances. They also help in the recycling of chemical elements and incorporate them into the soil, water and air. Certain autotrophic bacteria and algae play an important role in photosynthesis, which is a food and oxygen generating process. This process is critical to life on earth. Human intestine also contains bacteria that help in the digestion and synthesis of some vitamins required by our bodies (vitamin B and vitamin K). Microorganisms can be observed with a microscope ('micro' means small, 'scope' means to see). They can be observed with clarity as the present compound microscopes have magnification ten to thousand times greater than those of Robert Hooke's microscope with multiple lenses. In the year 1600, the first compound microscope was constructed by Zaccharias Janssen, a Dutch spectacle-maker. However, this microscope was of poor quality. Joseph Jackson Lister was responsible for considering a better microscope.

Microorganisms have commercial applications and play very important roles in our day-to-day life. Microorganisms are employed in the synthesis of products such as acetone, glycerine, organic acids, enzymes, alcohols, drugs and others. Industrial products also include vinegar, krait, pickles, alcoholic beverages, antibiotics, biopesticides, soy sauce, cheese, yogurt and bread. However, some microorganisms are pathogenic and cause diseases in humans, animals and plants. Such practical knowledge is of paramount importance in health and medicine.

Microorganisms are found almost everywhere (Fig.1.1). The invention of microscope unravelled the mysteries of this wonderful world of microbes. Diseases caused by bacteria such as smallpox, typhoid, diphtheria, tetanus, diarrhoea and others made the people to die in many devastating epidemics.

In ancient times, vaccination and antibiotics were not available to counter infectious diseases. Diseases like AIDS pose one of this century's formidable health threats, but it was not the first reported disease that is sexually transmitted. Syphilis caused many deaths around the world before the emergence of AIDS and it was produced by *Staphylococcus aureus*, a microorganism carried on human skin was responsible for many deaths.

Microorganisms have a great history and also have created history. In the year 1665, a crude microscope was invented by Robert Hooke. He was able to see individual cells with this microscope. Hooke's discovery marked the beginning of the cell theory, which states that all living things are composed of cells. Later discoveries have contributed largely to the understanding of the structure and functions based on this theory. Antony van Leeuwenhock was the first to observe the microorganisms through magnifying lenses that he had created.

Later scientists showed keen interest in knowing more about these tiny living things. Until the second half of 19th century, many scientists were philosophers who believed that life forms could arise spontaneously from non-living or *de novo* (nowhere). This theory was called spontaneous generation. In those days people believed that toads, mice, snakes etc. could arise from moist soils, while flies could emerge from manure.

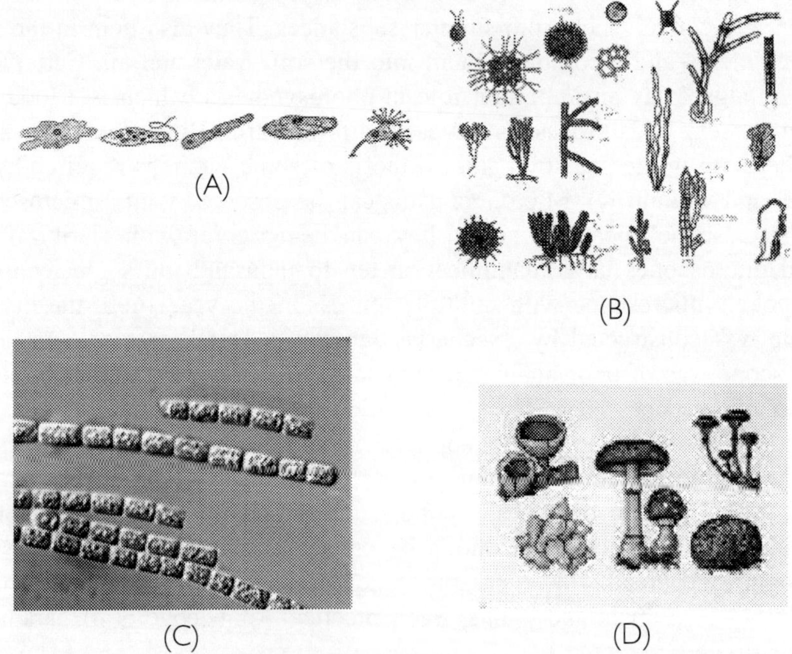


**Fig. 1.1.** An overview of microorganisms. A: protozoa; B: higher algae; C: cyanobacteria; D: forms of fungi.

Francesco Redi (1668) was of the opinion that maggots do not arise spontaneously from decaying meat. He demonstrated this by filling three jars with decaying meat and sealed them tightly, while three other jars having decaying meat were left open. Maggots appeared in open vessels but not in sealed vessels. This experimental result was a death blow to long held belief that life forms could arise from *de novo*.

John Needham (1745), a staunch supporter of spontaneous generation, stated that even after heating the nutrient fluids, the cooled solutions were supporting microorganisms. Twenty years later, Spallanzani, an Italian scientist, showed that nutrient fluids heated after being sealed in a flask did not develop microbial growth. In the year 1858, Rudolf Virchow, a German scientist challenged the theory of spontaneous generation by stating that living cells can arise only from pre-existing living cells (biogenesis).

Louis Pasteur (1861), the French scientist proved that microorganisms are indeed present in the air and they can contaminate seemingly, sterile solutions and argued that air itself does not give rise to microbial life. Pasteur conducted a series of experiments to prove this. He began by filling short necked flasks with beef broth and boiling them and few were left open. Within a few days, the open flasks were found contaminated with microbes. Flasks having boiled, cooled and sealed remained free of microorganisms. Pasteur proved that life can be destroyed by heat. Since then aseptic techniques were developed. Pasteur's work provided sufficient proof that microorganisms cannot originate spontaneously. After this there were numerous discoveries in microbiology. Actually the period, 1857–1914, is known as Golden Age of microbiology.

Pasteur later discovered that microorganisms called yeasts convert the sugars to alcohol in the absence of air. This process is now called fermentation and is employed in making wine, beer etc. The process of killing potentially harmful bacteria or microorganisms in milk or alcohol etc. is called pasteurization. It is a process to heat the milk etc. at a temperature just enough to kill most of the bacteria without affecting the nutrient levels and quality.

Microbes that cause diseases are called pathogens. The idea was referred as germ theory of disease. Pasteur (1865) gave a call to the people to fight silkworm disease which was responsible for the closure of silk industry in Europe. Late blight of potato, a fungal epidemic caused by *Phytophthora infestans* in Ireland (1845), made the people to starve and die.

Yeasts play an important role in fermentation. It is a link between activity of microorganisms, and physiological and chemical changes in organic materials. This discovery made the scientists to think that the microorganisms might have similar relationships with plants and animals, in particular that the microbes might cause disease. This was called the germ theory of disease.

The main proof that bacteria cause disease came from Robert Koch in 1876. Koch, a young German physician.[1] He was able to discover the bacterium (*Bacillus anthracis*), responsible for anthrax disease in cattle and sheep in Europe.[2] He cultured the bacterium artificially in nutrients and then injected into healthy animals.[3] Later he found that these animals became sick and died.[4] Koch isolated the same anthrax bacterium from the blood and compared it with original bacterial isolates.[5] Thus, Koch proved that a pathogenic bacterium was responsible for anthrax and it could be isolated artificially.[6] The artificially isolated bacterium on introduction into healthy animals produced anthrax disease.[7] The blood of the same, when examined, showed the anthrax bacteria, which was similar to that observed in the blood samples of animals died due to anthrax.[8] These steps are called Koch's postulates.

When numerous diseases appeared both on plants and animals, including humans, scientists started working on the prevention and control of diseases. Edward Jenner, a young British physician, conducted experiments to find protection from smallpox. Edward Jenner collected scrapings from smallpox blisters. Then he inoculated a healthy person with cowpox material by scratching his arm with smallpox contaminated needle. Within a few days the patient was mildly sick, but recovered and never had smallpox. This process was called vaccination.

The smallpox vaccine is an example of preventive medicine. The process of developing immunity is called vaccination. In Latin *Vacca* mean cow, and the protection from the disease, provided by vaccination (or by recovery from the disease itself), is called immunity. In this process, a living viral agent—the cowpox virus—was used to produce immunity. In China, the patients were immunized by removing scales from drying pustules of a person suffering from mild case of smallpox, grinding the scales to a fine powder and inserting the powder into the nose of the person to be protected. Some vaccines are still produced from virulent microbial strains that stimulate immunity to the related virulent strain. Other vaccines are made from killed virulent microbes or by genetic engineering techniques.

Scientists have started working on substance that could destroy pathogenic microorganisms without causing any damage to the infected animal or human. The treatment of disease by chemicals is called chemotherapy. Synthetic drugs were prepared and used to control the diseases. Chemicals produced by bacteria, actinomycetes and fungi that act against disease causing microorganisms or other microbes are called antibiotics. Paul Ehrlich was responsible for the discovery of *salvar san*, an arsenic derivate that was effective against syphilis. Similarly, quinine was extracted from a tree bark and was used to treat malaria. Many sulpha drugs were discovered later. Alexander Fleming, a Scottish physician and bacteriologist, observed that a fungus was able to control the growth of a bacterium. The mould was identified as *Penicillium notatum*. In 1928, Fleming named the active inhibitor from this mould as penicillin. This wonder drug has paved the way to the discovery of many new antibiotics. Although the synthetic drugs and antibiotics kill the pathogenic microbes, they have damaging side effects on infected host. Few successful antiviral drugs were also discovered.

New branches of Microbiology, such as Immunology and Virology, emerged soon. In recent times, recombinant DNA technology revolutionized research and could find practical applications in all areas of microbiology. Vaccines are now available for smallpox, measles, mumps, polio and Hepatitis B. The smallpox vaccine was so effective that not a single case was reported since 1977. Interferon, a substance generated by the body's own immune system, was discovered. This inhibits the replication of virus and cancerous growth also.

In 1892 Iwanowski reported that ultramicroscopic organism that caused tobacco mosaic disease could be passed through fine filters. This was called virus (poison). Stanley (1935) showed that this organism could be crystallized like a chemical. Since the discovery of electron microscopy and DNA / RNA material, scientists were able to observe the viruses in detail. Plant viruses contain RNA, and DNA is found in animal viruses. However, exceptions always exist.

Bacteria are easily cultivable on laboratory media and their life cycle is less than an hour. Thus, scientists have turned to unicellular organisms to unravel the mysteries of molecular biology. Oswald Avery and others have established that DNA is the hereditary material. Watson and Crick (1953) proposed model for the structure and replication of DNA. Jacob and Monod (1961) discovered messenger RNA. At the same time, scientists were able to understand genetic code and biochemical signals transmitted by DNA. Haeckel (1866), a German zoologist proposed a classification of three kingdoms (animalia, plantae and protista). The protista consisted of bacteria, some fungi, algae and protozoans.

Five kingdoms system was proposed by Robert Whittaker in 1969. Prokaryotes or Monera comprises eubacteria and archaebacteria. Protista includes slime moulds, protozoa and unicellular algae. Fungi has yeasts, multicellular moulds and mushrooms. Plantae includes algae, mosses, ferns, conifers and flowering plants. Lastly, animalia has sponges, worms, insects and vertebrata.

Few microorganisms are pathogenic, while others are useful. Microbes help in recycling vital elements such as nitrogen, carbon, phosphorus etc. Microbes are used in water treatment plants to remove pollutants and recycle water. Insects spread diseases and cause crop damage. Bacteria like *Bacillus thuringiensis* have been used extensively to control caterpillars, boll worms, tobacco bud worms and mosquitoes. Microbes are also used in the degradation of chemical compounds, which cause damage to the environment. Two important microbial products have attracted world attention, namely single cell protein and genetically engineered microbes (GEMS). Bacteria and fungi are used to produce single cell protein. Genetic engineering could find potential use in agriculture, medicine and others. We all live in a microbial world from birth until death. We have a variety of microbes on and inside our bodies. However, normal flora usually benefits us, but at times cause illness. Important natural resistance is provided by animals, plants and humans to infections.

## Anthony Van Leeuwenhock (1632–1723)

Leeuwenhock is called the father of modern bacteriology. He was the first to create microscope capable of seeing tiny bacteria and one-celled animals. Born in Deft, the Netherlands, he left school at 16 to become an apprentice dry goods merchant. At 21, he set up his own shop and also acted as Janitor to the town city hall. For 20 long years, he was busy grinding lenses and making microscopes far better than anything known in his days. His microscope consisted of a single, tiny, short focus lens, set between two pieces of metal. They could magnify 40 times, 100 times and even 300 times. Then he turned on everything within his grasp-ox eyes, rain water and tooth scrapings. One day he saw "little animalcules", later known as bacteria.

**Fig. 1.2.** Anthony Van Leeuwenhock.

In 1676, he wrote a letter to the Royal Society of England, about his discovery. The unschooled Dutchman was made a fellow along with other great scientists of his days. In later letters to the Royal Society, he recorded his many findings on bacteria, yeast, plants etc. These included the first drawing of bacteria and the discovery of the tiny capillaries, which join the veins and arteries. But most important was his discovery of bacteria, which led others to study bacteria and finally to control bacterial diseases.

## Edward Jenner (1749–1823)

Edward Jenner discovered a way to end the worldwide fear of smallpox. His discovery was nearly 100 years ahead of similar methods used to prevent others. Born in Berkeley, England, Jenner was prenticed at 12 to Daniel Ludlow, a surgeon in Sodbury. After studying in London under great anatomist, John Hunter, he returned to Berkeley to practise medicine.

**Fig. 1.3.** Edward Jenner.

Smallpox was then the deadliest disease known, ranging through most of the world. Injection of actual smallpox was practised to give healthy people a mild form of the disease in the hope of protecting them from its worst form. It was dangerous, however, and often spread smallpox still further. As an apprentice, Jenner had heard of the popular belief among the country people that a person could not get smallpox if he had already suffered from cowpox. Cowpox is a mild, weakened form of the disease caught from infected cows.

It was not until 1796 that he had a chance to check this belief. He took pus from a cowpox sore on a dairy maid's arm and scratched it into the skin of a small boy. Two months later the boy was injected with smallpox, but he did not develop the disease.

At first, no one believed Jenner's discovery. He was even advised by the Royal Society not to risk his reputation by telling about his findings. However, in 1798, he decided to publish a paper about his discovery, and soon vaccination was used throughout the world. Smallpox almost disappeared. Jenner received many awards from the Parliament, the Russian Czar, Napolean Bonaparte and Indian chiefs in the United States.

## Louis Pasteur (1822–1895)

Spontaneous generation would mean that life could arise from something non-living and many people could not imagine something so complex as a living cell arising spontaneously from dead materials The most powerful opponent of spontaneous generation was the French chemist Louis Pasteur (1822–1895), whose work on this problem was the most exciting and convincing. Pasteur first showed that structures were present in air that resemble closely the microorganisms seen in putrefying materials. He did this by passing air through guncotton filters, the fibres of which stop solid particles. After the guncotton was dissolved in a mixture of alcohol and ether, the particles that it had trapped fell to the bottom of the liquid and were examined on a microscope slide. Pasteur found that in ordinary air there exists constantly a variety of solid structures ranging in size from 0.01 mm to more than 1.0 mm. Many of these structures resemble the spores of common moulds, the cysts of protozoa and various other microbial cells. As many as 20–30 of them were found in 15 litres of ordinary air and they could not be distinguished from the organisms found in much larger numbers in putrefying materials. Pasteur concluded that the organisms found in purifying materials originated from the organized bodies present in the air. He postulated that these bodies are constantly being deposited on all objects. If this conclusion was correct, it would mean that, if food were treated to destroy all the living organisms contaminating it, then it should not purify.

Pasteur used heat to eliminate contaminants, since it had already been established that heat kills living organisms effectively. In fact, many workers had shown that if a nutrient solution was sealed in a glass flask and heated to boiling, it never putrefied. The proponents of spontaneous generation criticized such experiments by declaring that fresh air was necessary for spontaneous generation, and that the air inside the sealed flask was itself affected in some way by heating so that it would no longer support spontaneous generation. Pasteur skirted this objection simply and brilliantly by constructing a swan-necked flask, now called "Pasteur Flask" (Fig. 1.5). In such a flask, putrefying materials can be heated to boiling; after the flask is cooled, air can re-enter but the bends in the neck prevent particulate matter, bacteria or other microorganisms from getting into the flask. Pooled human gamma globulin may also be of value in the therapy of infectious hepatitis, if given early in the incubation period. The discovery of immunological phenomena has been at once scientific and practical achievement. Some of the most significant advances in public health and clinical medicine have come above via immunological manipulations.

The study of the mechanism of antibody formation has provided important insights into cellular and genetic phenomena in the animal's body.

Immunology arose out of the development of procedures for immunization against infectious diseases and two people are most closely identified with the early history, the British physician Edward Jenner and the French scientist Louis Pasteur. In 1798, Jenner first published his work showing that inoculation with material from cowpox lesions could induce immunity to smallpox infection. This was the first example of the use of an attenuated living agent to bring about immunity. The cowpox virus was sufficiently closely related to the smallpox virus to induce immunity, yet sufficiently low in virulence in humans so that fresh infection did not occur. Jenner's procedure came to be known as vaccination (from *vacca*, the Latin for cow). It was almost 20 years after Pasteur began his career that he started to work on infectious disease. After some initial work on anthrax (Koch had just proved that anthrax was caused by a bacterium), Pasteur turned to fowl cholera, a bacterial disease which at that time was affecting 10 percent of the chicken in France. He isolated the causal agent and was able to maintain its virulence by periodic passage through chickens, but found that if the organism was cultured for a while in the laboratory, it lost its virulence. Being aware of Jenner's

**Fig. 1.4.** Louis Pasteur.

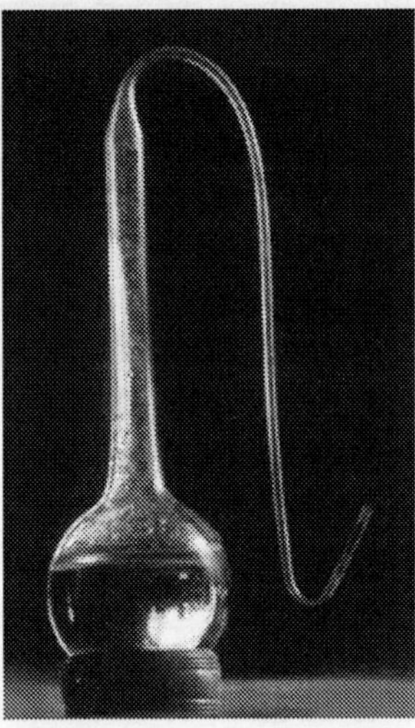

**Fig. 1.5.** Pasteur's gooseneck flask.

work, he concluded that the nonvirulent cultures were attenuated and had the same relationship to the virulent cultures that cowpox virus had to smallpox virus. He was able to show that such attenuated cultures could be used to vaccinate chicken and confer immunity to infection with the most virulent cultures. Pasteur realized that he had derived a general principle for immunization and he quickly turned to another infectious disease—anthrax—where immunization could be assumed to have more practical benefits. He developed an attenuated culture to anthrax, which could be used as vaccine, and he used this attenuated culture in a dramatic public experiment in 1881 at Pouilly-le-fort. On 5 May 1881, he vaccinated 24 sheep, 1 horse and 6 cows, and on 17 May they were revaccinated. Then on 31 May, all vaccinated animals plus 29 unvaccinated animals (24 sheep, 1 horse, 4 cows) were inoculated with a virulent culture. Boldly, Pasteur invited a large number of government officials, scientists, veterinarians and journalists (including a newspaper reporter from the *London Times*) to view the results. The visitors arrived on 2 June, 2 days after inoculation with virulent organisms. The vaccinated animals appeared completely healthy, whereas 20 of the unvaccinated sheep and the horse had died. Two other unvaccinated sheep died before the eyes of the spectators and the last died before the end of the day. The experiment was thus a complete success and demonstrated to the world the importance of this procedure. Pasteur's final work was with rabies virus and he developed attenuated strains, which could not only be used to induce immunity but could also be used prophylactically to protect individuals who has succumbed to bites from rabies infected animals. This was possible because the course of rabies virus infection is slow, and by infecting attenuated virus soon after a bite has occurred, immunity can be built up before the virus had time to induce its generally fatal symptoms.

## Iwanowski (1864–1920)

In 1892, D. Iwanowski of Russia had first shown that the causal agent of tobacco mosaic disease was filtrable, but Beijerinck went much further and provided strong evidence that causal agent was filtrable and it has many of the properties of living organism. He called the agent a *Contagium vivum fluidum,* a living germ that is soluble. He postulated that the agent must be incorporated into the living protoplasm of the cell in order to reproduce and that its reproduction must be brought about with the reproduction of the cell.

This is a postulate that comes very close to our current understanding of how viruses reproduce. Beijerinck also noted that there were other plant diseases for which causal agents had not been isolated and these might also be caused by filtrable agents. Soon a number of other filtrable agents were shown to be the cause of both plant and animal diseases. Such agents were called filtrable viruses, but as further work on these agents was carried out, the word, "filtrable" was gradually dropped.

**Fig. 1.6.** Iwanowski.

Today, the original meaning of 'virus' has been forgotten and the word is now used to refer to the kinds of contagious agents discussed in this chapter. Bacterial viruses were first discovered by the British scientist F.W. Twort in 1915 and independently by the French scientist F.D. Herelle in 1917, who called them bacteriophages (from the combining form *phago* meaning "to eat"). In 1935, the American scientist Wendell Stanley published work describing the crystallization of tobacco mosaic virus, providing the first demonstration that some of the properties associated with living organisms can be found in agents, which can be crystallized like chemicals. The crystals were shown to consist of only two components, protein and RNA. Stanley's work can well be said to mark the beginning of molecular biology, as it has provided the first insight into the chemical nature of virus itself.

## Robert Koch (1843–1910)

**Fig. 1.7.** Robert Koch.

Robert Koch is generally considered to be one of the two founders of modern bacteriology. The other is Louis Pasteur. Born on December, 11, 1843, in Klausthal, Germany, Koch was one of the thirteen children. His father was a mining official. He grew up to an avid chess player, an admirer of Johann Goethe and above all a brilliant scientific investigator. He obtained a medical degree from the University of Gottingen in 1866 after studying firstly natural science and then medicine. One of Koch's teachers at the university was a German pathologist Friedrich Henle, who has had a huge influence on the development of histology. Another teacher Friedrich Wohler was the chemist, who was the first to synthesize an organic compound (urea) from an inorganic substance in 1828. After graduation, Koch became a physician in various provincial towns and a surgeon during the 1870–72 Franco-Prussian war. The French parasitologist, Casimir-Joseph Davaine, inspired by the work of the French microbiologist Louis Pasteur, showed in 1863 that anthrax in sheep was due to the presence of rod-like bodies in the blood. However, the natural history of the disease was not complete. This required the painstaking observational work of Koch.

## Investigation of Anthrax

In 1872, Koch became District Medical Officer in Wollstem, where he began important investigation into anthrax despite the lack of research facilities. By using a suitable media, he cultivated the anthrax organisms on microscopic slides. It was a suitable medium that could help him to cultivate the anthrax organism on microscopic slides. He was able to demonstrate their growth into long filaments and discover oval translation bodies, the dormant spores within them.

The finding that dried spores remained viable for years explained the recurrence of the disease in pastures long unused for grazing. When conditions were favourable, the spores could develop into the rod-shaped bacilli of anthrax. On the invitation of an eminent botanist, Ferdinand Cohn, the life cycle of anthrax was demonstrated at Breslau in 1876. However, it was only after Pasteur's demonstration of an anthrax vaccine in 1882, that the Koch's findings were accepted.

After a brief period, in 1879, as Town Medical Officer in Breslau, he was appointed to the Imperial Health Office in Berlin to advise on hygiene and public health. With his collaborators, he devised new research methods. Koch had by now been began the daunting task of discriminating among bacteria, connecting microorganisms with particular effects.

He experimented with various dyes that stain bacteria and make them more visible under the microscope, he also devised an ingenious method of separating a mixture of bacteria. This involved inoculating an animal with the bacteria and passing the resulting infection from one animal to another until, at the end of the experimental chain, only one type of bacterium remained. Using this method he identified the bacteria responsible for several disorders, including septicaemia.

To obtain pure cultures outside the body, Koch mixed the organisms in melted gelatine. After solidification of the gelatine and growth of the organisms, portions of pure colonies were placed into separate tubes of broth or other media. This work offered the early formulation of Koch's postulates. These were formalized in 1882 and stated that to prove an organism was the cause of any disease, it was necessary to demonstrate:

- that the organisms could be discoverable in every instance of the disease;
- that, extracted from the body, the germ could be produced in a pure culture, maintainable over several microbial generations;
- that the disease could be reproduced in experimental animals through a pure culture isolated by numerous generations from the organisms initially isolated; and
- that the organisms could be retrieved from the inoculated animal and cultured a new.

Although some pathogenic entities, notably viruses, had to be accepted without meeting all the conditions as most conditions were able to be fulfilled. The applicability and thinking behind these rigorous postulates boosted the dogma of specific aetiology—the idea that a disease has a specific causative agent. The implication was that once the agent has been isolated, it will be possible to control the disease. Koch announced to the Physiological Society of Berlin on 24 March 1882 that he had isolated and grown the tubercle bacillus. He believed this to be the cause of all forms of tuberculosis. Koch formulated systematic method for biological research (Koch's Postulates, given above) which is still observed.

In 1905, Koch won the Noble Prize for Physiology and Medicine "for his investigation and discoveries in relation to tuberculosis".

## Martinus Beijerinck (1851–1931)

Martinus Beijerinck is known for his contributions in developing enriching culturing techniques in microbiology and introducing the concept of a virus. He was a professor at the Delft Polytechnic School in his later years of scientific career. A student of Botany, Beijerinck started his career studying the microbiology of plants. While studying the association of microorganisms with plants, he developed enrichment culturing techniques to isolate soil microorganisms of specific characters.

Beijerinck proposed selecting specific microorganism from a natural sample through the use of specific culture media and incubation conditions that favoured the growth of one type of organism while constraining the growth of others. This was a different approach by him to modify the practice of isolating microorganisms from nature by exposing nonselective culture media to some environment and allowing chance to dictate what grew. He called his techniques 'selective culture technique', which is known today as 'enrichment culture technique'. Beijerinck isolated the first pure culture of many microorganisms from soil and water environments, which included aerobic nitrogen-fixing bacteria, sulphate-reducing and sulphur-oxidizing bacteria, nitrogen-fixing root nodule bacteria, *Lactobacillius* sp., green algae and many others. His other contribution to the science of microbiology is the discovery that filtrable agent was the cause of tobacco mosaic disease, Beijerinck showed that the infectious agent (a virus) was not bacteria, but somehow became incorporated into the cells of the host plant and required the living plant to reproduce, in essence, describe the basic tenets of virology. Subsequently, the tobacco mosaic virus (TMV) was crystallized and characterized by others. The enrichment technique developed by Beijerinck is the most important tool for isolating and developing microbial pure cultures of interest today. Today's practice of enrichment culturing by making use of physical, chemical and biological environment for enriching the population of economically and biotechnologically important microorganisms in source sample is the technique introduced by Beijerinck.

**Fig. 1.8.** Beijerinck.

## Sergei Winogradsky (1856–1953)

Sergei Winogradsky was a Russian scientist with interests similar to that of Beijerinck. He was more interested in soil bacteria and their nutritional characterization with specific interest to study the cycling of nitrogen and sulphur compounds. Winogradsky isolated pure culture of nitrifying bacteria, showing clearly that the process of nitrification, the oxidation of ammonia to nitrate, was the result of bacterial action. He also studied the oxidation of hydrogen sulphide by sulphur-oxidizing bacteria directly in their natural habitats. To study the soil bacterial groups for their involvement in mineral cycling, he developed a technique of creating natural environment

**Fig. 1.9.** Sergei Winogradsky.

in a laboratory jar (Winogradsky's column) for observation. Growth of different microorganisms with different physiological processes from soil samples can be appreciated in a laboratory experiment by creating the required environment. With such experimentation, he explained that the bacteria can be geochemical agents. Winogradsky's observations on metabolic groups of soil microorganisms are of immense importance. He proposed the concept of 'chemolithotrophy' to explain the oxidation of inorganic compounds coupled to the release of energy for cell growth. With this source of energy, bacteria can live as autotrophs by assimilating $CO_2$ present in air. Today it is known as 'chemolithotrophic-autotrophy'. Such concept was explained by Winogradsky based on his observations on nitrifying bacteria. This concept was not readily accepted at that time, but today it is an important process and can support the growth of higher organisms. He isolated the free-living anaerobic nitrogen-fixing bacterium, *Clostridium pasteurianum*, by enrichment culturing method under anaerobic conditions. Winogradsky published many scientific papers, and his important monograph, Microbiologie du Soil (Soil Microbiology) is a milestone in microbiology, which contained many of his original drawings of organisms he had isolated and studied in his enrichment culturing or natural material during his scientific career.

## 1.2 SCOPE OF MICROBIOLOGY

Biology is the scientific study of life. It examines the structure, function, growth, origin, evolution and distribution of organisms. The unifying principles of biology are cell theory, evolution, gene theory and homoeostasis.

The total complement of genes in an organism or cell is known as the genome, which is stored in one or more chromosomes. DNA is a long strand on which thousands of genes are encoded. The rRNA gene data separates three domains: bacteria, archaea and eukaryotes. The exact relationships of the three domains are still being debated. New fields like genomics and proteomics are reversing the trend of evolution. Diversity is a variation in organisms with reference to morphology, physiology, biochemistry, ecology, habitat, genes, between species and genera etc.

India is rich in microbial and plant diversity and also in animal diversity. One third of global diversity exists in India. *Ex situ* and *in situ* conservation of microbes, flora and fauna is important. The diversity enables scientists to learn what important genes are actually contained in the millions of microbial and plant specimens housed in gene banks around the world.

There is no substance more important than DNA. It has provided basis for the evolutionary process that has generated millions of different life forms that have occupied the earth since the first living organism came into existence some 4 billion years ago. Genetic code is the language of living organisms.

**Table 1.1.** Discoveries in microbiology

| Scientist | Discovery | Year |
|---|---|---|
| Jensen | Magnification of object by hand lens | 1590 |
| Galileo | Telescope | 1608 |
| Francesco Redi | Opponent of spontaneous theory of generation | 1665 |
| Robert Hooke | Compound microscope | 1678 |
| Antony van Leeuwenhoek | Observation and description of bacteria | 1683 |
| Joblot | Extraneous origin of microorganisms | 1711 |
| John Needham | Proponent of spontaneous theory of generation (abiogenesis) | 1745 |
| Spallanzani | Opponent of spontaneous theory of generation | 1796 |
| Edward Jenner | Immunisation in smallpox by cowpox virus | 1796 |
| Appert | Preservation of foods (appertization) | 1805 |
| Augustino Bassi | Pathogenic role of microorganisms | 1835 |
| Franz Schulze | Microorganisms in the air as causative agents | 1836 |
| Theodor Schwann & Schroeder von Dusch | Boiled infusions remain sterile even after passing through air | 1837 |
| Cogniard Larour, Schwann & KutzingSchwann | Yeast as source of alcoholic fermentation | 1837 |
| Louis Pasteur | Disproved spontaneous theory of generation of microorganisms, microorganisms as agents for chemical changes, pasteurization of milk, fermentations, discovery of anaerobic life | 1822–1895 |
| Lister | Antiseptic surgery | 1864 |
| John Tyndall | Method of sterilization by repeated heating (tyndallization) | 1877 |

**Table 1.1.** Contd.

| Scientist | Discovery | Year |
|---|---|---|
| Robert Koch | Father of medical microbiology; Koch's postulates, isolation of pure cultures of bacteria | 1843–1910 |
| Hansen | Leprosy *Bacillus* | 1874 |
| Neisser | *Gonococcus* | 1881 |
| Metchnikoff | Developed phagocytic phenomenon | 1883 |
| Christian Gram | Staining of bacteria | 1884 |
| Frankel | *Pneumococcus* | 1886 |
| Adolf Mayer | Mosaic disease caused by virus | 1886 |
| Weichselbum | *Meningococcus* | 1887 |
| Rous and Yersin | Diphtheria toxin | 1888 |
| Kitasato | Isolation of tetanus *Bacillus* | 1889 |
| Yersin | Plague *Bacillus* | 1890 |
| Iwanowski | Reproduction of tobacco mosaic virus | 1892 |
| S. Winogradski | Chemoautotrophic bacteria | 1856–1953 |
| M.W. Beijerinck | Nitrogen fixation by microorganisms and enrichment culture techniques (in association with Winogradski) | 1851–1931 |
| H. Buchner | Discovery of cell-free alcoholic fermentation | 1897 |
| Loeffler and Frosch | Transmission of foot and mouth viral disease of cattle | 1898 |
| Sir Ronald Ross | Sexual cycle in malarial parasite | 1897 |
| M.W. Beijerinck | Tobacco mosaic disease caused by virus | 1898 |
| Walter Reed | Human disease of yellow fever (viral origin) transmitted through mosquito bite | 1902 |
| Theobald Smith | Rickettsiae | 1907 |
| Paul Ehrlich | Chemotherapy | 1909 |
| Landsteiner and Popper | Poliomyelitis due to filtrable virus | 1909 |
| Ruska | Morphological study of viral particles by electron microscope | 1934 |

**Table 1.1.** Contd.

| Scientist | Discovery | Year |
|---|---|---|
| Twort and d'Herelle | Bacteriophages | 1915–1917 |
| Alexander Fleming | Discovery of penicillin | 1929 |
| Stanley | Crystallization of virus particles | 1935 |
| Delbruck and Luria | Mutation in bacteria | 1943 |
| Avery, Mc Leod and McCarty | Bacterial genetic transfer (transformation) | 1944 |
| J.D. Watson and F.H.C. Crick | DNA structure | 1953 |
| S.A. Waksman | Discovery of streptomycin | 1952 |
| Isaacs and Lindenmann | Discovery of interferon | 1957 |
| Beadle, Tatum and Lederberg | Microbial genetics | 1958 |
| Centre for Disease Control, Atlanta, USA | First report of AIDS HIV | 1981 |
| S.B. Prussiner | Discovery of prions | 1997 |
| Barry J. Marshall & J. Robin | *Helicobacter pylori* and its role. Warren in gastritis and peptic ulcer disease | 2005 |
| Andrew Z. Fire & Craig C. Mello | RNA interference gene silencing by double-stranded RNA | 2006 |

Biotechnology is the science of manipulation or altering biological systems for the benefit of mankind.

The elucidation of structure of DNA, discovery of restriction endonucleases and ligase enzymes and related discoveries led to the establishment of the array of new tools and techniques used in biotechnology. The practical implication in the society includes a number of life saving drugs, transgenic microorganisms, transgenic animals and plants with improved efficiency.

Microbial technology is important as it elaborates antibiotics, enzymes, proteins etc. The tools and techniques used involve the synthesis, isolation, cloning, sequencing, and transfer of genes. Recent developments in genetic engineering such as PCR, reverse transcriptase, gene sequencer, gene synthesizer, microchips etc. have opened new opportunities to manipulate genes across the genetic barriers for obtaining useful commercial products or the development of transgenic organisms. Achievements of biotechnology in agriculture, medicine and industry have proved useful, and it is now predicted that the coming centuries would be entirely dependent on it. Genetically engineered plants, animals and microbes capable of withstanding hostile environmental stress like high temperature, toxic gases, low oxygen and low gravity would accompany humans in long distance voyages. Developments in biotechnology, nanotechnology, environmental biotechnology (bioremediation) and others will blur the distinction between man and machine. Cloning is a means to generate genetically identical copies. Research in stem cells could serve as the source of cells and tissues for transplantation or replace defective or damaged tissues and organs. The recombinant DNA technique allows for the manipulation of DNA and the subsequent genetic modification of a living organism on demand. Already scientists have identified a gene conferring.

More effective vaccines, antibiotics and drugs would be available as unravelling the genomes of infective organisms is continued. It is projected that within the next 25 years, human body shops would be available to deliver organs longevity, the Methuselah gene, in Icelandic population.

## Cell Theory

There are three points to modern cell theory and these points were derived by two early cell biologists, Schleiden and Schwann, in the 1840s.

Cells are the basic unit of life. All living organisms are composed of one or more cells. All cells are derived from previously existing cells.

Living cells consist of a cell membrane surrounding cell fluid, called cytoplasm. Within the cytoplasm are structures that carry out cell functions. These structures include chromosomes, where the information needed to construct the cell is stored. Ribosomes make the information stored in the chromosomes into proteins. Enzymes, which are proteins that allow the cell to carry out chemical reactions and structural proteins, give shape to cells.

Cells obtain energy from their environment, and use it to build and maintain their internal structure. Cells use the energy to concentrate nutrients and build the components they need to grow (increase in mass) and reproduce (divide)

Prokaryotes and eukaryotes have cells, are known to produce exact copies of themselves and contain genetic material (DNA), obtain and use energy, are surrounded by a membrane and maintain internal homoeostasis.

Microbiology is the science that deals with microscopic organisms and their scope. Based on their internal structure, cells can be divided into two major groups: prokaryotic cells and eukaryotic cells. Prokaryotic cells have a single chromosome and no internal membrane bounded structures. These cells are structurally simple, and are represented in the earliest fossils known. Eukaryotic cells have many linear chromosomes and a vast array of internal specialized membrane bounded structures, including a nucleus, mitochondria, and a complex internal membrane system that is involved in synthesizing and processing proteins.

Organisms alive today are classified according to cell and gene structure into three large groups. Two of these groups are prokaryotic: the archaea, which are believed to represent the earliest organisms on the earth: and the bacteria, which includes most of the microorganisms we will study in this course—viruses, mycoplasma and others.

The eukarya, which includes all eukaryotes, include most of the organisms with which we are familiar: plants, animals, protists, and fungi.

Molecular analysis of the genes encoding a very conservative kind of RNA, called ribosomal RNA (rRNA) indicate that the archaea are as different from the bacteria as eukaryotes.

All organisms have a binomial nomenclature. There are a few rules about naming organisms. First, no two bacteria, for example, can have the same genus and species name. Only one bacterium can have the name *Escherichia coli.* Second, The genus name, which comes first, is always capitalized, and the species name always begins with a lower case letter. *Escherichia Coli* and *escherichia coli* are both incorrect. Third, since the names are considered to be Latin (a foreign language), they must be italicized or underlined when written. Escherichia coli is incorrect, but *Escherichia coli* is correct.

### Groups of prokaryotes

**Eubacteria:**
Typical bacteria
appendaged or budding bacteria
Actinomycetes
sheathed bacteria
Mycoplasmas
gliding bacteria
Myxobacteria
Spirochetes
Rickettsia
Chlamydia
Cyanobacteria

**Archaea:**
extreme thermophiles
halophiles
methanogens

In ancient times, microorganisms have played their roles not only in spoilage of foods but also in fermenting them to a more palatable, nutritious and easily digestible ones. With the discovery of microscope, the science of microbiology has become an important branch of science like other branches. Further the upcoming and front line areas of genetics, molecular biology and bioengineering have not only added strength to the discovery of specific microbes but also led to the identification of specific functions. Today we know that selected bacteria, yeast and moulds have been employed in industrial microbial technology, which has resulted in products and processes that benefited mankind.

There are nitrogen-fixing bacteria (*Rhizobium Azotobacter* and others) that are applied to soil to enrich the nitrogen supply to the plant growth and plant growth hormones like gibberillins produced by *Gibberella fujikorai* (fungus) that aid in improved crop yield are the examples that benefited agriculture. Food and industrial enzymes such as pectinases, amylase and cellulases elaborated by *Aspergillus, Bacillus, Cellulomonas* and others have greatly benefited food, feed, brewing and textile industries. Industrially produced microbial products including lactic acid, vitamins, colours, flavours, single cell proteins and others have contributed to the

**Table 1.2.** Comparison of bacteria, archae, and eukaryotes

| Characteristic | Bacteria | Archae | Eukaryote |
|---|---|---|---|
| Cell membrane | ester-linked hydrocarbons | ether-linked hydrocarbons | ester-linked hydrocarbons |
| Cell wall | murein | no pseudomurein | no murein |
| Cellular organelles | No | No | Yes |
| Chromosomes | one, circular | one circular | linear |
| Membrane-bounded nucleus | No | No | Yes |
| Protein synthesis sensitive | Yes (Chloramphenicol, erythromycin, streptomycin) | No | No |
| Ribosomes | 70S | 70S | 80S |
| RNA polymerase | One type | single RNA polymerase. sub-unit structure has different types (5–11) | 3 different types |

nutrition and health of humans. Alcohol and baker's yeast used in bread making are the products of yeast fermentation. Penicillin, Streptomycin produced by *Penicillium* and *Streptomyces*, respectively are the first generation antibiotics. Thus it is clear that industrial microbiology has contributed a great deal to the economy and mankind at large. Organic farming due to the role of microorganisms would provide organic nutrients, hormones, maintain soil texture and pH favouring for plant growth and yield of food grains. There is ample scope and opportunity for microbiologists to identify the microorganisms and develop scientific industrial processes for compost having the above properties. Production of starter cultures for fermentation of foods and beverages is necessary and helps in the production of quality foods having nutritional and therapeutic values.

Selection of suitable mould, yeast and bacteria not only makes the food easily digestible but also imparts better nutrition like vitamin B 12, flavour and therapeutic value besides detoxifying the food from aflatoxin contamination. This kind of technology would help the entrepreneurship with economic benefits.

Antibiotic resistance developed by disease causing bacteria poses several problems in the medical field. Hence there is a need to discover more potent antibiotics to deal with the above problem. Development of therapeutic food is another approach to combat the diseases. Microbiology has opened up new avenues like vaccine technology development of anti-viral antibiotics and anti-cancer antibiotics. Approach like virosomal drug delivery is new and is applied to prevent cancer, neuro-degenerative disorders and infectious diseases.

Virosomes are spherical, unilamellar vesicles with a mean diameter of 150 nm. They represent reconstituted empty influenza virus envelops, devoid of the nucleocapsid including the genetic material of the source virus. DNA barcoding has an edge over traditional morphological identification in being able to identify cryptic species. Cryptic species are a group of species that satisfy the biological definition of species—they are reproductively isolated from each other—but are not morphologically distinguishable.

More than 500 bacterial species and *Escherichia coli* present in the normal gut and colon are generally beneficial as they synthesize folic acid, vitamin K and biotin. *Lactobacillus* species is a beneficial bacterium, which converts lactose and other sugars to lactic acid.

# Microscopy

## 2.1 INTRODUCTION

The microscope is an optical instrument and it can be used similar to the optical lens. When man learnt to grind lenses during the Middle Ages, he used them only as reading glasses. The first microscope was developed in 1590 by Hans and Zaccharias Janssen in Holland. There was little further development in microscopy for over a century. By about 1673, Antony van Leeuwenhoek in Holland started grinding lenses and making simple microscopes with which he examined various objects. In the course of the next few years, he discovered and described with the help of drawings, 'the animalcules'. During his lifetime he constructed more than 200 microscopes, and the approximate magnification of his microscopes was around 200 times. The first compound microscope was designed by Galileo in 1610 and later by Robert Hooke in 1665. In the course of the next century, several agencies in Europe designed and manufactured new types of compound microscope. Significant progress in this field was made only in the second half of the nineteenth century and this was mainly due to various discoveries in Physics. The development of water immersion lenses in 1840, the substage or Abbe condenser in 1872 and the apochromatic lenses in 1886 contributed to the construction of the modern compound microscope. During the past two or three decades, several improvements have been made in the compound microscope and new ones like stereozoom, ultraviolet, phase-contrast and electron microscopes have been developed, which have gone a long way in understanding the microorganisms.

For a microbiologist, a microscope is an essential instrument, and depending on the specific field of interest, one may use different types of microscope. A microscope is an optical instrument consisting of one or more lenses in order to enlarge and/or magnify images of minute objects. In a simple microscope, only a single optical lens is provided, whereas in a compound microscope, at least two lenses are provided, of which one is known as the objective and the other as eyepiece. The construction of compound microscopes, though based on the same scientific principle, varies to some extent with the manufacturer. The general principles involved in the use of a microscope are briefly outlined below.

The compound bright-field microscope (Fig. 2.1), which is most widely used by students of microbiology, consists of two parts, mechanical and optical. The mechanical parts are only secondary, but are necessary for operating the instrument. There is a *stand* or framework for all parts, which rests on a *base* or foot that is mostly or a horse-shoe shape; from the base in *pillar* rises, at the top of which, through an inclination joint, an *arm* or *limb* is attached. Attached at the top of the pillar is the *stage* with central circular opening called *stage aperture*. There are also two stage or spring clips, which hold the microscope slide carrying the object of examination. A *substage* beneath the stage carries the *condenser*, which focuses the light rays on the specimen. Attached to the condenser is an *iris diaphragm*, which helps in regulating the light, passing through the mirror, into the condenser and then to the object. There is a mirror below the condenser, which is made of a lens with one plane surface and another concave surface. The mirror is mounted on a frame and attached to the pillar in such a manner that it can be focused in three different directions. Hinged up at the top of the arm, in line with the stage aperture, is a body tube, which is worked up and down by two separate arrangements, one called coarse adjustment worked with pinion head, and the other fine adjustment worked with micrometer head. The tube is a hollow cylinder, usually of a standard diameter. At the lower end of the tube is an arrangement for screwing in the objective lens. In most microscopes, there are two or more such objectives of different powers, which are screwed to a revolving nosepiece. One of these objectives carry an oil-immersion lens used in examining microorganisms at high magnifications. Each one of the objectives, which swung into line, operates with the eyepiece lens, which is fixed at the top of the body tube. The eyepieces are mostly of standard dimensions and made in different power lenses. They are slided into the body tube and are replaceable with different powered lenses. An improvement over the old type of microscope is the binocular or double-ocular system with two eyepieces so that both eyes could be used simultaneously to observe the object and thus reduce eye strain. The stands of modern compound microscopes are so made that they could be changed to either monocular or binocular types without much difficulty.

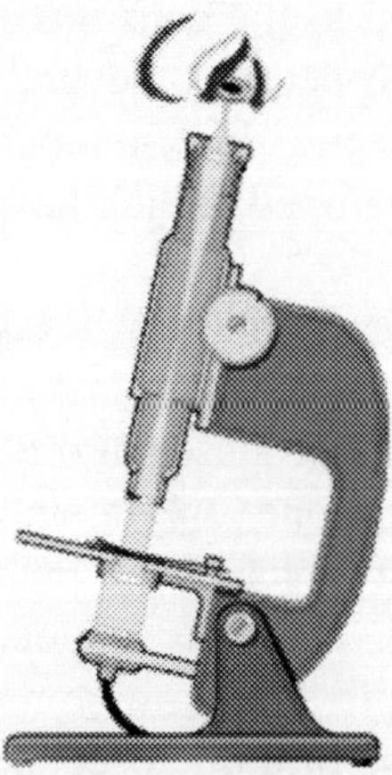

**Fig. 2.1.** Compound microscope.

The optical parts of the microscope include the mirror, condenser, objective and ocular. All of these should be kept in perfect optical axis for best results. The specimen or object is placed on the microscope slide, so that the beam of light passing from the mirror through the condenser and stage aperture should fall on the specimen and then into the objective and ocular. The path of light through the compound microscope is illustrated in Fig. 2.2. The light passes through the object and forms a real and inverted image, which is called primary image. The ocular intercepts the ray coming from the object and magnifies and forms the virtual image, which is again inverted, i.e., it comes back to the original position of the object, and this virtual image is then examined by the eye lens. The distance between the eye and the virtual image is known as the projection distance.

The magnification or the relative size of the image is determined by multiplying the magnifying power of the eyepiece by that of the objective. Thus, when an eyepiece of 10 X and objective of 43 X are used in the system, the magnification would be 430. The magnification could be changed by using different oculars as well as by varying the length of the body tube. But the extent of magnification is limited by the resolving power of the system. (The resolving power is the ability to show, separately and distinctly, two closely adjacent points of the object.) This limitation is determined by the wavelength of the light used. Light is transmitted from a luminous source to the eye by a wavy or vibrational movement, the velocity of this transmission being 298,080 km per second and its wavelength varies from 385 to 760 nm, 1 nm being equivalent to 1 / 1,000 or a micron, i.e., the length of light wave is the distance from the crest of one wave to the crest of the next. The ability of a lens to form distinct images of two separate object points is proportional to the angle of light and also to the refractive index of the space in front of the lens. The two factors are usually combined and expressed as numerical aperture.

Using light of 0.5640 μm wavelength and oil immersion lens in a compound microscope with a numerical aperture of 1.25, we get a resolution power of:

$$\text{Resolving power} = \frac{\text{Wavelength}}{\text{Numerical aperture}}$$

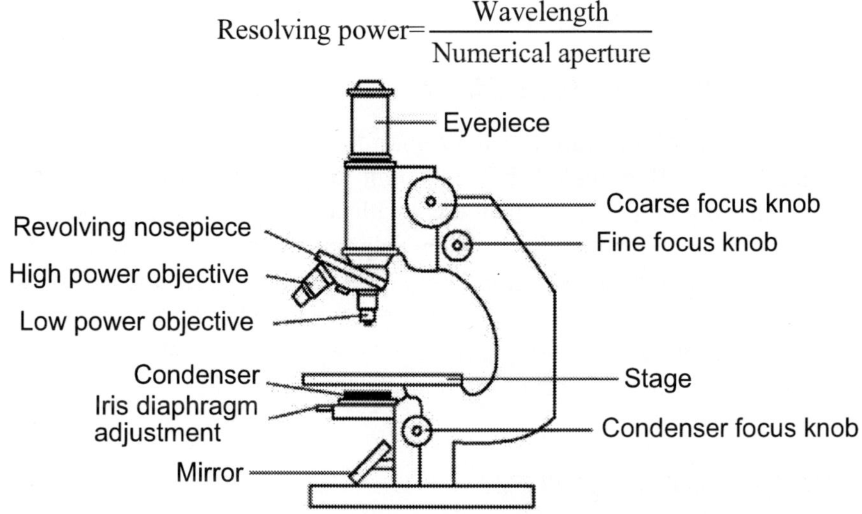


**Fig. 2.2.** Parts of compound microscope.

For a compound microscope, operated with bright light and fitted with a condenser, the equation is

$$\text{Resolving power} = \frac{\text{Wavelength}}{2 \times \text{Numerical aperture}}$$

$$\frac{0.5640}{2 \times 1.25} = \frac{0.5640}{2.5} = 0.223\ \mu\text{m}$$

i.e., two points placed at a distance of 0.223 μm could be resolved through the microscope. Magnifications beyond the resolving power of the system would result in large images, less distinct in their details and appearance. But by using a colour filter, the wavelength of the transmitted light could be adjusted, and with shorter wavelength there will be better resolution of the object though the system, and the finer details of the object could be seen.

When a dry objective lens is used to view a specimen through the microscope, the air space between the condenser and the objective lens scatters away the light and thus the resolution is reduced. If the air space between the coverslip and the objective is filled with a fluid having a higher refractive index than air (refractive index of the air is approximately 1.0), there is lesser wastage of light as the liquid medium permits more light to pass into the objective lens. By using a liquid medium such as cedar wood oil, whose refractive index is 1.51, almost equal to that of glass (1.56), better resolution is obtained. There are still other ways of improving the resolution power. In the lenses used in microscopes, there are certain defects known as aberrations, which may be chromatic aberrations or spherical aberrations. If these defects are rectified by the use of apochromatic lenses, there will be better resolution in the microscope.

In bright-field microscopy, the object absorbs and refracts the light passing through it and thus it contrasts with the background in which it is placed.

## 2.2 PHASE-CONTRAST MICROSCOPY

Since the structural elements constituting a microbial cell do not differ much in their optical quality, they can hardly be differentiated in bright-light microscopy. In phase-contrast microscopy, the unstained microorganisms are examined with a phase-contrast condenser, a phase-contrast objective and a special magnifier, which help in accentuating the difference in the transmission of light rays through the structural details and thus enhance their resolution by the eyepiece. In this system, even slight variations in the refractive index of the component parts of the object were well differentiated, as the phase-contrast assembly transforms the slight wave-front irregularities into more distinct variations (Fig. 2.3).

## 2.3 ULTRAVIOLET MICROSCOPY

The ultraviolet light has a shorter wavelength and this helps in better resolution of the object than bright light. Since glass lenses are opaque to ultraviolet light, quartz lenses are used as

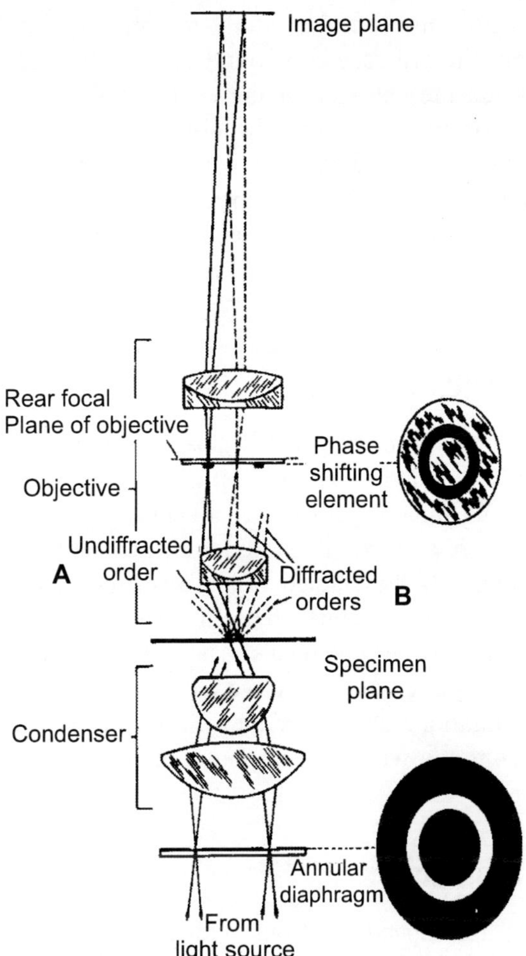


**Fig. 2.3.** Phase contrast microscope.

ocular and objective and as the mirror and condenser of the microscope. The ultraviolet light is invisible, so a sensitive photographic plate must be used to record the image, and then the picture on the plate develops in the usual manner. Ultraviolet microscopy is used extensively to study larger viruses.

## 2.4 DARK-FIELD MICROSCOPY

In dark-field microscopy, the object under observation is brightly illuminated against a dark background. The dark-field microscope is used for viewing living bacteria, particularly those so thin that they approach the limit of resolution of the compound microscope. Many pathogenic spirochaetes fall into this category, particularly the syphilis spirochaetes, *Treponema pallidum*.

Dark-field microscopy is not only useful to study morphology but also motility. The dark-field microscope differs from the ordinary compound light microscope only in that it has special condenser, which prevents the entrance of the central ray of light, passing directly from the mirror to the front lens of the objective. The oblique rays do not enter the tube of the microscope, making the field of vision dark. If the slide contains objects that differ from it in refractive index, the oblique rays are scattered by reflection and refraction. The scattered light enters the objective and thus the object becomes visible as a brightly illuminated speck against a dark background.

## 2.5 FLUORESCENCE MICROSCOPY

Some chemical substances in the microorganisms absorb ultraviolet light and thus alter their inherent colour and appear luminous by exciting remissible or visible light. This phenomenon is also observed in a few living microorganisms. If a bacterial smear is stained with a fluorescent chemical, such as auromine, and examined under a microscope, the bacteria would appear bright against dark background. If the stained organisms are irradiated with ultraviolet rays, the luminescence is brought out more clearly. The limitation of this method is that not many bacteria take the stain with fluorescent chemicals. The auromine-O stain is more selective for tubercle bacilli and is commonly used as a selective technique to detect them. This is a specialized field of scientific investigation. The techniques employed are so specialized that they do not have many features in common with the compound microscope.

## 2.6 ELECTRON MICROSCOPY

In the electron microscope, high speed electrons with an extremely short wavelength, which is about one in 100,000th of the visible light, are used. The short wavelength of the electrons helps in better resolution of the object. A hot tungsten filament forms the source of electron, while the object is interposed in the path of moving electrons and the magnetic condenser lens causes the primary magnification. A second magnetic lens amplifies the primary image and this image is viewed on a fluorescent screen in an air-tight window and recorded on photographic plates. The pathway of electrons in an electron microscope compared with that of light in ordinary light microscope and phase contrast microscope is given in Fig. 2.4.

Two types of electron microscopes are in use today. The first type, the transmission electron microscope (TEM), is used to see the fine structural details within cells. Ultra-thin sections of the object must be prepared because electrons can penetrate matter only a short distance. This is accomplished by embedding or freezing the specimens and sectioning it with a diamond or a glass knife.

The sections are then floated in water and picked up on a wire grid. They are stained with a heavy metal, such as gold or palladium to make certain parts dense and inserted in the vacuum chamber of the microscope. A photograph prepared from the image may be enlarged with enough resolutions to achieve a total magnification of over 20 million times. Objects as small

**Fig. 2.4.** Electron microscope.

as 1.0 nm may be observed. The second type, the scanning electron microscope (SEM), developed in the late 1960s allows surface topography of objects to be seen without sectioning. The objects are placed into the vacuum chamber and covered with a thin coating of gold to increase electrical conductivity. The electron beam then sweeps across the objects, building an image line by line as in a TV camera. Magnifications with the SEM are limited to about 75,000 to 100,000 times. Electron microscopy has extended man's vision into the smallest of living and non-living things, including viruses and has revealed several cytological details of smaller and bigger forms of organisms. However, there are certain limiting factors in electron microscopy:

1. The specimens examined become dry and hence living organisms cannot be examined.
2. The objects are exposed to high vacuum and electron rays, which may even distort the structures of microorganisms; and
3. The high cost of construction and the specialized techniques involved in its use prevent many microbiologists from using electron microscopes.

# Cell and Cell Organization

## 3.1 INTRODUCTION

The earliest fossil evidence of animals dates from the Vendian Period (650 to 544 million years ago), with coelenterate-type creatures that left traces of their soft bodies in shallow-water sediments. The first mass extinction ended that period, but during the Cambrian Period which followed, an explosion of new forms began the evolutionary radiation that produced most of the major groups, or phyla, known today. Vertebrates (animals with backbones) are not known to have occurred until the Ordovician Period (50 to 438 million years ago).

Plants and animals are composed of eukaryotic cell. In the microbial world, bacteria are prokaryotes and fungi, protozoans and true algae are eukaryotes. Prokaryotes and eukaryotes are chemically similar as they contain nucleic acids, proteins, lipids and carbohydrates. Their chemical reactions to metabolize food, proteins and stored energies are also same. The differences exist primarily in the structure of cell walls, membranes and organelles.

Viruses are non-cellular elements with some cell-like properties, which do not fit between living and non-living cells. These are genetic particles that replicate but are unable to perform the usual chemical activities of living cells.

## 3.2 ANIMAL CELL

Animal cells are typical of the eukaryotic cell, enclosed by a plasma membrane and containing a membrane-bound nucleus and organelles. Unlike the cells of the two other eukaryotic kingdoms, plants and fungi, animal cells do not have a cell wall. This feature was lost in the distant past by the single-celled organisms that gave rise to the kingdom Animalia.

The lack of a rigid cell wall allowed animals to develop a greater diversity of cell types, tissues, and organs. Specialized cells that formed nerves and muscles—tissues impossible for plants to evolve—gave these organisms mobility. The ability to move about by the use of specialized muscle tissues is the hallmark of the animal world. (Protozoans locomote, but by nonmuscular means, i.e., cilia, flagella, pseudopodia.) The animal kingdom is unique amongst eukaryotic organisms because animal tissues are bound together by a triple helix of protein called collagen. Plant and fungal cells are bound together in tissues or aggregations by other molecules, such as pectin. The fact that no other organisms utilize collagen in this manner is one of the indications that all animals arose from a common unicellular ancestor. Animals are a large and incredibly diverse group of organisms. Making up about three-quarters of the species on the earth, they run the gamut from sponges and jellyfish to ants, whales, elephants, and—of course—human beings. Being mobile has given animals the flexibility to adopt many different modes of feeding, defence, and reproduction.

The following are the organelles of animal cell (Fig. 3.1).

**Centrioles**—Centrioles are self-replicating organelles made up of nine bundles of microtubules and are found only in animal cells. They appear to help in organizing cell division, but are not essential to the process.

**Cilia and Flagella**—For single-celled eukaryotes, cilia and flagella are essential for the locomotion of individual organisms. In multicellular organisms, cilia function to move fluid or materials from an immobile cell as well as moving a cell or group of cells.

**Endoplasmic Reticulum**—The endoplasmic reticulum is a network of sacs that manufactures, processes, and transports chemical compounds for use inside and outside of the cell. It is connected to the double-layered nuclear envelope, providing a connection between the nucleus and the cytoplasm.

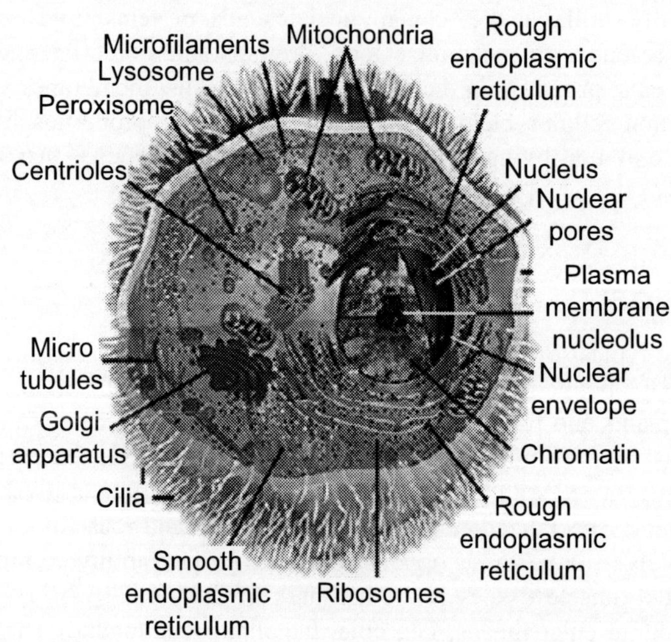


**Fig. 3.1.** Anatomy of animal cell.

**Golgi Apparatus**—The Golgi apparatus is the distribution and shipping department for the chemical products of the cell. It modifies proteins and fats built in the endoplasmic reticulum and prepares them for export to the outside of the cell.

**Lysosomes**—The main function of these microbodies is digestion. Lysosomes break down cellular waste products and debris from outside the cell into simple compounds, which are transferred to the cytoplasm as new cell-building materials.

**Microfilaments**—Microfilaments are solid rods made of globular proteins called actin. These filaments are primarily structural in function and are an important component of the cytoskeleton.

**Microtubules**—These straight, hollow cylinders, composed of tubulin protein, are found throughout the cytoplasm of all eukaryotic cells and perform a number of functions.

**Mitochondria**—Mitochondria are oblong-shaped organelles that are found in the cytoplasm of every eukaryotic cell. In the animal cell, they are the main power generators, converting oxygen and nutrients into energy.

**Nucleus**—The nucleus is a highly specialized organelle that serves as the information and administrative centre of the cell.

**Peroxisomes**—Microbodies are a diverse group of organelles that are found in the cytoplasm. These are roughly spherical and are bound by a single membrane. There are several types of microbodies, but peroxisomes are the most common.

**Plasma Membrane**—All living cells have a plasma membrane that encloses their contents. In prokaryotes, the membrane is the inner layer of protection surrounded by a rigid cell wall. Eukaryotic animal cells have only the membrane to contain and protect their contents. These membranes also regulate the passage of molecules in and out of the cells.

**Ribosomes**—All living cells contain ribosomes—tiny organelles composed of approximately 60 percent RNA and 40 percent protein. In eukaryotes, ribosomes are made of four strands of RNA. In prokaryotes, they consist of three strands of RNA.

The plant cell differs from the animal cell in that it possesses a cell wall made up of cellulose which is external to its outer cellular (plasma) membrane. Cytoplasmic continuity is maintained between neighbouring cells via plasmodesmata (small channels through the cell wall). The vacuole often occupies a large volume within the cell. A plant is dependent for its turgidity upon the effect (if the plant is adequately watered) of the turgor pressure of the vacuole pushing out the cytoplasm against the rigid cell wall. Photosynthetic parts of a plant contain chloroplasts, which have a double membrane enclosing the stroma in which stacks of thylakoids form grana (these contain the photosynthetic apparatus of the organelle).

Otherwise, at the ultrastructural level, plant and animal cells appear quite similar. Each cell is surrounded by an outer plasma membrane. The nucleus is surrounded by a double membrane (the nuclear envelope), which is perforated in places by nuclear pores. Within the nucleus is the chromatin, which may appear densely (heterochromatin) or lightly (euchromatin) stained. Nucleoli (the sites of ribosomal RNA synthesis) are visible usually as spherical regions, which are granular and fibrillar in nature.

The rest of the cell is occupied by cytoplasm, which contains a range of organelles. The latter are visibly identifiable structures, which have different functions within the cell. The outer nuclear membrane may be continuous with the rough endoplasmic reticulum (RER) in places. It is termed 'rough' because it has ribosomes (often in spiral aggregates called polyribosomes or polysomes) on the outer surface of its membrane. Ribosomes (and polysomes) may also be 'free' within the cytoplasm. Smooth ER has no ribosomes associated with it and is involved in lipid and sterol synthesis. The golgi body (or apparatus) within a cell is thought to be an interconnected system, but in thin sections appear as discrete units (termed dictyosomes in plant cells). These units may be compared to a stack of saucers, with each saucer being one sac or cisterna (membrane-enclosed space) of the Golgi. At the periphery of these cisternae, secretory vesicles bud off via tubular extensions. The mitochondria have two membranes, the inner one of which invaginates to form the cristae. Within this inner membrane is the matrix of the mitochondrion.

All the above organelles of the cell are surrounded by the 'cytoskeleton', an extremely fine network of proteinaceous strands (microfilaments and microtubules), which, along with forming a structural 'skeleton' for the cell, are also involved in morphogenesis and motility (e.g., targeting of secretory vesicles to the cell membrane).

In general terms, the organelles exist to compartmentalize away discrete biochemical functions within the cell. For example, within the confines of the mitochondria, conditions are maintained that are optimal for the enzymes responsible for oxidative phosphorylation and ATP synthesis. Organelles also provide a structural framework for the specific biochemical processes that occur within them (e.g., enzyme attachment to organized membranes in the mitochondria and chloroplast).

The organelles should in no sense, however, be imagined to operate in isolation as there is a continuous exchange of metabolites, etc. between those which are functionally-related. Indeed, some of the organelles (such as the ER and Golgi) are functionally and physically related. Returning to the idea of the dynamic nature of the cell, there is what is termed the 'endomembrane concept' of membrane 'flow' or turnover. The ER is generally considered to be the major site of membrane biogenesis (formation), and from here (and the outer nuclear membrane, to which the ER contacts in places) there is a functional continuum involving the physical transfer of membrane to the Golgi and thence to Golgi-derived secretory vesicles and to the plasma membrane (and/or the tonoplast [vacuolar membrane] in plants). During this process, for example, specific proteins may be assembled in the (rough) ER, and then various enzyme activities through the ER and Golgi will result in the required end-product being packaged ultimately into secretory vesicles for export.

## 3.3 PLANT CELL

Like the fungi, another kingdom of eukaryotes, plant cells, has retained the protective cell wall structure of their prokaryotic ancestors. The plant cell has the basic construction of a typical eukaryotic cell, but does not have centrioles, lysosomes, cilia, or flagella as does the animal cell. Plant cells do have specialized structures: a rigid cell wall, made up of cellulose, central vacuole, plasmodesmata, and chloroplasts (Fig. 3.2).

It is estimated that there are at least 260,000 species of plants in the world today. They range in size and complexity from small, non-vascular mosses to giant Sequoia trees, the largest living

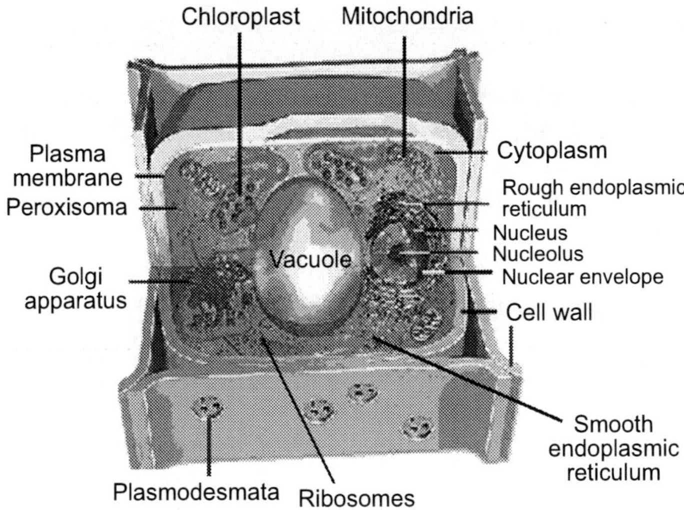


**Fig. 3.2.** Anatomy of plant cell.

organisms as tall as 330 feet (100 meters). Only a tiny percentage of these species is directly used by people for food, shelter, fibre, and medicine. Nonetheless, plants are the basis for the earth's ecosystem and food web, and without them, the complex animal life forms (such as humans) could never have evolved.

Thought to have evolved from the green algae, plants have been around since the early Paleozoic era, more than 500 million years ago. The earliest fossil evidence of land plants dates to the Ordovician Period (505 to 438 million years ago). By the Carboniferous Period, about 355 million years ago, most of the earth was covered by forests of primitive vascular plants, such as lycopods (scale trees) and gymnosperms (pine trees, ginkgos). Angiosperms, the flowering plants, did not develop until the end of the Cretaceous Period, about 65 million years ago.

## 3.4 PROKARYOTIC CELL

Prokaryotes are single-celled organisms without a nucleus, i.e., bacteria and archaea. This name comes from the Greek root *karyon* meaning nut, combined with the prefix pro-, meaning before. Cells with true nucleus are called eukaryotes, where the prefix eu- means good or true.

They are as unrelated to human beings as living things can be, but bacteria are essential to human life and life on the earth. Although they are known for their role in causing human diseases, from tooth decay to the Black Plague, there are beneficial species that are essential to good health. For example, one species, (*Escherichia coli*), which lives symbiotically in the large intestine manufactures vitamin K, an essential blood clotting factor. Other species are beneficial indirectly. Bacteria help in the preparation of yogurt, its tangy flavor and sourdough bread for its sour taste. They make it possible for ruminant animals (cows, sheep, goats) to digest plant cellulose, and for some plants (soybean, peas, alfalfa) to convert nitrogen to a more usable form.

Bacteria are prokaryotes, lacking well-defined nuclei and membrane-bound organelles, and with chromosomes composed of a single closed DNA circle. They are in many shapes and sizes, from minute spheres, cylindrical and spiral threads, to flagellated rods and filamentous chains. They are found practically on the earth and live on diversified habitats and seemingly inhospitable places (extreme habitats).

Evidence shows that bacteria were in existence as long as 3.5 billion years ago, making them one of the oldest living organisms on the earth. Even older than the bacteria are the archeans—(also called archaebacteria) tiny prokaryotic organisms that live only in extreme environments: boiling water, super-salty pools, sulphur spring, volcanic vents, acidic water, and deep in the Antarctic ice. Many scientists now believe that the archaea and bacteria developed separately from a common ancestor nearly four billion years ago. Millions of years later, the ancestors of today's eukaryotes split off from the archaea. Despite the superficial resemblance to bacteria, biochemically and genetically, the archea are as different from bacteria as bacteria are from humans.

In the late 1600s, Antony van Leeuwenhoek became the first to study bacteria under the microscope. During the nineteenth century, the French scientist Louis Pasteur and the German physician Robert Koch demonstrated the role of bacteria as pathogens (causing disease). The twentieth century saw numerous advances in bacteriology, indicating their diversity, ancient lineage, and general importance. Most notably, a number of scientists around the world made contributions to the field of microbial ecology, showing that bacteria were essential to food webs and for the overall health of the earth's ecosystems. The discovery that some bacteria produced compounds lethal to other bacteria led to the development of antibiotics, which revolutionized the field of medicine.

There are two different ways of grouping bacteria. They can be divided into three types based on their response to gaseous oxygen. Aerobic bacteria require oxygen for their health and existence and die without it. Anaerobic bacteria cannot tolerate gaseous oxygen at all and die when exposed to it. Facultative anaerobes prefer oxygen, but can live without it.

The second way of grouping them is by how they obtain their energy. Bacteria that have to consume and break down complex organic compounds are heterotrophs. This includes species that are found in decaying material as well as those that utilize fermentation or respiration. Bacteria that create their own energy, fueled by light or through chemical reactions, are autotrophs.

The following are the details of the bacterial cell structure:

(a) **Capsule**—Some species of bacteria have a third protective covering, a capsule made up of polysaccharides (complex carbohydrates). Capsules play a number of roles, but the most important are to keep the bacterium from drying out and to protect it from phagocytosis (engulfing) by larger microorganisms. The capsule is a major virulence factor in the major disease-causing bacteria, such as *Escherichia coli* and *Streptococcus pneumoniae*. Non-encapsulated mutants of these organisms are avirulent, i.e., they do not cause disease.

(b) **Cell Envelope**—The cell envelope is made up of two to three layers: the interior cytoplasmic membrane, the cell wall, and in some species of bacteria, an outer capsule.

(c) **Cell Wall**—Each bacterium is enclosed by a rigid cell wall composed of peptidoglycan, a protein-sugar (polysaccharide) molecule. The wall gives the cell its shape and surrounds the cytoplasmic membrane, protecting it from the environment. It also helps to anchor appendages like the pili and flagella, which originate in the cytoplasm membrane and protrude through the wall to the outside. The strength of the wall is responsible for keeping the cell from bursting when there are large differences in osmotic pressure between the cytoplasm and the environment.

Cell wall composition varies widely amongst bacteria and is one of the most important factors in the analysis and differentiation of bacterial species. For example, a relatively thick, mesh-like structure that makes it possible to distinguish two basic types of bacteria. A technique devised by Danish physician Hans Christian Gram in 1884 uses a staining and washing technique to differentiate between the two forms. When exposed to a gram stain, gram-positive bacteria retain the purple colour of the stain because the structure of their cell walls traps the dye. In gram-negative bacteria, the cell wall is thin and releases the dye readily when washed with an alcohol or acetone solution.

(d) **Cytoplasm**—The cytoplasm, or protoplasm of bacterial cells is the structure where the functions for cell growth, metabolism, and replication are carried out. It is a gel-like matrix composed of water, enzymes, nutrients, wastes, and gases and contains cell organelles such as ribosomes, a chromosome, and plasmids. The cell envelope encases the cytoplasm and all its components. Unlike the eukaryotic (true) cells, bacteria do not have a membrane enclosed nucleus. The chromosome, a single, continuous strand of DNA, is localized, but not contained, in a region of the cell called the nucleoid. All the other cellular components are scattered throughout the cytoplasm.

(e) **Cytoplasmic Membrane**—A layer of phospholipids and proteins, called the cytoplasmic membrane, encloses the interior of the bacterium, regulating the flow of materials

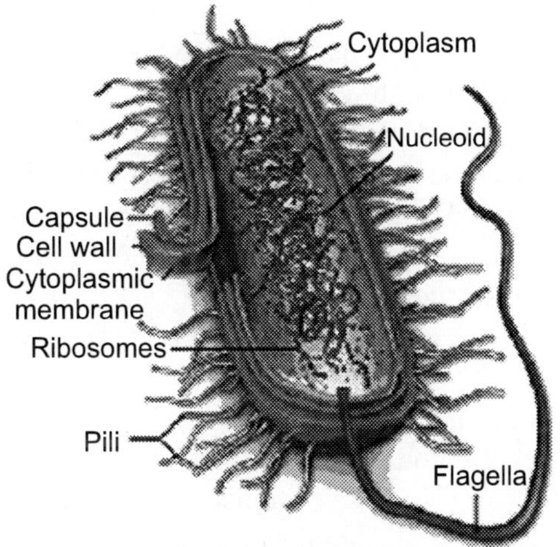


**Fig. 3.3.** Structure of prokaryotic cell.

in and out of the cell. This is a structural trait bacteria share with all other living cells; a barrier that allows them to selectively interact with their environment. Membranes are highly organized and asymmetric having two sides, each side with a different surface and functions. Membranes are also dynamic, constantly adapting to different conditions.

One of those components, plasmids, are small, extrachromosomal genetic structures carried by many strains of bacteria. Like the chromosome, plasmids are made of a circular piece of DNA(plasmids). Unlike the chromosome, they are not involved in reproduction. Only the chromosome has the genetic instructions for initiating and carrying out cell division, or binary fission, the primary means of reproduction in bacteria. Plasmids replicate independently of the chromosome and, while not essential for survival, appear to give bacteria a selective advantage.

Plasmids are passed on to other bacteria through two means. For most plasmid types, copies in the cytoplasm are passed on to daughter cells during binary fission. Other types of plasmids, however, form a tube-like structure at the surface called a pilus that passes copies of the plasmid to other bacteria during conjugation, a process by which bacteria exchange genetic information. Plasmids have been shown to be instrumental in the transmission of special properties, such as antibiotic drug resistance, resistance to heavy metals, and virulence factors necessary for infection of animal or plant hosts. The ability to insert specific genes into plasmids have made them extremely useful tools in the fields of molecular biology and genetics, specifically in the area of genetic engineering.

(f) **Flagella**—Flagella (singular, flagellum) are hair-like structures that provide a means of locomotion for those bacteria that have them. They can be found at either or both ends of a bacterium or all over its surface. The flagella beat in a propeller-like motion to help the bacterium move toward nutrients; away from toxic chemicals; or, in the case of the photosynthetic cyanobacteria; toward the light.

(g) **Nucleoid**—The nucleoid is a region of cytoplasm where the chromosomal DNA is located. It is not a membrane bound nucleus, but simply an area of the cytoplasm where the strands of DNA are found. Most bacteria have a single, circular chromosome that is responsible for replication, although a few species do have two or more. Smaller circular auxiliary DNA strands, called plasmids, are also found in the cytoplasm.

(h) **Pili**—Many species of bacteria have pili (singular, pilus), small hair-like projections emerging from the outside cell surface. These outgrowths assist the bacteria in attaching to other cells and surfaces, such as teeth, intestines, and others. Without pili, many disease causing bacteria lose their ability to infect because they're unable to attach to host tissue. Specialized pili are used for conjugation, during which two bacteria exchange fragments of plasmid DNA.

(i) **Ribosomes**—Ribosomes are microscopic "factories" found in all cells, including bacteria. They translate the genetic code from the molecular language of nucleic acid to that of amino acids—the building blocks of proteins. Proteins are the molecules that perform all the functions of cells and living organisms. Bacterial ribosomes are similar to those of eukaryotes, but are smaller and have a slightly different composition and molecular structure. Bacterial ribosomes are never bound to other organelles as they sometimes

**Table 3.1.** The major differences between prokaryote and eukaryote genetic and cellular organization.

| Prokaryotes | Eukaryotes |
|---|---|
| DNA free in the cytoplasm. | DNA is in a membrane bound nucleus. |
| Only one chromosome | >1 chromosome. Two copies of each chromosome may be present (diploid) |
| DNA associated with histone-like proteins | DNA completed with histone proteins |
| May contain extrachromosomal elements called plasmids | Plasmids only found in chromosomes |
| Introns not found in mRNA | Introns found in all genes |
| Cell division by binary fission—asexual replication only | Cells divide by mitosis |
| Transfer of genetic information occurs by conjugation, transduction and transformation | Exchange of genetic information occurs during sexual reproduction. Meiosis leads to the production of haploid cells (gametes) |

are (bound to the endoplasmic reticulum) in eukaryotes, but are free-standing structures distributed throughout the cytoplasm. There are sufficient differences between bacterial ribosomes and eukaryotic ribosomes that some antibiotics will inhibit the functioning of bacterial ribosomes, but not in eukaryote's, thus killing bacteria but not the eukaryotic organisms they are infecting.

Microbes have been used extensively as model organisms for the investigation of biochemical and genetical processes as they are much easier to work with than more complex animals and plants. Millions of copies of the same single cell can be produced in large numbers very quickly and at a low cost to give plenty of homogeneous experimental material. An additional advantage is that most people have no ethical objections to experiments with these microorganisms.

# Diversity of Microbes

## 4.1 INTRODUCTION

Microbiology is a relatively new subject in biology and deals with prokaryotic biology and its importance. Though the art of making several fermentation products was practised, the science behind such practices was understood very late. The subjects like Botany and Zoology were known much prior to microbiology because the plants and animals were visible for observation. The science of microbiology began with the observations of Antony van Leeuwenhoek who reported the living (moving) objects under microscope. Prior to him, microscope was used for observing objects of non-living nature. Knowledge on microorganisms increased slowly with refinement in microscopy. Initially some microorganisms were interpreted to be as plant-like, some as animal-like and some were neither like plants nor like animals. Therefore, it was proposed to have a separate kingdom to include such organisms. The earliest proposal for this was made in 1866 by E.H. Haeckel, a German zoologist. He suggested that the third kingdom called 'Protista', be formed to include those unicellular microorganisms that are typically neither plants nor animals. These organisms, the protists, include bacteria, algae, fungi and protozoa. Of these, bacteria were referred to as lower protists and others as higher protists. Viruses are not included in protists, because they are not cellular organisms and therefore made a separate kingdom.

Until 1940s, several unicellular organisms like bacteria, yeasts, micro-algae etc., were not clearly demarcated and classified under protista. With the refinement in light microscopy and development of electron microscope, more cellular details were understood for intracellular organization of living cells. It was discovered that in some cells, e.g., typical bacteria and blue-green algae, the nuclear substance was not enclosed by nuclear membrane. In other cells, such as typical algae, yeasts and fungi, the nucleus was enclosed in a membrane. This discovery, the absence of membrane-bound internal cell structures in one group of protists (bacteria) and the presence of membrane-bound internal cell structures in all other microscopic organisms (protists), was of fundamental significance. The concept of bacteria and blue-green algae as prokaryotic organisms was established in 1962 by Stanier and Van Niel. With this the concept of 'Prokaryotes' and 'Eukaryotes' has come into use for all living organisms. All such categorizations were based on the microscopic (light or electron) observations of living organisms for their cellular organization. Prokaryotic microorganisms were called 'lower protists' and eukaryotic microorganisms were called 'higher protists'.

## 4.2 WHITTAKER'S FIVE KINGDOM CONCEPT

The characterization and taxonomical grouping of living organisms till early 1960s was based on the cellular organization as revealed by microscopic (light microscope and/or electron microscope) studies. A comprehensive system of classification based on the nutrition and absorption of nutrients by the living organisms was proposed by R.H. Whittaker in 1969. This is called Whittaker's Five Kingdom concept. He proposed the classification of all living organisms into five kingdoms (Fig. 4.1). This is based on three levels of cellular organization which have evolved to accommodate three principal modes of nutrition: photosynthesis, absorption and ingestion. Whittaker explained that some organisms obtain their nutrients (mainly carbon) by photosynthetic assimilation, some organisms obtain (organic carbon) nutrients by absorption from their living environment and some others obtain (organic carbon) nutrients by ingestion. He grouped plants and other photosynthetic algae, fungi and the animals in the above three groups (kingdoms), respectively. The prokaryotic organisms are included into a new kingdom, Monera (lower protists). These lack the ingestive mode of nutrition. The unicellular eukaryotic microorganisms are placed in the kingdom Protista (higher protists). All three nutritional types are represented in Monera and Protista. This becomes the basis to explain evolutionary group to start with (1) Kingdom-Monera, (2) Kingdom-Protista, (3) Kingdom-Plantae, (4) Kingdom-Fungi and (5) Kingdom-Animalia. The nutritional modes are continuous, where the mode of nutrition of microalgae is photosynthetic, the mode of nutrition of the protozoa is ingestive, and the mode of nutrition in some other protists is absorptive, with some overlap to the photosynthetic and ingestive modes.

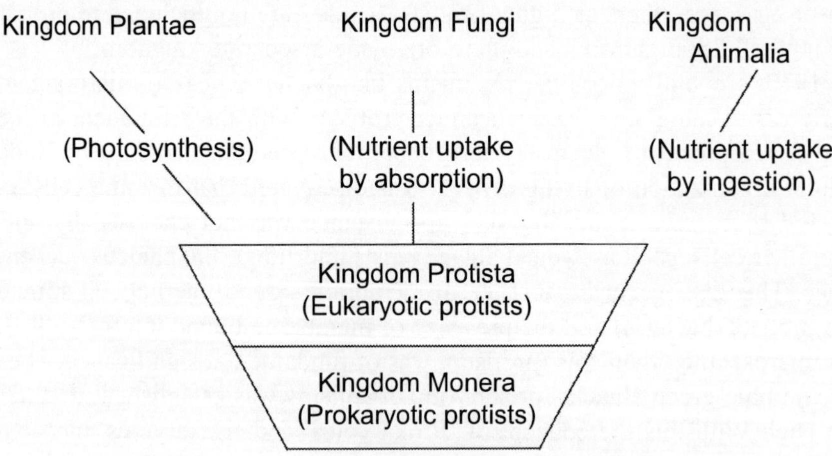


**Fig. 4.1.** Simplified Whittaker's five kingdom system.

The multicellular and multinucleate eukaryotic organisms are found in the kingdom Plantae (multicellular green plants and higher algae), Animalia (multicellular animals), and Fungi (multinucleate higher fungi). Their diversified nutritional modes lead to a more diversified cellular organization. Microorganisms are found in three of the five kingdoms: Monera (bacteria and blue-green algae or cyanobacteria), Protista (microalgae and protozoa) and Fungi (yeasts and moulds).

## 4.3 CARL WOESE CONCEPT

All explanations and grouping given by Whittaker were based on microscopic studies on living organisms till 1960s. Subsequently, developments in molecular biology techniques lead to more deeper understanding of living organisms at molecular level. These studies were done mainly in terms of type of genetic materials, the DNA and RNA and their sequence, structure and function of biomacromolecules in different living organisms and their molecular structure, stability of genetic structures and genes has come into understanding. Scientists have studied the genomic structures and nucleotide sequence of genetic materials in different organisms. The DNA base composition, G + C percentage ratio, DNA homology and gene sequence of more stable genes of prokaryotic and eukaryotic organisms were analyzed. The 16S rRNA and 18S rRNA gene structures and sequence of prokaryotes and eukaryotes respectively, were proposed to be taken into the major criteria for classification and identification of living organisms. This

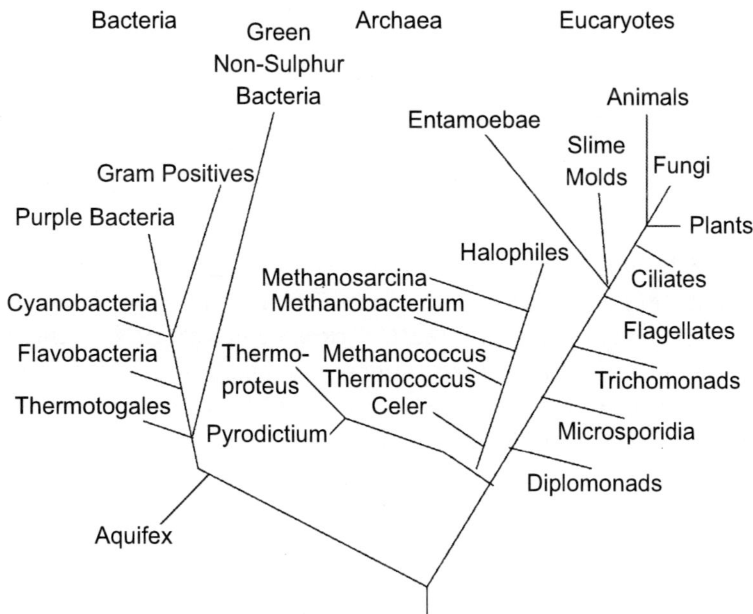


**Fig. 4.2.** Carl Woese concept for grouping of all living organisms into three domains (kingdom) based on 16S or 18S rRNA gene sequence.

**Table 4.1.** Phenotypic properties of bacteria and archaea compared with eukarya

| Property | Eukaryotes | Bacteria | Archaea |
|---|---|---|---|
| Cell configuration | Eukaryotic | Prokaryotic | Prokaryotic |
| Nuclear membrane | Present | Absent | Absent |
| Number of chromosomes | >1 | 1 | 1 |
| Chromosome organization | Linear | Circular | Circular |
| Murein in cell wall | – | + | – (pseudomurein) |
| Cell membrane lipids | Ester-linked Glyceroldiesters Unbranched; Polyunsaturated | Ester-linked Glyceroldiesters Unbranched; Saturated or Monounsaturated | Ether-linked Glyceroldiethers Branched Saturated |
| Cell membrane sterols | Present | Absent | Absent |
| Organelles (mitochondria and chloroplasts) | Present | Absent | Absent |
| Ribosome size | 80S (cytoplasmic) | 70S | 70S |
| Cytoplasmic streaming | + | – | – |
| Meiosis and mitosis | Present | Absent | Absent |
| Transcription and Translation coupled | – | + | + |
| Amino formyl acid initiating | Methionyl | Methionyl | Methionine Methionyl |
| Protein synthesis Inhibited by streptomycin and chloramphenicol | – | + | – |
| Inhibited by diphtheria | + | – | + |

concept was proposed by Carl Woese. Carl Woese and his group in 1990 noted that bacteria are distant from plants and animals and, by contrast, plants and animals are not so far from each other (Fig. 4.2). All this is based on the 16S rRNA and 18S rRNA gene sequencing and homology studies, DNA-DNA hybridization studies, and the DNA base analysis or DNA finger printing by restriction fragment length polymorphism (RFLP) analysis.

They proposed radical change in classification of all living organisms into three important domains and called them Bacteria, Archaea and Eukarya to explain that the three domains (domain is taken as a kingdom) are evolved from common ancestors (Fig. 4.2). There are certain common characteristics between bacteria and archaea and also between archaea and eukarya (Table 4.1). These include the cell wall and membrane structures, chemical composition of membranes, ribosomal function, RNA polymerase structure, and sensitivity to certain antibiotics. The latest classification and taxonomy of bacteria is based mainly on these characteristics. The latest edition of Bergey's Manual of Systematic Bacteriology (second edition) explaining the grouping of bacteria and their identification gives more emphasis on their genetic relatedness and phylogenetic identification. Bacteria identified and grouped earlier based on non-genetic criteria have changed their taxonomic position based on their phylogenetic relatedness.

## 4.4 MYCOPLASMA

In 1898, the French scientist, E. Nocard and E.R. Roux, studying pleural fluids of cattle suffering from bovine pleuropneumonia discovered the organisms that were entirely different from any other microorganisms known then. When cultivated on rich organic media containing about 20% of animal serum, the organisms were found in different forms as spheroid, thin, branching filaments, stellate or asteroid structures and other irregular forms. Similar pleomorphic (Gr. *Pleo* = many; *morphe* = forms) organisms were later isolated from other animals such as sheep, goat, dogs, rats, mice, and human beings. Similar forms were also found growing as saprophytes in decaying organic matter. These were named as pleuropneumonia-like organisms (PPLO). The species discovered by Nocard and Roux was given the first binomial as *Asterococcus mycoides* by Borrel et al. (1910), meaning rounded and stellate forms with radial, mould-like filaments. It was later on put under the genus *Mycoplasma* by Nowak (1929) and these organisms are now commonly called mycoplasmas. Characteristic features of the mycoplasmas are as follows:

1. Unicellular, prokaryotic, usually non-motile and form fried egg-shaped colonies
2. Highly pleomorphic, form varying with culture conditions. Under different microscopy, they appear as small coccoid bodies, ring forms and fine filaments, some of which are branched.
3. Filtrable through bacterial filters.
4. Cell wall absent. Cells delimited by a triple layered lipoproteinaceous unit membrane, the plasma membrane.
5. Both DNA and RNA present. DNA base composition ranges from 23 to 36 (39 to 41 in *Mycoplasma pneumoniae*) mole per cent GC (cf L-forms).
6. Resistant to antibiotics such as penicillins that act on cell walls.
7. Inhibited by tetracyclines and similar antibiotics that act on metabolic pathways.
8. Mostly free-living; parasites and saprophytes.

They are true cells, and like animal cells they have however, no demonstrable cell walls. The sole retaining structure is the cytoplasmic membrane, which like most other cell membranes is phospholipid-protein bilayer or unit type, continuous, flexible and 7–11 nm thick.

Mycoplasmas are simpler than the cells of higher plants and animals. They possess all the necessary biochemical machinery to grow and multiply in the absence of other cells. The genetic machinery is in the form of DNA, RNA and ribosomes.

They vary in size mostly from 300 nm to about 0.2 μm in diameter. The volume of the smallest PPLO is about $1 \times 10^{-3}$. The internal structure is typical of prokaryons, generally and closely resembles that of true bacteria. The cell membrane encloses the cytoplasm, which contains numerous ribosomes, a network of fibrillar DNA, electron dense areas, soluble RNA and soluble protein. Nuclear structures, however, are less evident than typical bacteria. Mesosomes are absent.

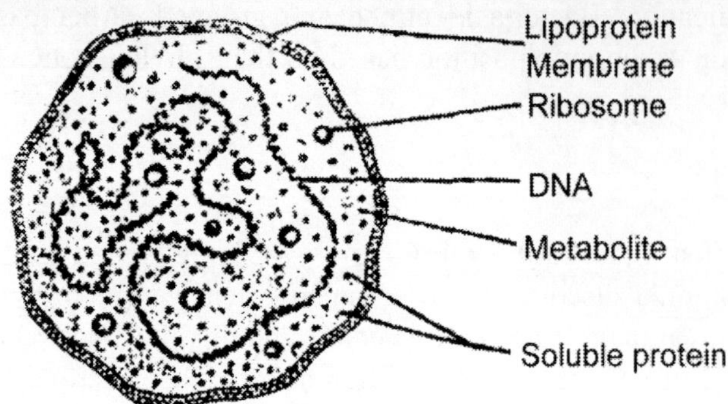


**Fig. 4.3.** Mycoplasma (0.5 μm).

The cells of mycoplasma divide unevenly into very minute bodies called elementary bodies or minimal reproductive units. They are commonly formed inside the large bodies or mature cells. These elementary bodies range in size from about 330 nm to 450 nm. These can pass through bacteria retaining filters like viruses, but are viable on ordinary media. These have bacteria-like structure. These bodies are often cited as the smallest independently living entities (Fig. 4.3). These represent a stage in the life cycle. They enlarge to form long filaments and mycelia and chains of minute spheres like conidia, but much smaller in size. It is thought by some that these conidia-like bodies are liberated and each increases in size to become larger body several μm in diameter, inside which new elementary bodies are formed. These are released by rupture of the membrane of larger body. Growth rate of mycoplasma is very rapid, and generation time is 1–3 hrs. Like viruses and animals cells, mycoplasmas are resistant to penicillin.

### 4.4.1 Plant Pathogenic Mycoplasmas (Spiroplasma)

For many years the yellows of plants were thought to be caused by viruses. In 1967, however, the Japanese workers, Doi et al., discovered pleomorphic mycoplasma-like organisms (MLO), called spiroplasma in the phloem cells of plants, affected by different yellows-type diseases (Fig. 4.4). Since then MLOs have been found to be associated with over 70 plant diseases, which infect over 300 genera of plants. They occur from temperate to tropical regions, but it is in the warmer areas that serious losses occur in crops such as coconuts, citrus, rice, maize, cotton and potatoes. They also occur in insects and as saprophytes in soil and sewage. A number of mycoplasmal diseases of plants are listed by Ghosh and Raychoudhuri (1972). During 1940s, there occurred an unusual increase in the cases of human pneumonia. Monroe Eaton isolated a tiny virus-like agent from the respiratory tract of patients. This was named *Mycoplasma pneumoniae*, the Eaton agent. The disease came to be known as primary atypical pneumonia (PAP). *Mycoplasma hominis* causes a sexually-transmitted disease, mycoplasmal urethritis. Some of the plant diseases caused by mycoplasma are given in Table 4.2.

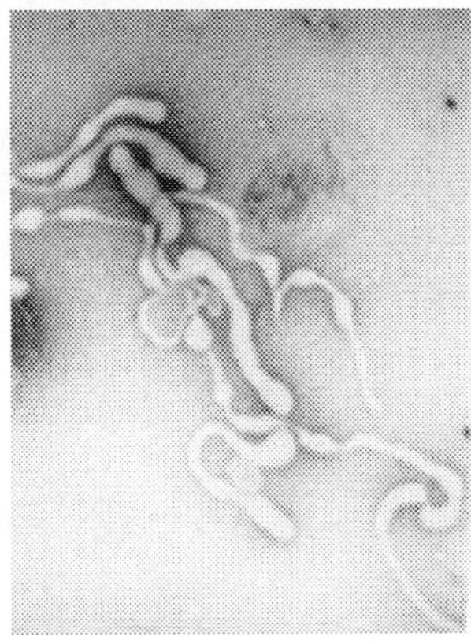

**Fig. 4.4.** Spiroplasma, plant pathogenic mycoplasma.

**Table 4.2.** Mycoplasmal plant diseases

| | |
|---|---|
| Alfalfa witche's broom | Brinjal little leaf |
| Apple proliferation | Citrus greening and stubborn |
| Aster yellows | Clover dwarf |
| Clover phyllody | Peanut witche's broom |
| Clover stolbur | Pear decline |
| Clover virescence | Phloem necrosis of elm |
| Corn stunt | Phormium yellow leaf |
| Cotton virescence | Potato purple top |
| Cranberry false blossom | Potato stolbur |
| Crimean yellows | Potato witche's broom |
| Cryptomeria witche's broom | Rice stripe |
| Flavescence doriee | Rice yellow dwarf |
| Giallume yellow of rice | Rubus stunt |
| Grassy stunt of rice | Safflower phyllody |
| Legume little leaf | Sandal spike |
| Legume witche's broom | Sesame phyllody |
| Little cherry | Strawberry green petal |
| Little peach | Sugarcane white leaf |
| Lucerne witche's broom | Sweet pòtato little leaf |
| Mal Azul (blue) disease of tomato | Sweet potato witche's broom |
| Mulberry dwarf | Tobacco yellow dwarf |
| *Opunita tuna* witche's broom | Tomato bid bud |
| Papaya bunchy top | *Vicia faba* virenscence |
| Para stolbur | *Vinca rosea* yellows |
| Paulownia witche's broom | Western X-disease of peaches and cherries |
| Pea and green pea yellow dwarf | Yellow wilt of sugarbeet. |

## 4.5 L-FORMS

Klieneberger in 1935 first studied a form of *Streptobacillus moniliformis*, what was thought to be a symbiotic species of PPLO. He, while working at the Lister Institute in London, called this form $L_1$, L being the initial of the Lister Institute named for Lord Lister, pioneer in aseptic surgery. The $L_1$, organism was later shown by Klieneberger and others to be a stable protoplast of L-form of *Streptobacillus moniliformis*. The two forms interchange under proper growth conditions. L-forms of bacteria are entirely different from L cells, a clone of animal tissue culture cells (Fig. 4.5).

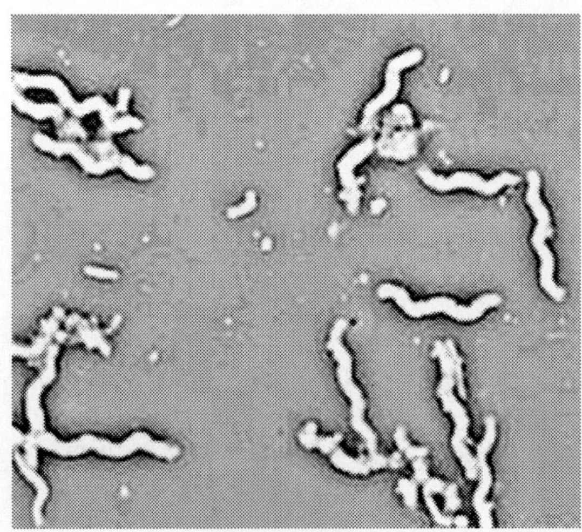

**Fig. 4.5.** L-Forms.

## 4.6 PROTOPLASTS

Some microorganisms, both eukaryotic and prokaryotic (e.g., yeasts, moulds, bacteria, some higher plants) that have a distinct cell wall can at times exist without a cell wall in what is designated as wall-less state. The cell membrane and its intact contents are then called a protoplast. Unless the solute concentration of the suspending fluid is osmotically protective i.e., high enough (e.g., 3–20% glucose; 2–5% NaCl; 10–20% serum) to balance the intracellular osmotic pressure, protoplasts no longer retained by their thick walls, usually burst. Protoplast are therefore said to be osmotically fragile.

An important property of bacterial cell wall is sensitivity to antibiotic penicillin. Penicillin inhibits the synthesis of peptidoglycans. Young actively growing Gram-positive bacteria are, therefore, sensitive to penicillin. Thus, these bacteria can be made into protoplasts by treating them with penicillin. Animal cells, including human cells, are not susceptible to penicillin, because there are no peptidoglycans.

## 4.7 SPHEROPLASTS

Gram-negative bacteria, where the bulk of cell wall is made by lipopolysaccharides and lipoprotein, are usually partly or wholly resistant to penicillin (as it inhibits only peptidoglycan) (Fig. 4.6.)

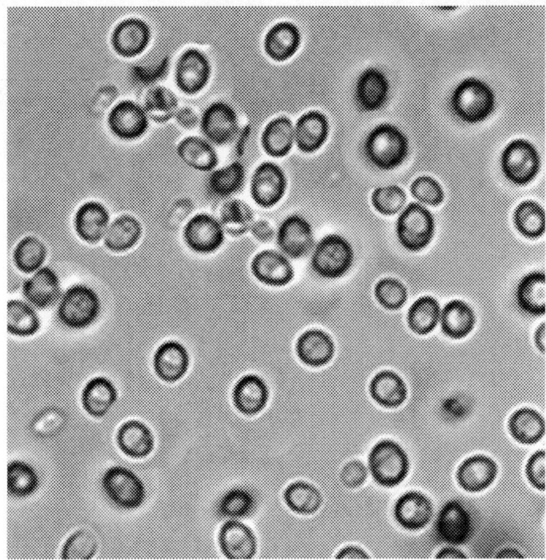

**Fig. 4.6.** Spheroplasts.

These Gram-negative bacteria are also protected by their lipopolysaccharides and lipoproteins of cell wall (in outer layers) from the action of a group of enzymes, lysozyme, discovered by Fleming. Lysozyme is found in egg white secretions of skin and mucous membranes, tears and elsewhere. It attacks superficially the glycosidic bonds in the polysaccharide backbone of peptidoglycan of bacterial cell walls. When Gram-positive bacteria are treated with lysozyme, they are rapidly denuded of their cell walls and become naked protoplasts. In contrast, to make Gram-negative bacteria (though peptidoglycan present in their cell wall, but it is protected by outer layers of lipocomplexes) vulnerable to lysozyme, it is necessary first to remove the lipocomplex of cell wall. These are removed with lipid solvents such as NaOH or ethylenediaminetetraacetate (EDTA). Even then the wall is not completely removed and the result is osmotically fragile cells, still retaining some remnants of cell wall. Such partially denuded cells are called spheroplasts.

Thus protoplasts or spheroplasts may be prepared from various bacteria either by removing the cell wall or by metabolically blocking the synthesis of cell wall. Lysozyme, which dissolves the cell wall, and penicillin, which inhibits cell wall synthesis are most frequently used in preparation of protoplasts. Such protoplasts in hypertonic or isotonic fluid are often able to grow and multiply on agar media. The colonies resemble the mycoplasmas. These cells without cell walls have been designated as stable L-forms. L-forms can be selected by cultivation of bacteria on an osmotically buffered penicillin containing medium. L-forms are, therefore, bacteria in which the primer for peptidoglycan has either been eliminated or modified by penicillin treatment. L-form can be obtained from both Gram-positive and Gram-negative bacteria. Stable L-forms obtained from animal and man are different from mycoplasmas. Recent genetic studies have shown that L-forms are different from mycoplasma. For example, DNA base composition of *Mycoplasma* species ranges from 23 to 36 (39–41 in *M. pneumoniae*) mole percent GC, whereas most other bacteria and the L-forms usually have higher per cent GC content (63-64 mole per cent GC).

L-forms are formed in *Mycobacterium tuberculosis*, *Streptococcus* and others. Some bacteria undergo spontaneous mutation due to loss of cell walls. For example, if DAP is withheld from the medium, which is osmotically protective, the bacteria grow without a cell wall i.e., they are L-forms of the bacteria.

## 4.8 ARCHAEBACTERIA

The archaebacteria are predominantly terrestrial and aquatic microbes occurring in anaerobic or hypher or hydrothermally and geothermally heated environments. Some also occur as symbionts in the digestive tracts of animals. They consist of aerobes, anaerobes and facultative anaerobes that grow chemolitho-autotrophically, organotrophically, or facultatively organotrophically. Archaebacteria may be mesophiles or thermophiles, with some species growing even above 100°C (Fig. 4.7).

A unique biochemical feature of archaebacteria is the presence of glycerol isopranyl ether lipids. The lack of murein (peptidoglycan containing muramic acid) in cell walls make archaebacteria insensitive to alpha-lactam antibiotics. The "common arm" of the tRNA contains pseudouridine or 1-methylpseudouridine instead of ribothymidine. The sequences of 5S, 16S and 23S rRNAs are very different from the corresponding ones in eubacteria and eukaryotes.

Archaebacteria share some molecular features with eukaryotes: (a) the elongation factor 2(EF-2) contains the amino acid diphthamide and is therefore ADP-ribosylable by diphtheria toxin, (b) amino acid sequences of the ribosomal "A" protein exhibit sequence homologies with the corresponding eukaryotic protein, (c) the methionyl initiator tRNA is not formylated,

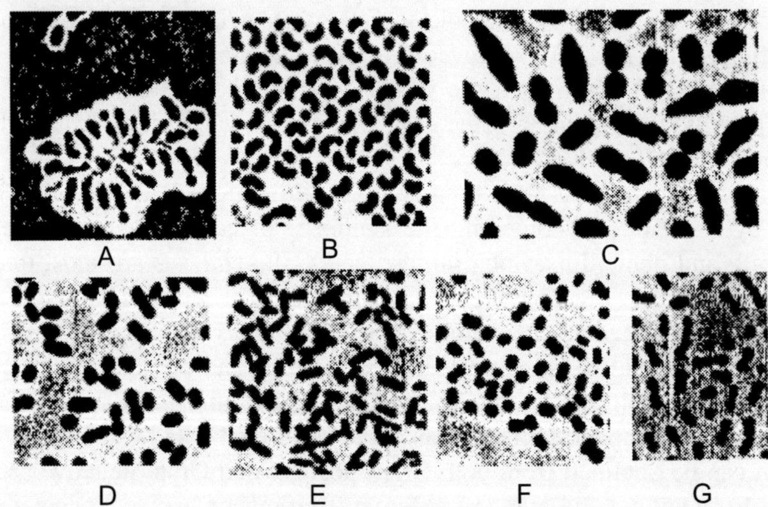


**Fig. 4.7.** Methanotrophs (A: *Methylosinus;* B: *Methylocystis;* C: *Methylobacter;* D, E: *Methylomonas;* F, G: *Methylococcus.*)

(d) some tRNA genes contain introns, (e) the aminoacyl stem of the initiator tRNA terminates with the base pair "AU", (f) the DNA dependent RNA polymerases are multicomponent enzymes and are insensitive to the antiobiotic rifampicin and streptolydigin, (g) like the alpha-DNA polymerases of eukaryotes, the replicating, archaeobacteria DNA polymerases are not inhibited by aphidicolin or butylphenyl-dGTP, and (h) protein synthesis is inhibited by anisomycin but not by chloramphenicol.

Autotrophic archaebacteria do not assimilate carbon dioxide via the Calvin cycle. In *Methanobacterium*, $CO_2$ is fixed via an acety-CoA pathway, whereas in *Acidianus* and *Thermoproteus*, autotrophic $CO_2$ is fixed via a reductive tricarboxylic acid pathway. Fixation of $N_2$ has been demonstrated by some methanogens.

Gram stain results may be positive or negative within the same order because of very different types of cell envelopes. Gram-positive species possess pseudomurein, methanochondroitin and heteropolysaccharide cell walls; Gram-negative cells have (glyco-) protein surface layers. The cells may have a diversity of shapes, including spherical, spiral, plate- or rod-shaped; unicellular and multicellular forms in filaments or aggregates also occur. The diameter of an individual cell may be 0.1 – >15 μm and the length of the filaments can be upto 200 μm. Multiplication is by binary fission, budding, constriction, fragmentation or unknown mechanisms. Colour of cell masses may be red, purple, pink, orange-brown, yellow, green, greenish black, grey or white.

The major groups of archaebacteria include (a) the methanogenic archaebacteria, (b) the archaebacterial sulfate reducers, (c) the extremely halophilic archaebacteria, (d) the cell-wall-less archaebacteria and (e) the extremely thermophilic S-metabolizers (Fig. 4.7).

## 4.9 CYANOBACTERIA

The cyanobacteria are the largest and most diverse group of photosynthetic bacteria. Older classification has as many as 2000 or more species. In one recent system this has been reduced to 62 species and 24 genera. Section 19 of the current edition of *Bergey's Manual of Determinative Bacteriology* describes 34 genera. Although cyanobacteria are true prokaryotes, their photosynthetic system resembles that of the eukaryotes because they have chlorophyll and a photosystem II and carry out oxygenic photosynthesis.

These organisms use phycobili proteins as accessory pigments. Photosynthetic pigments and electron transport chain compounds are located in thylakoid membrane lined with particles called phycobilisomes. These contain phycobilin pigments, particularly phycocyanin and transfer energy to photosystem II. Carbon dioxide is assimilated through the Calvin cycle, and the reserve carbohydrate is glycogen. Since cyanobacteria lack the enzyme alpha-ketoglutarate dehydrogenases, they do not have a fully functional citric acid cycle. Cyanobacteria vary greatly in shape (1-10 mm dia) and exist as cellular forms, as colonies of many shapes or form filaments called trichomes.

A trichome is a row of bacterial cells that are in close contact with one another over a large area. Although most appear blue-green because of phycocyanin, a few are red or brown in colour because of the red pigment phycoerythrin. Despite this variety, Cyanobacteria have typical prokaryotic cell structures and a nodal Gram-negative type cell wall. They often use gas vesicles to move vertically in the water and many filamentous species have gliding motility.

Cyanobacteria reproduce by binary fission, budding, fragmentation and multiple fission. In this a cell enlarges and then divides several times to produce many smaller progeny, which are released upon the rupture of the parental cell. Fragmentation of filamentous cyanobacteria can generate small, motile filaments called hormogonia. Some species develop akinetes, specialized, dormant, thick-walled resting cells that are resistant to desiccation. These often germinate to form new filaments (Fig. 4.8). Many filamentous cyanobacteria fix atmospheric nitrogen by means of special cells called heterocysts. Around 5 to 10% of the cells can develop into heterocysts when preferred nitrogen sources, and while transforming themselves into heterocysts, cyanobacterial cells synthesize a very thick new wall, reorganize their photosystem II, and synthesize the nitrogen-fixing enzyme nitrogenase. Photosystem II is absent. This inability to generate $O_2$ is critical because the nitrogenase is extremely oxygen sensitive. It obtains nutrients from adjacent vegetative cells and contributes to fixed nitrogen in the form of the amino acid glutamine.

Cyanobacteria are highly tolerant of environmental extremes and are present in almost all waters and soils. Thermophilic species may grow at temperatures up to 75°C in neutral to alkaline hot springs. Some unicellular forms even grow in the rocks. In nutrient-rich warm ponds and lakes, surface cyanobacteria such as *Anacystis* and *Anabaena* can reproduce rapidly to form blooms. The release of large amounts of organic matter upon the death of the bloom, microorganisms stimulate the growth of chemo-heterotrophic bacteria that subsequently depletes the availabe oxygen.

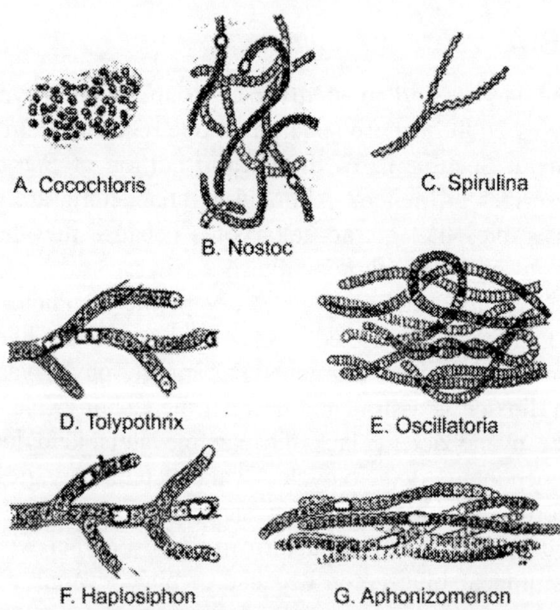


**Fig. 4.8.** Some examples of cyanobacteria. (A: *Cocochloris;* B: *Nostoc;* C: *Spirulina;* D: *Tolypothrix;* E: *Oscillatoria;* F: *Haplosiphon;* G: *Aphonizomenon*).

**Table 4.3.** Characteristics of the cyanobacterial orders

| Order | Structure | Reproduction | Heterocysts | G+C % | Motility | Examples |
|---|---|---|---|---|---|---|
| *Chroococcales* | Unicellular rods or cocci; nonfilamentous aggregates | Binary fission, budding | – | 40–49 | Almost always non-motile | *Chamaesiphon Gleocapsa* |
| *Pleurocapsales* | Unicellular rod or cocci; May be held together in aggregates | Multiple fission to form baeocytes | – | 40–46 | Only some baeocytes are motile | *Pleurocapsa Dermocarpa* |
| *Oscillatoriales* | Filamentous, unbranched trichome may contain specialized structures | Binary fission in a single plane, fragmentation | – | 40–67 | Usually motile | *Lyngbya, Oscillatoria Spirulina* |
| *Nostocales* | Filamentous, unbranched trichome with only vegetative cells | Binary fission in a single plane fragments of form hormogonia | + | 38–47 | Often motile, may produce akinetes | *Anabaena, Cylindrospermum, Nostoc* |
| *Siigonematales* | Filamentous trichomes either with branches or composed of more than one row of cells | Binary fission in more than one plane, hormogonia formed | + | 42–46 | May produce akinetes complexity and differentiation in cyanobacteria | *Fischerella, Stigonema* |

Some species that grow in water such as *Microcystis aeruginosa,* produce toxins which may be toxic to animals that drink such water. Some algae like *Oscillatoria* species are resistant to pollution.

Cyanobacteria are particularly successful in establishing symbiotic relationships with other organisms. They are photosynthetic partners in most lichen formations. Cyanobacteria are symbionts with fungi and nitrogen-fixer species that form associations with a variety of plants: (liver-wort) mosses, gymnosperms and angiosperms.

## 4.10 PROTOZOA

The protozoa are a diverse group of invertebrate (non-chordata) organisms. There are over 30,000 known species and they are found in all environments where usually water is present. Each protozoan functions as an independent unit and is able to perform effectively all the activities necessary for life. However, protozoa are sometimes separated from animal kingdom and are placed in (protoctista–Protista) with some algae. Protozoans possess numerous animal features, hence they are included in animal kingdom.

The general characteristics of Protozoa are as follows (Table 4.4):

1. Small, usually one-celled, some in colonies of few to many similar individuals, symmetry none, bilateral, radial or spherical.
2. Cell form usually constant, oval, elongate spherical, or otherwise varied in some species and changing with environment or age in many.
3. Nucleus distinct, single or multiple, other structural parts are such as organelles, no organs or tissues.
4. Locomotion by flagella, pseudopodia, cilia or movements of cell itself.
5. Some species have protective housing or tests; many species produce resistant cysts or spores to survive unfavourable conditions or for dispersal.

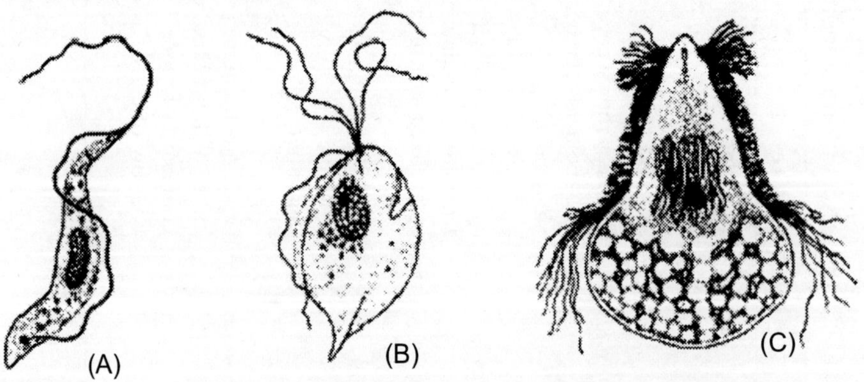


**Fig. 4.9.** Protozoans (A: *Trypanosome;* B: *Trichomonas;* C: *Trichonympha*).

**Table 4.4.** Characteristics of different Classes of Protozoa

| Mastigophora | Sarcodina | Sporozoa | Ciliata | Suctoria |
|---|---|---|---|---|
| Some possess chromotophores, others do not. | With pseudopodia for locomotion | No locomotor organs or contractile vacuoles | Cilia usually present throughout | Cilia found only in adult attached by stalk |
| Semi-rigid cell, covering is a pellicle | No chromatophore | Asexual reproduction by multiple fission or sexual phases, usually producing spores | Macro- and micro-nucleus | With tentacles for feeding |
| Adult movement by one or more flagella | Variable shape, one nucleus | All internal parasites usually with intracellular stages. | | No cytosome |
| One nucleus | Asexual reproduction by binary fission | | | Mostly sessile in salt or fresh water |
| Asexual reproduction by longitudinal or binary fission, multiple fission, cyst e.g., *Euglena, Trypanosoma* | Many sporulate e.g., *Amoeba Arcella, Polystomella* | e.g., *Monocystis, Plasmodium, Noisema* | e.g., *Opalina, Paramaecium,* | e.g., *Podophyra, Trichophyla* |

6. Mode of life: free-living, commensal, mutualistic or parasitic.
7. Nutrition: *Holozoic* – living on other organisms, *saprophytic*, living on dissolved substances in their surroundings, *saprozoic* – living on dead animal matter and *holophytic or autotrophic* – producing food by photosynthesis.
8. Asexual reproduction is by binary fission, multiple fission or budding in some and sexual reproduction is by fusion of gametes or by conjugation.

The protozoa are divided into five classes according to the locomotion structures they possess.

1. *Mastigophora or Flagellates* with one or more whip-like flagella.
2. *Sarcodina or Rhizopoda* with pseudopodia.
3. *Sporozoa* with no locomotor organelles.
4. *Ciliata or ciliates* with cilia throughout life.
5. *Suctoria* with cilia in young and tentacles in adult stages.

The extreme age of phylum is evident by finding the hard remains of *Radiolaria* and *Foraminifera* in precambrian rock. Many of the flagellates of mastigophora have cells embedded in a common gelatinous matrix and show physiological coordination between the individuals. Some are connected to one another by protoplasmic threads, and in *Volvox* there is differentiation into vegetative and reproductive cells. This condition is parallel to the *formation of tissues* and segregation of somatic and germ cells in Metazoa. Certain protozoans are ordinarily free-living and are occasionally found living within the bodies of other animals, showing derivations of parasitic forms. Some chlorophyll bearing flagellates resemble the green algae in structure and physiology and suggest common origin for plants and animals. The mastigophora are probably the most primitive, the ciliate most specialized and sporozoa are probably simplified as a result of their strictly parasitic manner of life.

## 4.11 ACTINOMYCETES

The members of this large and important group are mainly soil inhabitants. They were regarded as intermediate forms between bacteria and fungi on account of their mycelial structure, but are now grouped with eubacteria because of their prokaryotic nature and cell wall composition. When growing on agar media, they produce mycelium, which is partly aerial and partly submerged in the medium, formed from much branched but thin hyphae. Most members also produce single-celled spores. Thus they resemble fungal colonies in culture. In liquid media also they form individual colonies and do not make the medium turbid like other bacteria. The spores are formed either externally as conidia or inside sporangia as sporangiospores. The hyphae usually divide into long cells with more than one nucleoid in them. The spores are easily detached and disbursed. Many actinomycetes produce pigments of various shades. The cell walls are chemically distinguishable into four types. While most actinomycetes are saprophytes, a few of them are pathogens also. They are classified on the basis of mycelial and spore morphology, pigments, secondary metabolites and cell wall features. Representative genera are illustrated in Figs. 4.10, and 4.11.

Actinomycetes, particularly the genus *Streptomyces,* are economically important. Majority of the clinically useful antibiotics are produced by actinomycetes. Taxonomically, the following six groups are recognized:

*Nocardioforms* (e.g. *Nocardia, Faenia*). They develop substrate mycelium, which easily breaks up into individual cells. They are mostly saprophytes in soil and water but *Nocardia asteroids* causes a lung disease in humans and animals of low resistance.

*Actinomycetes with multilocular sporangia.* This section includes *Frankia* and *Geodermatophilas*. The hypha divides in two planes to produce clusters of spores. *Frankia* forms root nodules in non-leguminous trees, in symbiotic association. The spores of *Geodermatophilus* are motile.

*Actinoplanates.* In this group, extensive substrate mycelia are formed but aerial mycelium is absent or rudimentary, with sporangiospores rising above the agar surface. The spores are arranged in coils in parallel rows in the sporangium. (e.g. *Actinoplanes*). *Micromonospora* produces single sporangiospores on branched sporophores.

*Streptomyces and related forms. Streptomyces* comprises a large genus with nearly 380 species. They are aerobic, produce aerial and subtrate mycelia and chains of non-motile spores (conidia) enclosed in a thin fibrous sheath. Different species are distinguished on the basis of morphological and biochemical features such as colour of mycelium, arrangement of conidiophores and conidia, spore surface characters, utilization of various sugars, nitrate reduction, melanin production, antibiotic production etc. Soil is their natural habitat and they occur in large numbers, especially in tropical soils as they can decompose a large variety of organic substrates and also withstand desiccation. The characteristic earthy smell that soil gives out after a shower is due to volatile substances (geosmin, etc.) that they produce. Waksman (1952) was awarded Nobel Prize for his discovery of streptomycin from *Streptomyces.*

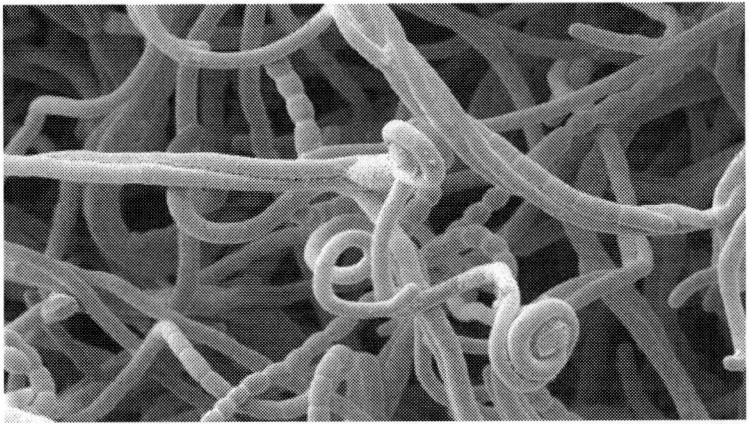

**Fig. 4.10.** Aerial mycelium of *Streptomyces* illustrating the arrangement of hyphae (scaning image).

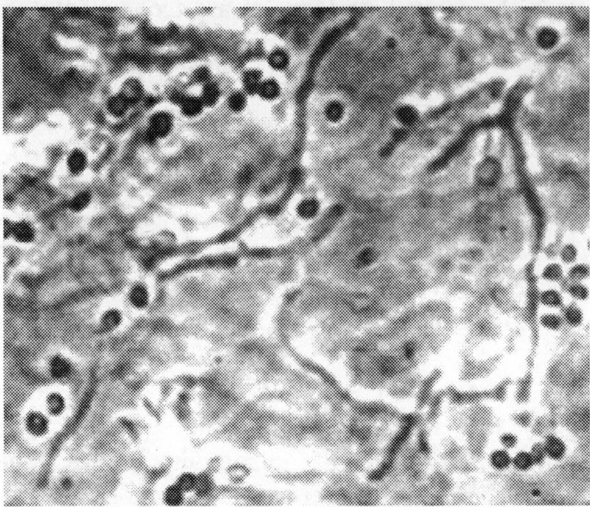

**Fig. 4.11.** Spores of Streptomyces sp.

*Maduromycetes.* These are sporangia-forming genera producing a characteristic sugar madurose in the cell. *Streptosporangium* produces an antibiotic and *Actinomadura* causes a disease, actinomycetoma.

*Thermomonospora and related genera.* These genera are morphologically variable but are heat tolerant. Conidia are produced either on single hyphae or on compact clusters of hyphae. *Thermomonospora* occurs in dung heaps, compost pits etc. where high temperatures develop. *Thermoactinomycetes* develop heat-resistant endospores.

## 4.12 RICKETTSIAE

These are spherical or rod-shaped gram negative cells occurring as intracellular parasites. They have a cell and plasma membrane. The cells are small and range from 0.3 to 5 mm in diameter and 0.8 to 2.0 mm in length. They have both DNA and RNA. Cells divide by binary-fission like in bacteria. Rickettsias are intracellular prarasites that replicate in the host cytoplasm. Rocky mountain spotted fever and Q-fever are the examples of the diseases caused in humans and are transmitted by arthropod vectors.

## 4.13 PRIONS

Prions are the simplest living entities made up of only a small protein but are capable of infection and replication. Prions modify a host protein to replicate the protein molecule. The prions cause few diseases in animals and man. Scrapie disease in sheep, kuru disease and mad cow disease in humans are examples (Fig. 4.12).

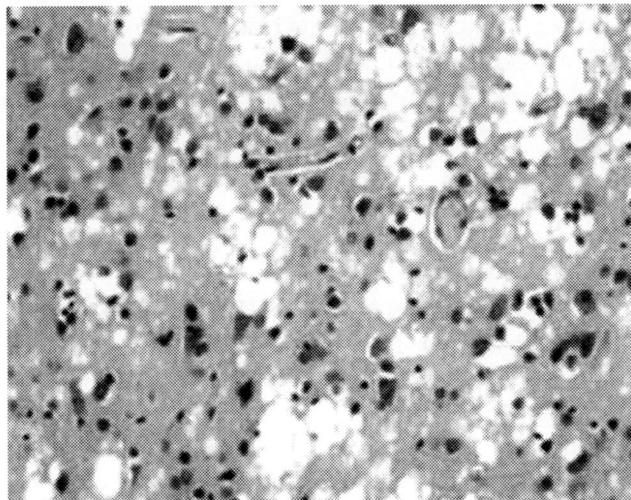

**Fig. 4.12.** Structure of prion, causative agent of mad cow disease.

## 4.14 VIROIDS

Viroids are infectious agents containing single-stranded RNA molecule and are without a protein coat. They do not have a definite shape. Cadang-Cadang disease is caused by them. The RNA molecule is in the range of 240–375 nucleotides and shows a circular configuration. The viroid molecule is known to replicate in the host nucleus utilizing the host RNA polymerase.

## 4.15 CHLAMYDIAS

The psittacosis group, which comprises the genus *Chlamydia*, probably represents a further stage in degenerate evolution. The chlamydias are obligate parasites in which there has been an even greater loss of metabolic functions. In fact, for many years the chlamydias were considered to be large viruses rather than bacteria, and only since the nature of virus replication has been well understood that the bacterial nature of the chlamydias has been firmly established. Many chlamydias are smaller than some of the true viruses, such as the smallpox virus, but the chlamydias divide by binary fission and do not replicate in the manner viruses do. Of the diseases caused by chlamydias, pisttacosis, an epidemic disease of birds that is transmitted occasionally to humans, is the most important, although a number of other diseases of birds and humans and other mammals are known. The chlamydias are of most interest, however, they are intriguing because of the biological and evolutionary problems they pose. The bacterial nature of the chlamydias was first suspected when it was discovered that, unlike viruses, they were susceptible to penicillin and other antibiotics whose action is restricted to the bacteria. When the specificity of penicillin for the bacterial cell wall was understood, the apparent viral nature of the bacterial cell wall was understood and the apparent viral nature of the chlamydias was further refuted. Biochemical studies showed that the chlamydias have typical bacterial cell walls as they have both DNA and RNA. Electron microscopy of this section of infected cells show forms that are clearly undergoing binary fission. The biosynthetic capacities of the chlamydias are very restricted, however, even more so than the rickettsias. This raises the interesting question of the limits to which evolutionary loss of function is still retained.

Two cellular types are seen in a typical life cycle—a small dense cell, which is relatively resistant to drying and is the means of dispersal of the agent, and a larger less dense cell, which divides by binary fission and is the vegetative form. The chlamydias are not transmitted by arthropods but are primarily air-borne invaders of the respiratory system—hence the significance of resistance to frying of the small cells. When a virus infects a cell, it loses its structural integrity and liberates nucleic acid. When chlamydia enters a cell, however, it changes in form, it remains as a structural unit and enlarges and undergoes binary fission. After a number of divisions, the vegetative cells are converted into small, dense cells that are released when the host cell disintegrates and then infect other cells. Generation times of 2 to 3 hours have been reported, which are considerably faster than those found for the rickettsias (Fig. 4.13).

Chlamydia cells have a chemical composition similar to that of other bacteria. Both RNA and DNA are present in approximately equal proportions. The DNA content of a chlamydia corresponds to a molecular weight of approximately $4 \times 10^8$, twice that of vaccinia virus and one-tenth that of *Escherichia coli.* The base composition is about 29 percent G–C. At least

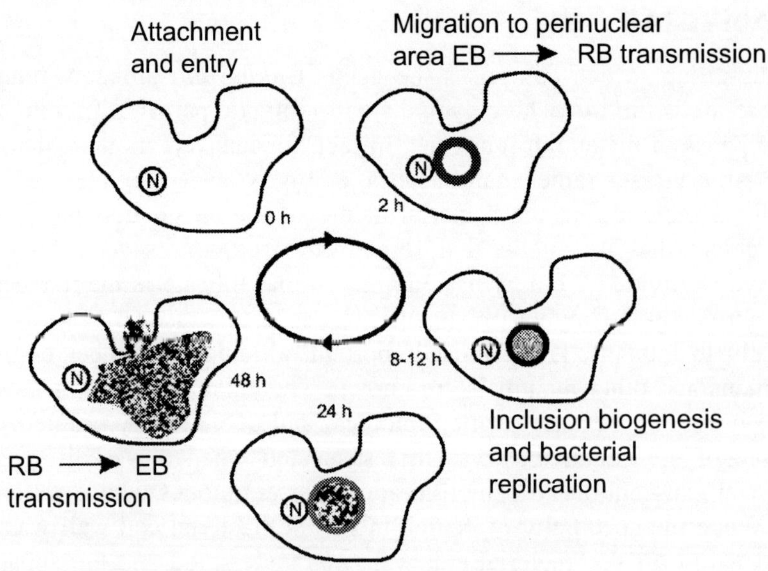


**Fig. 4.13.** Infectious cycle of Chlamydia.

some of the RNA is in the form of ribosomes, and like those of other prokaryotes, the ribosomes are 70-S particles composed of one 50-S and one 30-S unit.

The small, dense chlamydial form has a typical rigid bacterial cell wall and muramic acid has been shown to be present. When the small form reorganizes to the large form, it loses its cell-wall rigidity and there is some evidence that the cross-linking of the peptidoglycan is reduced in extent.

# Fungi

## 5.1 INTRODUCTION

Fungi are achlorophyllous living organisms. These are heterotrophic and absorptive in nutrition. The absence of chlorophyll has enforced fungi either to become parasitic or saprophytic. Parasitic fungi like rusts, smuts and powdery mildews live on living hosts. Thallus is holocarpic (single celled) or eucarpic (multicellular). Cell wall is typically chitinised. Life cycle is simple to complex. Reproduction is by means of asexual and sexual methods. Fungi occur in diversified habitats and are cosmopolitan in distribution. Some of the skin disease causing fungi are called dermatophytes, which are parasitic on human beings and animals. In the absence of particular host, the fungus specific to that cannot live and multiply. There are some fungi like *Phytophthora, Fusarium,* which cause late blight of potato disease and wilt disease on crop plants, respectively and also *Pythium* causing damping off disease in castor seedlings. Many other such fungi live for some time as parasites, and as soon as host plant dies, the fungus becomes saprophyte. Such fungi are known as facultative parasites. Some of these fungi can be grown artificially under laboratory conditions in test tubes using specific media. However, there are some fungi that cannot be grown artificially in test tubes and always need living hosts. These are called obligate parasites (e.g., *Albugo, Puccinia, Erysiphae*). Fungi that grow luxuriantly on dead organic matter of plants and animal origin are called saprophytes (e.g., *Aspergillus, Penicillium, Mucor, Rhizopus*).

Some fungi whose mycelium gets associated with plant roots are called mycorrhiza (*Myco*-Fungus, *rrhiza*-root). Fungi cannot prepare their own food, hence they get carbohydrates from plants and in turn fungi mobilize phosphorus and nutrients through their hyphae and transport them to the host tissue. Such mutualistic association is called symbiosis. They are broadly divided into ectomycorrhiza and endomycorrhiza. Ectomycorrhiza form sheath-like structures on roots of woody trees (e.g., *Pisolithus* on *Casuarina, Eucalyptus*). These fungi are of *Agaricus* type and are grown on agar media. They are grown in fermenters and multiplied. Endomycorrhiza are those that form hyphae, vesicles and arbuscules inside the host root cortex and some hyphae bearing spores and sporocarps are formed outside the root. These fungi are also called Vesicular Arbuscular Mycorrhizal Fungi (VAM) or Arbuscular Mycorrhizal Fungi (AM). These fungi are obligate symbionts and cannot be multiplied in axenic culture, however, they can be grown in soil-pot culture. Mycorrhizal fungi are used as biofertilizers, which increases plant growth and productivity.

Fungi degrade any complex organic matter and ultimately convert it into glucose. This is called absorptive nutrition. Such kind of nutrition is not present in any form of life. Further ingestion is very rare.

The body or thallus of fungi is either holocarpic or eucarpic. The thallus does not show differentiation into vegetative and reproductive stages. The thallus is usually unicellular, which does both the functions—vegetative and reproductive e.g., yeast. In some fungi, there may be rhizoids at the base of single-celled thallus e.g., chytrids. Thalli of this kind are called holocarpic (holo—complete, carpic—body). Fungi such as *Rhizopus, Aspergillus Penicillium* and *Agaricus* are multicellular and different parts of thallus perform different functions. For example, in *Rhizopus*, rhizoids get hold on the substratum, hyphae absorb nutrients and sporangia release sporangiospores, which multiply the fungus asexually, and zygospores are produced sexually either through homothallic or heterothallic gametangial copulation. In *Aspergillus* and other such fungi the hyphae are septate, while in *Rhizopus*, the hyphae are tube-like without cross walls and multinucleate. Such thalli are termed eucarpic (eu—true, carpic—body). In some fungi like yeasts, which are also pathogenic on animals, both unicellular and multicellular thalli are observed. Such thalli are called dimorphic.

The most important character that distinguishes several groups of fungi is the septum. In lower fungi such as chytrids, the thallus is unicellular. In *Rhizopus* the hyphae are tubular, multinucleate and without septa, called coenocytic. In *Aspergillus*, the cross walls are present. Such hyphae are called septate hyphae (Fig. 5.1). The members of Ascomycetes, Basidiomycetes and Deuteromycetes have septate hyphae. In septate hyphal forms, the hyphal segments may contain one, two or more nuclei. When the nuclei are genetically identical, it is said to be homokaryotic. If mycelium contain nuclei of different genotypes, it is said to be heterokaryotic. Sometimes in certain hyphal cells of higher fungi, genetically distinct nuclei are present, hence they are called dikaryotic.

Cell wall is well defined, typically chitinised. The hyphae assumes toughness due to the turgor of protoplast it contains. The changes in thallus form are due to wall structure. Chitin is the

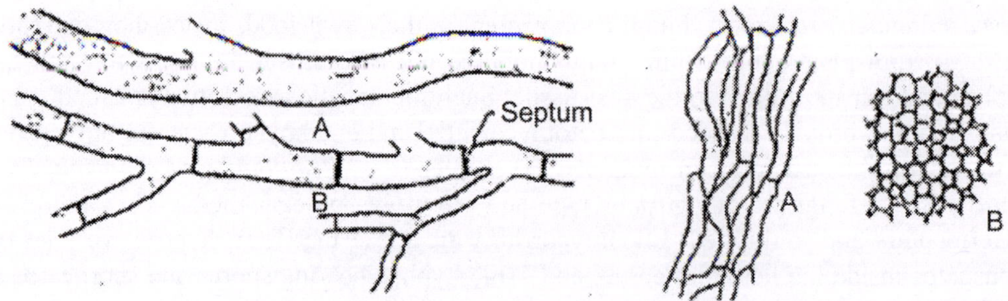


**Fig. 5.1.** A: Coenocytic hyphae; B: Septate hyphae.

most usual cell wall component in fungi along with glucans. Chitin is a polymer of N-acetyl-glucosamine. In *Pythium* and related fungi of Oomycetes, the cell wall is made up of cellulose along with glucans. Mannan and glucans are present in the cell wall of yeast (*Saccharomyces*). In fungi like *Agaricus,* dolipore septum is present. Through this septum only cytoplasm can flow, but not nuclei. The dolipore septum has small pore with parenthosomes (swollen structures) on either side.

Many fungi are not seen with naked eye but can be seen with the help of a microscope. These are bigger than bacteria and are called microfungi. Fungi like mushrooms, toad stools, earth stars etc. are seen with naked eye and are called macroscopic fungi.

The cells of the fungi are true nucleate (Eukaryote) and contain all familiar cell organelles except chloroplast. The nucleus is surrounded by a double membrane and is in continuity with endoplasmic reticulum. Numerous pores are present on the nuclear membrane, which allow interchange of materials from cytoplasm to nucleus and vice-versa. During nuclear division nuclear membrane remains intact—a peculiar character of fungi. This is called Karyochoresis. Mitochondria are usually elongate, large and move in cytoplasm. Endoplasmic reticulum is smooth or rough and is concerned with protein synthesis. Cytoplasmic microtubules are reported, which help in the movement of cytoplasm and maintenance of cell shape. Among eukaryotes, fungi have small nuclei (Fig. 5.2).

Golgi apparatus or dictyosomes, which have stacks of folded membranes, are present and help in the secretion of lipid droplets and glycogen (storage product). The surface of the fungal protoplast is the plasmalemma containing lipoprotein. Vacuoles are present and are delimited by tonoplast.

Growth of the hyphae is apical, and as hyphae extends, rapid synthesis of new wall components and membrane components takes place. Vacuolation is more in growing apices. Chito some has been identified as the precursor of chitin. Lomasomes are reported in some fungi. Normally, the growth of the fungal hyphae is monopodial. Apical growth and chemotropic growth are reported in fungi.

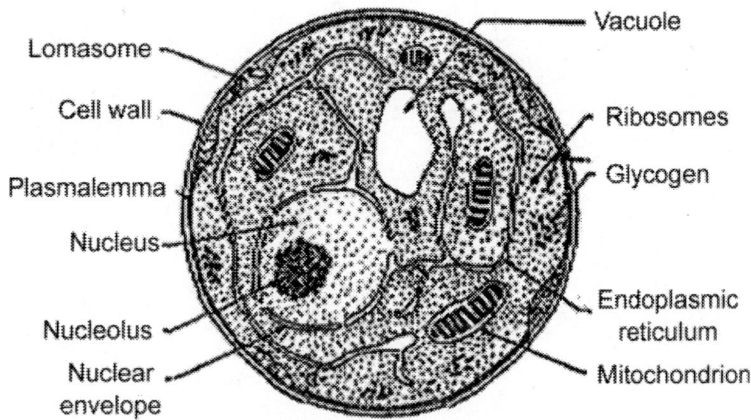


**Fig. 5.2.** Fungal cell.

Some macroscopic fungal structures are formed due to hyphal aggregation/grouping. Aggregation of vegetative hyphae in the form of strands is called rhizomorph. Thick-walled vegetative fruit bodies of pin head size containing reserved food materials called sclerotia are formed. The fruit bodies formed in larger ascomycetes and basidiomycetes are made up of pseudoparenchymatous and prosenchymatous hyphal aggregation.

Reproduction is a means of multiplication and continued sustenance of hereditary material from one generation to other. Fungi reproduce vegetatively, asexually and sexually. Vegetative reproduction is by means of rhizomorphs, sclerotia, thick-walled spores (chlamydospores) and by the fragmentation of hyphae. There are no definite sex organs.

Asexual reproduction is by means of motile spores (zoospores) and non-motile spores (sporangiospores and conidia) (Fig. 5.3).

All these reproductive units are produced through mitosis, hence called mitospores. These spores are produced in large number, but many of them vanish or die. The survived ones germinate and continue the generation. Asexual reproduction is a means of multiplication where nuclear fusion and meiosis occur; reproductive propagules are haploid.

Motile mitospores are called zoospores. Zoospores are produced inside a sac-like structure called zoosporangium. Zoosporangia are of different shapes and sizes. They are mostly cylindrical or oval. Many sporangia are produced inside one sporangium called sporangial proliferation e.g., *Saprolegnia*.

The zoospores may also show more than one encystment and motile stage, hence known as polyplanetism. Zoospores that are produced in lower fungi are of three kinds:

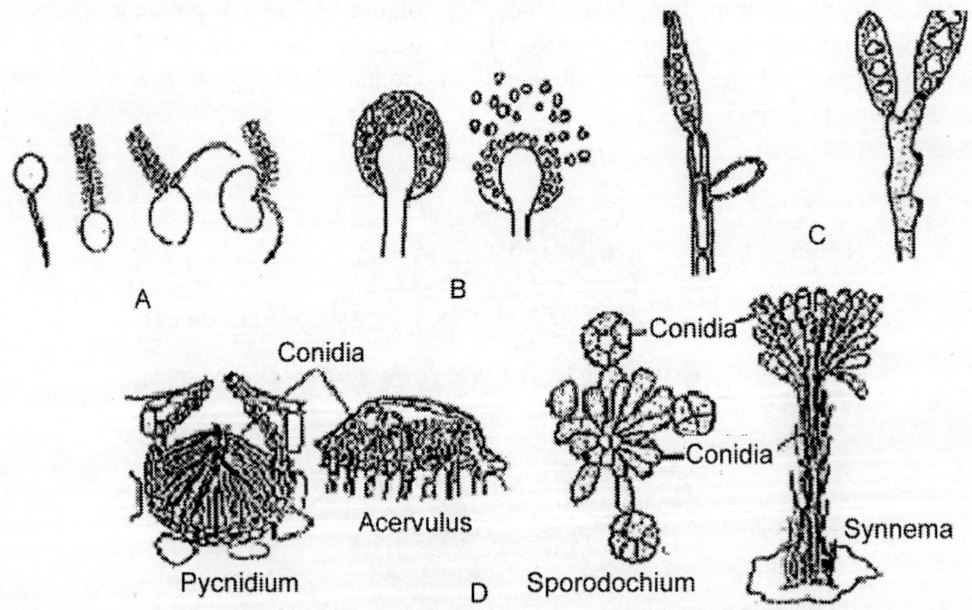


**Fig. 5.3.** A: Zoospores (1: Posteriorly uniflagellate (whiplash); 2: Anteriorly uniflagellate (tinsel); 3: Apically biflagellate; 4: Laterally biflagellate); B: Sporangiospores; C: Conidiophores and conidia.

1. Posteriorly uniflagellate zoospores with whiplash type flagella, characteristic of Chytridiomycetes.
2. Anteriorly uniflagellate zoospores with tinsel type flagella, characteristic of Hyphochytridiomycetes.
3. Biflagellate zoospores with anteriorly or laterally attached flagella, one is whiplash and the other is of tinsel type.

Zoospores are best studied by means of electron microscope in *Allomyces*. The zoospore is tadpole-like with a pear-shaped head, 17 × 9 μm, having a single trailing flagellum, 20 μm long. The conspicuous internal structure is the dense crescent-shaped nuclear cap, which surrounds the more transparent nucleus. The nuclear cap is rich in ribonucleic acid and protein and is filled with ribosome. In some fungi, a large mitochondrion may be present besides other structures like gamma particles, lipid granules, microtubules etc. The flagellum usually orginates from a basal body or kinetosome, which is called blepharoplast. Each flagellum has a main shaft called axoneme, which continues as a ring of nine doublet microtubules surrounding the central pair of single microtubules. The whole 9 + 2 microtubule is surrounded by a flagellar sheath. The flagellum helps in forward and backward movement. Usually the lower fungi such as *Allomyces, Achlya, Saprolegnia, Chytrids, Pythium, Phytophthora,* which live in water or humid soils, reproduce through zoospores. In case of tinsel type, the axoneme of the flagellum bears a series of fine cilia called mastigonemes along the whole of its length.

The second type of asexual spores that are non-motile but are produced inside sporangia as mitospores are termed sporangiospores e.g. *Rhizopus, Mucor.*

The third type of asexual spores which are not motile but are produced on fertile hypha-like structures called conidiophores. These are mitospores produced exogenously termed conidia. Conidia are produced either by the breaking of thallus (thallic) or as blown out structures like balloon (blastospores). Conidia are oval, spiral, cylindrical, elliptical, lunar, crescent-shaped, etc. Conidia may be coloured or colourless (Fig. 5.4). Conidia may be uniseptate or multiseptate. Transversely or longitudinally/obliquely arranged septa are present.

Sexual reproduction in fungi involves:

1. Plasmogamy—Fusion of cytoplasm
2. Karyogamy—Fusion of nuclei
3. Meiosis—Reduction division

In fungi like *Pythium* and *Phytophthora,* the reduction division takes place immediately after nuclear fusion. In some fungi like *Emericella,* belonging to Ascomycotina and in Basidiomycotina (*Puccinia, Agaricus*), the plasmogamy is followed by pairing of nuclei. The pairing of nuclei leads to dikaryotization. Karyogamy and meiosis gets delayed.

Compatibility is the main factor in fungi. Some fungi are homothallic in which the same thallus produces sexual spores and zygote. In heterothallic species, mating of two different (+ and –) strains is essential. The sex hormones like serinene in *Allomyces*, oogonial and antheridial in *Achlya* play an important role. Trisporic acid in *Rhizopus* has been very much confirmed with reference to heterothallic species.

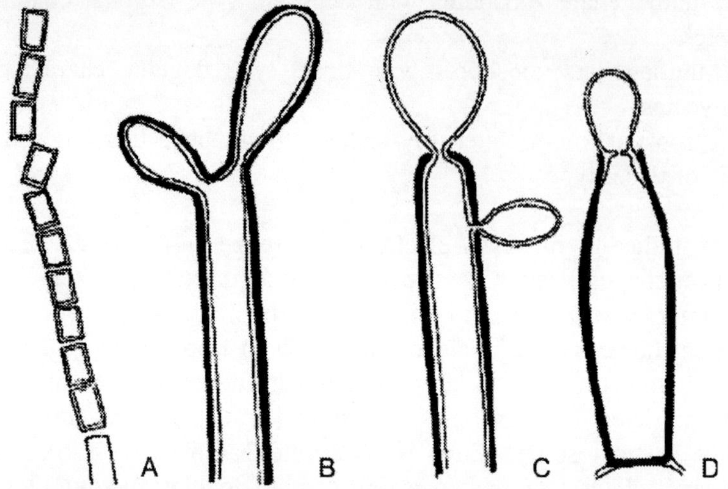


**Fig. 5.4.** Development of conidium. A: Thallic; B, C: Blastic; D: Phialidic.

As a result of sexual reproduction, oospores, zygospores, ascospores and basidiospores are produced. Oospores and zygospores are diploid spores, which undergo meiosis followed by mitosis and release zoospores or sporangiospores on germination. Ascospores and basidiospores are the haploid spores.

The following are the modes of sexual reproduction that occur in fungi (Fig. 5.5):

1. Planogametic copulation: The motile gametes of +/– strain are released and fuse to form motile zygote, e.g., chytrids.
2. Gametangial contact: Definite sex organs, namely oogonium and antheridium (Ascogonium and Antheridium) are produced. Sex organs come closer and plasmogamy occurs through a pore or fertilization tube followed by nuclear fusion and oospore formation (*Pythium*). In some, plasmogamy is followed by dikaryotization and delayed karyogamy and meiosis resulting in ascospore or basidiospore production, which are haploid.
3. Gametangial copulation: There are no definite sex organs. Hyphae get differentiated into + or – strains. Gametangia are formed followed by their fusion. Zygospore formation is by the fusion of a part of gamentangium of + and – strains, e.g., *Rhizopus*.
4. Spermatization: In this method, sterile hyphae act as trichogynes or receptive (+ strain), while microconidia act as male reproductive units (- strain). Thus male reproductive unit falls on receptive hyphae of trichogyne, not of the same thallus but of other thallus resulting in plasmogamy and dikaryotization, and at last delayed karyogamy takes place e.g., *Neurospora, Puccinia*.
5. Somatogamy: There is no involvement of sex organs nor the production. Sexual reproduction takes place by the fusion of two somatic cells, which become reproductively potential cells, e.g., *Saccharomyces, Peziza, Agaricus*.

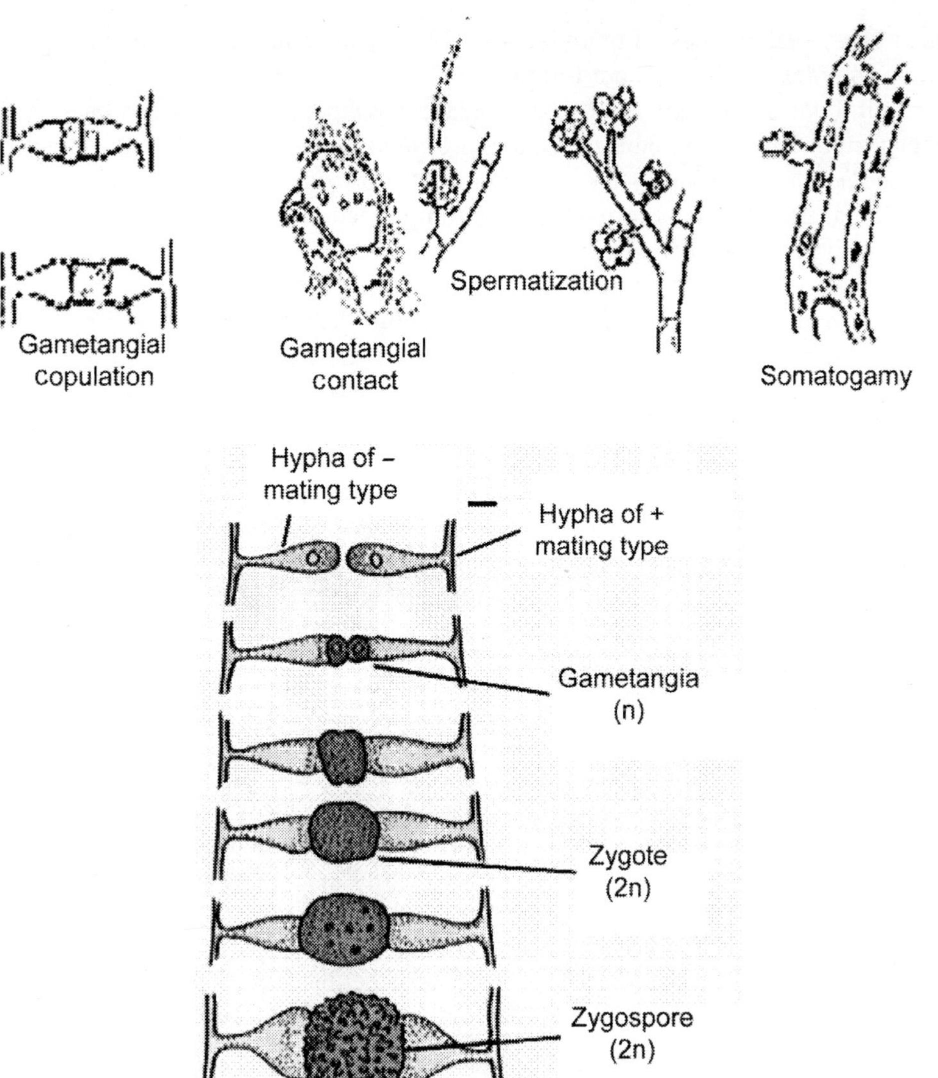


**Fig. 5.5.** Sexual reproduction in fungi.

Besides the above, there is an alternative sexual mechanism called parasexuality. In this method, partial genetic recombination takes place. In this method, heterokaryosis takes place followed by haplodization, diplodization and mitotic crossing over. Thus, new variants are produced in *Emericella nidulans*.

Further heterokaryosis, homothallism and hetrothallism add much more basic information in understanding fungal genetics. Fungal genetics are well understood through bipolar and tetrapolar heterothallism also.

As a result of asexual and sexual reproduction, not only the reproductive units are generated but also many complex fruit bodies are formed.

Asexual fruit bodies, namely Sporodochia (saucer shaped) in *Myrothecium,* Acervulus (saucer shaped fruit body with stroma and setae) in *Colletotricum,* Pycnidia (flask shaped body–cavity like) in *Phoma* and broom stick like synnemata formed in *Graphium* are some of the noteworthy asexual fructifications formed in Fungi imperfecti producing mitospores/conidia.

Sexual fruit bodies, like in *Gymnoascus*, the asci are covered with loosely interwoven hyphae. Closed, non-ostiolate, covered with peridium and having randomly distributed asci called cleistothecia are formed in *Eurotium* and *Emericella*. Hulle cells may be present or absent. In *Chaetomium*, the fruit body is a flask-shaped structure having palisadely arranged basal fasicle of asci with or without paraphyses and periphyse at the apical region of the ostiole.

Stromatic perithecia are produced in *Claviceps*. Non-ostiolate but rigid fruit bodies having thick peridium and bitunicate asci are produced in Loculoascomycetes (Pseudothecia), e.g., *Pleospora*. Open and saucer-shaped fruit body with stipe, excipulum, hymenium bearing uni- or mostly bitunicate asci are produced in Discomycetes (*Peziza*). Perithecia are produced by some fungi as exoparasites on insects in labouleniomycetes. Basidiospores are produced on basidia which may be non-septate or septate single celled basidiospores are produced on basidia located on fertile hyphae of the basidiospores.

In fungi, there are two kinds of life cycles:

1. *Sexual Cycle:* The events in the life cycle of a sexually reproducing fungus are nuclear fusion and meiosis (reduction division). In this there are haplobiontic 'A' life cycle (nuclear fusion–reduction-mitosis), haplobiontic 'B' life cycle (nuclear fusion–mitosis-reduction) and diplobiontic life cycle (Fusion-mitosis-reduction-mitosis).
2. *Non-Sexual Cycle:* In many fungi the sexual reproduction has not been observed and mitospores are produced.

**Classification:** Fungi do not have matching characters, neither with plants nor with animals, hence they are raised to a kingdom level—Mycota. Mycota is divided into two divisions—Myxomycota and Eumycota. The Eumycota has the following sub-divisions:

1. Motile spores present, perfect state spores typically oospores, motile cells absent *Mastigomycotina*
2. Perfect state present, Perfect state absent *Deuteromycotina*
3. Perfect state, Zygospores, Zygospores absent *Zygomycotina*
4. Perfect state, ascocarp present, unit of sexual reproduction, Ascospores *Ascomycotina*
5. Perfect state, basidiocarp and basidiospores present *Basidiomycotina*

# Eukaryotic Algae

## 6.1 INTRODUCTION

The algae are the simple chlorophyllous living lower plants, colonizing diversified habitats (water, rocks, humid soils, tree trunks etc). The presence of chlorophyll helps them in synthesizing their own food by means of photosynthesis. The thallus range from unicellular to multicellular organism. The study of algae is known as algalogy or phycology. The word *phykos* in Greek means seaweed.

Algae are known to occur in stagnant waters, lakes, rivers, sea water, thermal springs, ice, damp soil, rocks, tree trunks, old walls etc. Free floating algae in water are known as phytoplankton. Some of them live inside the tissue of other plants. These are known as endophytes. Few algae live on other plant parts, molluscan shells etc. These are called epiphytes. Algae like *Cephaleureus* is an angiospermic parasite. Some algae called symbionts live in mutualistic association with others.

Algae like *Chlamydomonas* and *Chlorella* are unicellular and generally spherical, oblong or pear-shaped. *Chlamydomonas* has two flagella, while *Chlorella* is non-motile. Alage, e.g., *Volvox* is a colonial form, ball-like with indefinite number of motile cells covered by a gelatinous sheath. In forms like *Gonium*, definite number of cells are embedded in a gelatinous matrix and are held together. This thallus is called coenobium. Filamentous habit is represented in many alga. Unbranched (*Zygnemia*), branched (*Cladophora*) and false branched (*Scytonema*) types are known. Thallus with parenchymatous structure or pseudoparenchymatous tissue are reported in *Enteromorpha* (Tubular parenchymatous); ribbon like-foliose parenchymatous thallus is observed in *Ulva*. Prostrate creeping system with a projecting or erecting system of the thallus, usually branched is exhibited in *Coleochaete* and *Stigeoclonium*. Coenocytic tubular multinucleate thallus, which is differentiated into rhizoidal system, rhizomatic system and erect system (e.g., *Caulerpa*) is reported.

Complex thallus with root-like, stem-like and leaf-like organs are present in thalli of *Laminaria* and *Polysiphonia* (Fig. 6.1). All algal cells are eukaryotic, but blue green algae (cyanophyceae) are prokaryotic.

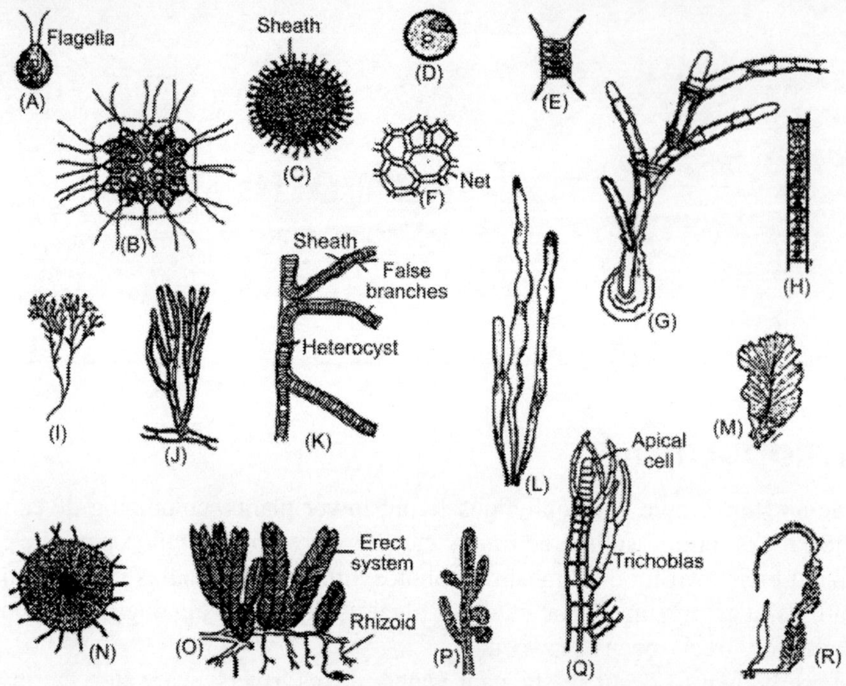


**Fig. 6.1.** Thallus structure of algae. A: Unicellular–*Chlamydomonas;* B: *Coenobium–Gonium;* C: Colonial–*Volvox;* D: *Chlorococcum;* E: *Scenedesmus;* F: *Hydrodictyon;* G: *Ecballocystopsis;* H: *Spirogyra;* I and J: *Cladophora;* K: *Scytonema;* L: *Enteromorpha;* M: *Ulva;* N: *Coleochaete;* O: *Caulerpa;* P: *Vaucheria;* Q: *Polysiphonia;* R: *Laminaria.*

Cell contains a gelationous sheath. Cell wall is a multilayered structure external to plasmalemma and internal to the sheath. The cell wall is made up of monopolymers containing muramic acid, glucoseamine, diaminopimelic acid, glutamic acid and alanine. Besides this, peptidoglycan, lipopolysaccharides and proteins are also present. Cyanophycean food granules, protein bodies, lipid, photosynthesis lamellae and lamellosome are suspended in the cytoplasm. Phycocyanins are the pigments present. DNA fibrils are located in the centre without nuclear membrane (Fig. 6.2).

In all other algae, cell wall is thin, transparent and consists of outer cellulose and inner pectic layer. The cytoplasm is granuled containing mitochondria, volutin granules, starch, golgi complex, chloroplast and pyreniods etc. The nucleus is eukaryotic (Fig. 6.2).

Reproduction is mainly of three types (Figs. 6.3 and 6.4).

### 6.1.1. Vegetative Reproduction

It is carried out by fragmentation in which the thallus breaks accidentally or by means of separation disc-like cells (*Oscillatoria*). Cell division is a common mode of reproduction.

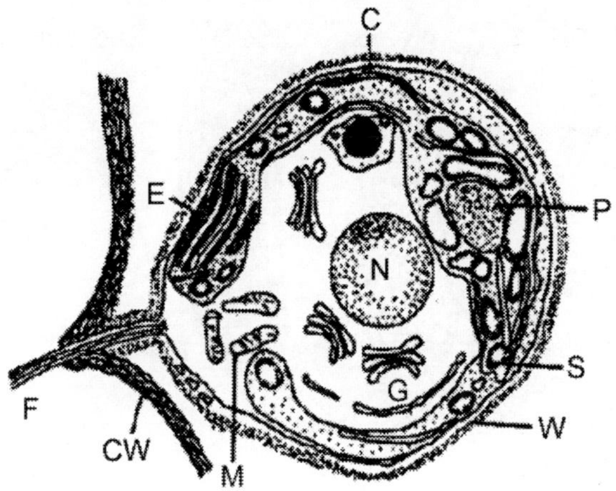


**Fig. 6.2.** Structure of algal vegetative cell (*Volvox*). F: Flagellum; CW: Colony wall; W: Cell wall; S: Starch; P: Pyrenoid; C: Chloroplast; E: Eyespot; N: Nucleus; M: Mitochondria; G: Golgi body.

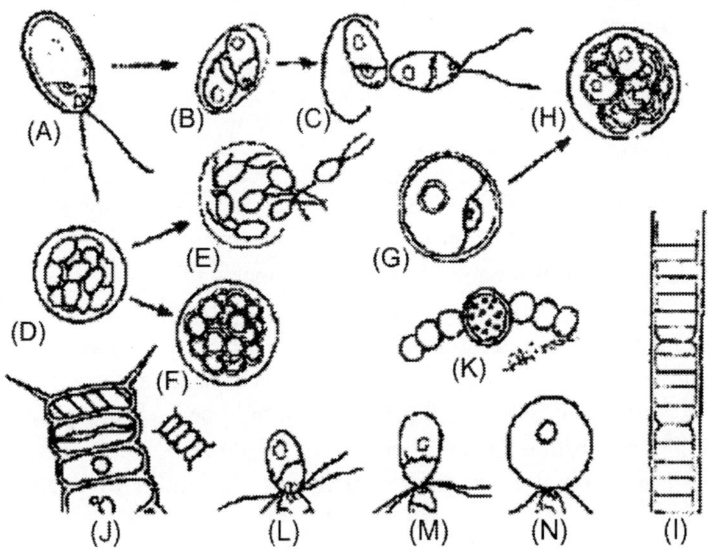


**Fig. 6.3.** Methods of reproduction in algae. A–C: Binary fission; D–E: Zoospore formation; F: Aplanospore formation; G–H: Autospore formation; I: Fragmentation or formation of hormogonium; J: Autocolony formation; K: Akinete formation; L: Isogamy; M: Anisogamy; N: Oogamy.

**(a) Asexual reproduction** takes place by means of Akinetes (thick-walled)—cells resistant to unfavourable conditions, which germinate with the onset of favourable conditions, e.g., *Pithophora*. The flagellated zoospores are produced inside the zoosporangium. Zoospores

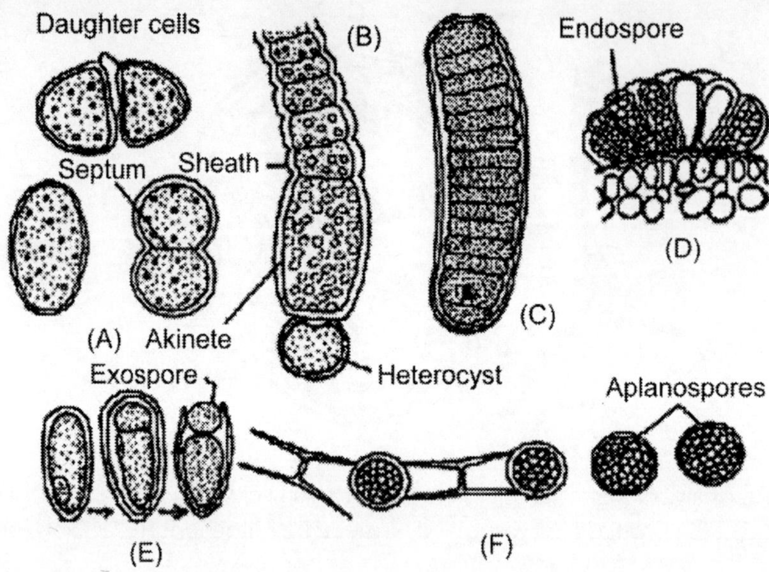


**Fig. 6.4.** Spore types and asexual reproduction in algae. A: Simple division in unicellular *Synecococcus;* B: Filament of *Gleotrichia natans* with heterocyst and akinete; C: Homospore of *Westiella;* D: Endosporess of *Thermocapsa prasina;* E: Exospore of *Chamaesiphon incrustuns;* F: Aplanospores of *Microspora willeane.*

may be biflagellate or multiflagellate. Zoospores swarm in water for some time and then come to rest, shed their flagella and germinate into new individuals. Non-flagellate spores called aplanospores are also produced. Autospores superficially look like aplanospores but lack the motilities. These are miniature forms of the parent cells that produce them. Homocysts are the fragments of filamentous form, which develop thick sheath around them. Heterocysts also help in reproduction and are the larger vegetative cells, transparent and covered with thick layer. These are produced in Cyanophyceae. Endospores, exospores and other spore types are also produced in other groups of algae.

**(b) Sexual reproduction:** Sexual reproduction involves the union of protoplasm, union of their nuclei and meiosis.

The process of Isogamy involves the fusion of motile and non-motile gametes resulting in zygote formation. In *Oedogonium*, gametangia are morphologically differentiated and the male gamete fuses with egg nucleus of the female gametangium. In some the motile and non-motile gametes are morphologically dissimilar. In Oogamy, there is fusion of a large non-motile egg with a smaller motile sperm, except red algae.

**(c)** In phaeophyceae and rhodophyceae, complex fruit bodies are formed after sexual reproduction, e.g., Cystocarp.

## 6.2 LIFE CYCLE

The sequence of number of stages in morphological or cytological stages during the growth and development of an alga are important and is called life cycle. On the basis of a number of distinct morphological phases, different kinds of life cycle are named:

1. Monophasic type
2. Diphasic type
3. Triphasic type

**Classification:** Round (1965) has classified the algae into:

| | | | | |
|---|---|---|---|---|
| Procaryota | - | Cyanophyta | - | Cyanophyceae |
| Eucaryota | - | Chrysophyta | - | Xanthophyceae |
| | | Bacillariophyta | - | Bacillariophyceae |
| | | Chlorophyta | - | Chlorophyceae |
| | | Phaeophyta | - | Phaeophyceae |
| | | Rhodophyta | - | Rhodophyceae |

# Bacteria

## 7.1 INTRODUCTION

Bacteria are very small, simple and single-celled living organisms. The hereditary material is not a true nucleus and is a double-stranded DNA. Genetic material is not enclosed by nuclear membrane. Such organisms are called prokaryotes (pre-nucleus). Bacteria are included in kingdom Monera proposed by Whittaker. Some microbiologists call it *prokaryotae*.

Bacteria are considered homogenous group of primitive forms of microorganisms. However, the morphological, physiological and biochemical characteristics are found to be varied and distinct from those of other forms under the plant kingdom. This group of organisms deserves separate kingdom. Bacteria have different shapes ranging from spherical (coccus), cylindrical rod-like (*bacillus*) forms to ellipsoidal, spiral, (cork screw) and helicoidal. They are classified on the basis of their shape and other morphological and physiological characteristics. Individual bacteria may form pairs, chains, cluster or other groupings.

Bacteria are enclosed in cell walls made up of a substance called peptidoglycan. Bacteria reproduce by dividing into two equal cells (binary fission) and also through endospore formation. Bacteria use organic chemicals, which can be derived from dead organisms or a living host. There are bacteria, which manufacture their own food by photosynthesis. These are called autotrophs or photosynthetic bacteria. Some bacteria can derive nutrition from inorganic substances. Many bacteria or photosynthetic bacteria swim through flagella.

## 7.2 SHAPE AND SIZE

Cocci are round, elongated or flattened, and on division the cells remain attached to one another. Paired bacteria are called Diplococci—chain-like—are streptococci, in group of four are known as tetrads, cube-like groups of eight are called sarcinae, and grape-like clusters are called staphylococci. Bacilli are rod-shaped bacteria while spiral bacteria have

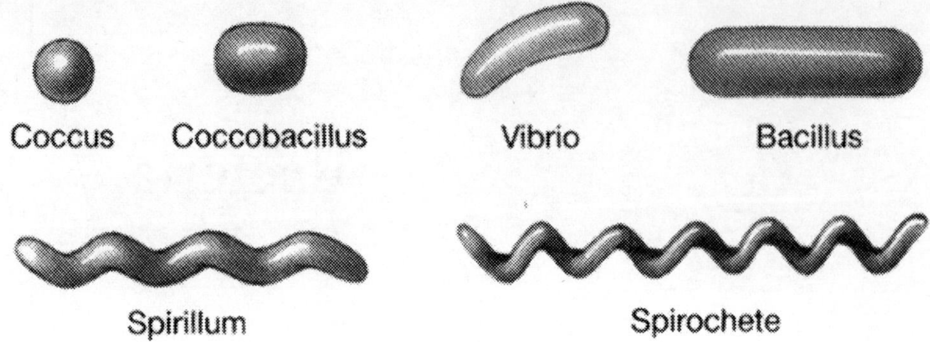


**Fig. 7.1.** Shapes of bacteria.

**Table 7.1.** Bacterial cell size

| Name of bacterium | Length (μm) |
|---|---|
| *Escherichia coli* | 2.0 |
| *Clostridium botulinum* | 3.8 |
| *Clostridium tetani* | 2.0–5.0 |
| *Mycobacterium tuberculosis* | 0.5–4.0 |
| *Neisseria meningitis* | 1.0 |
| *Pasturella pestis* | 1.0–2.0 |
| *Salmonella typhi* | 0.5–4.0 |
| *Staphylococcus* sp. | 0.8 |
| *Streptococcus pneumoniae* | 1.0–2.5 |
| *Treponema pallidum* | 6.0–14.0 |

one or more twists (commas, vibrios, helical) and spirochetes (Fig. 7.1). The shape of a bacterium is determined by heredity besides the influence of environmental parameters. Some bacteria like *Rhizobium* and *Corynebacterium* are genetically pleomorphic. Size ranges between 1–14 μm (Table 7.1).

## 7.3 MORPHOLOGY AND ULTRA-STRUCTURE

The bacterial cells are surrounded by a slime layer or capsule, flagella, axial filaments and pili. Slime layer is also called glycocalyx and is an extracellular polymeric substance. It is composed of polysaccharide, polypeptide or both and is viscous. This material is firmly attached to the cell wall, hence described as capsule, while loosely attached one is known as slime layer. Capsules protect pathogenic bacteria from phagocytosis besides getting adhesion to various surfaces in order to survive in its natural environment.

### 7.3.1 Flagella

Flagellum, singular—meaning whip, is a long filamentous appendage that helps in the movement of bacteria. Four types of flagellar attachment are observed in bacteria (Fig. 7.2).

1. Monotrichous—Single polar flagellum.
2. Amphitrichous—Single polar flagellum at both ends.
3. Lophotrichous—Two or more flagella at one or both poles of the cell.
4. Peritrichous—Flagella distributed over the entire cell.

Flagellum is made up of a protein called flagellin, which helps in the identification of some pathogenic bacteria. Bacteria with flagella are motile and they can move on their own.

The movement because of stimulus towards chemical is called chemotaxis, and towards light is known as phototaxis. Receptors are present either on the cell wall or under the cell wall. Spirochaetes move by the means of axial filaments, the bundles of fibrils that arise at the ends of cell beneath the outer sheath and spiral around the cell. The flagella beat in a propeller-like motion to help the bacterium move towards nutrients away from toxic materials, and towards light in cyanobacteria.

### 7.3.2 Cell Wall

Cell wall is a complex and semi-rigid structure offering not only protection but also characteristic shape of the cell (Figs. 7.3 and 7.4). Cell wall is composed of macromolecular network called peptidoglycan (murein), which is a mucopolysaccharide having repeated disaccharide attached to chains of four or five amino acids. The component monosaccharides called N-acetyl glucosamine and N-acetyl muramic acid are related to glucose and have amino acid groups attached. Diaminopimelic acid is present in some of the bacteria.

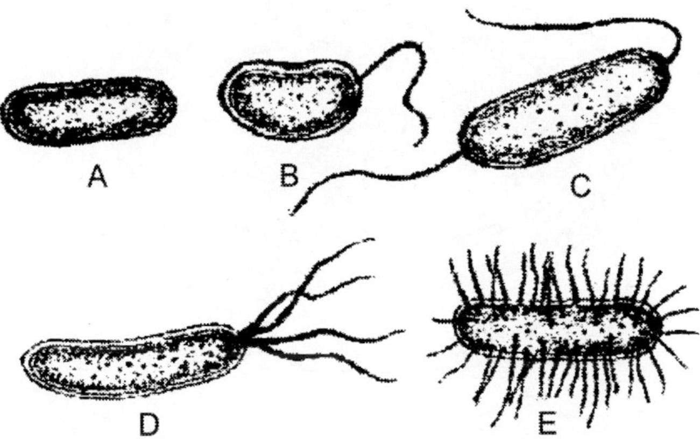


**Fig. 7.2.** Types of bacterial flagella. A: Atrichous; B: Monotrichous; C: Amphitrichous; D: Lophotrichous; E: Peritrichous.

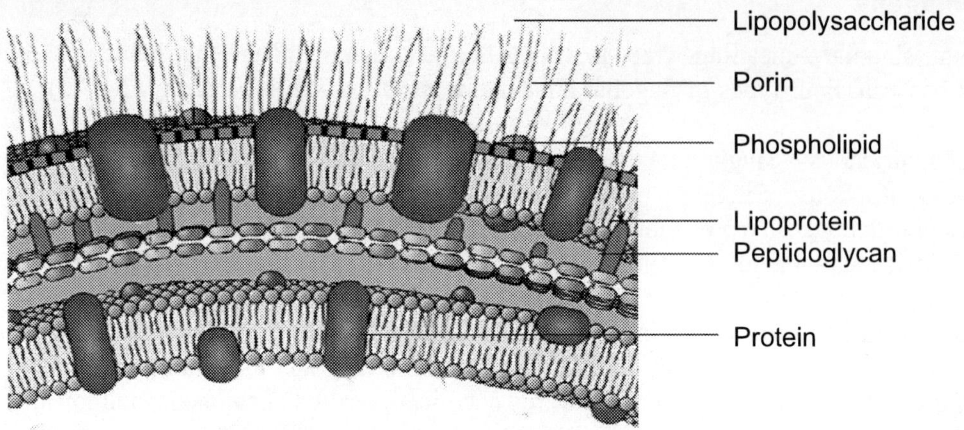


**Fig. 7.3.** Structure of bacterial cell wall.

In gram-positive bacteria, the cell wall consists of several layers of peptidoglycan connected by peptide side chains and cross bridges. The cell wall is thick in gram positive bacteria and contains teichoic acids, which help in storing phosphorus. Gram negative bacteria contain small amounts of peptidoglycans and teichoic acids are absent. In gram negative bacteria, peptidoglycans are surrounded by an outer membrane having lipoproteins, lipopolysaccharides and phospholipids. The major differences between Gram positive and Gram negative bacteria are given in Table 7.2.

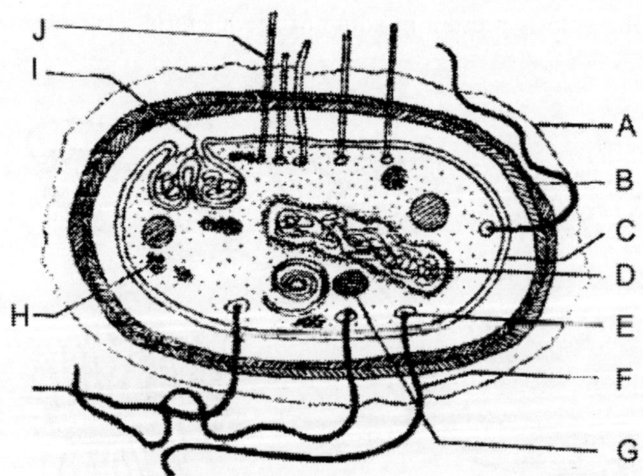


**Fig. 7.4.** Structure of bacterial cell. A: Mucilaginous sheath; B: Cell wall; C: Plasma membrane; D: Nuclear material with nuclear membrane. E: Basal granule; F: Flagellum; G: Stored food granule; H: Polyribosome; I: Mesosome; J: Pili.

**Table 7.2.** Differences between gram-positive and gram-negative bacteria

| Character | Gram-positive | Gram-negative |
|---|---|---|
| Retention of crystal violet colour | | |
| Acceptance of counterstain on (saffranine) | – | + |
| outer membrane | – | + |
| Appearance of peptidoglycan | Thick layers | Single thin layer |
| Lipids and lipoproteins | Low | High |
| Lipopolysaccharides | Absent | Present |
| Teichoic acid | Present | Absent |
| Periplasmic space | Absent | Present |
| Flagellar rings in basal body | 2 | 4 |
| Presence of toxin | Exotoxin | Endotoxin |
| Wall strength | High | Low |
| Resistance to drying | High | Low |
| Antibiotic sensitivity(penicillin) & sulfonamide | High | Low |
| Sensitivity to chloramphenicol, streptomycin and tetracycline | Low | High |
| Inhibition to basic dye | Significant | Not much |

## 7.3.3 Plasma Membrane

It is a thin phospholipid bilayer lying inside the cell wall and enclosing the cytoplasm. This membrane is less rigid besides having protein also. The function of the plasma membrane is to serve as a selective barrier through which materials enter and exit the cell. Plasma membranes are selectively permeable. Plasma membrane helps in the breakdown of nutrients and production of energy. Plasma membrane often contains one or more large and irregular folds known as mesosomes, which may play a role in reproduction and metabolism.

Materials move into plasma membrane through passive and active processes besides having simple diffusion. While performing active transport, the cell uses energy in the form of adenosin triphosphate to move substances across the plasma membrane. Membranes are dynamic, constantly adapting to different conditions.

## 7.3.4 Cytoplasm

It refers to the internal matrix of the cell present inside the plasma membrane. It has 80% water and contains primarily proteins, carbohydrates, lipids, inorganic ions and compounds of low molecular weight. Inorganic ions may also be present in higher concentration. It is thick, semi-transparent and elastic and also contains ribosomes, DNA, plasmids and reserve deposits known as inclusions. Metachromatic granules are called volutin and they stain red with certain blue dyes. Volutin is a polyphosphate that can be used in the synthesis of ATP. Polysaccharide granules consist of glycogen and starch. These granules appear reddish-brown, and starch granules appear blue in the presence of iodine. Lipid inclusions appear in various species while sulfur granules are found in iron bacteria. Carboxysomes are polyhedral and hexagonal inclusions that contain ribulose 1, 5 diphosphate carboxylase. There are hollow cavities in many aquatic bacteria called gas vesicles. Their function is to maintain buoyancy. Cytoplasm is a gel-like matrix composed of water, enzymes, nutrients, wastes and gases. Chromosome, the single, continuous strand of DNA is localized; all the other cellular components are scattered throughout the cytoplasm.

### 7.3.5 Endospores

Under stress and unfavourable conditions, certain gram-positive bacteria like *Clostridium* and *Bacillus* form specialized resting cells called endospore (Fig. 7.5). These are formed inside the bacterial cell membrane. Endospores can survive extreme heat, lack of water, radiation and exposure to toxic chemicals due to a chemical substance, dipicolinic acid, which is characteristic of spores but not of vegetative cells.

### 7.3.6 Nuclear Material

Bacterial cell contains nucleoid, in which a single, long, circular bacterial chromosome is present. DNA carries all information required for the structure and function of the cells. Bacterial chromosomes do not include histones and are not surrounded by a nuclear envelope. Besides, the bacterial chromosome contains additional small circular and double-stranded DNA molecules called plasmids. Plasmid is an extra-chromosomal genetic element and is not connected to the main bacterial chromosomes. This plasmid also replicates independently of chromosomal DNA. Plasmids do carry genes, which offer resistance to antibiotics, toxic metals, toxins and also help in the synthesis of enzymes. Plasmids are transferable from one bacterium to another. For most plasmid types, copies in the cytoplasm are passed on to daughter cells during binary fission. Other types of plasmids form a tube-like structure at the surface called pili, which passes copies of the plasmid to other bacteria during conjugation, a process by which bacteria exchange

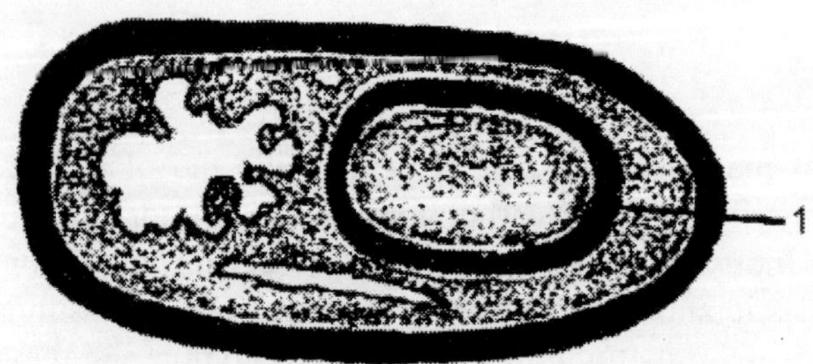


**Fig. 7.5.** Endospore in bacterial cell (1).

genetic information. Plasmids are instrumental in the transmission of special properties, such as antibiotic drug resistance, resistance to heavy metals and virulence factors for the infection of plant or animal hosts. The ability to insert specific genes through plasmids have made them extremely helpful tools in the field of molecular biology and genetic engineering. Nucleolus is absent. DNA base ratios as moles percentage of guanine and cytosine range from 28 to 73.

### 7.3.7 Pili

Many bacterial species have pili, small hair-like projections emerging from the outside of cell surface. These outgrowths assist the bacteria in attaching to other cells and surfaces, such as teeth, intestines and rocks. Without pili, many disease causing bacteria lose their ability to infect because they are unable to attach to host tissue. Specialized pili are used in conjugation, during which two bacteria exchange fragments of plasmid DNA.

### 7.3.8 Ribosomes

Ribosomes are present in cytoplasm and serve as the site of protein synthesis. Each ribosome has two subunits and each subunit is composed of a protein and a type of RNA called ribosomal RNA. Prokaryotic ribosomes are called 70 S ribosomes as they are smaller and less dense than eukaryotic, which are 80 S type. The letter S refers to Svedberg units, which is related to sedimentation rate. The subunit of 70 S ribosomes is a small 30 S subunit containing one molecule of rRNA and larger 50 S subunit containing two molecules of rRNA. Bacterial ribosomes are smaller and are never bound to other organelles.

## 7.4 CELL DIVISION

Cell division is one of the multiplication methods by which the population increases. It is a simpler process in prokaryotes and is by binary fission. The cell along with constituents increases in size. The DNA gets attached to a part of the membrane that replicates. A wall is laid down separating the cell contents (Fig. 7.6). Cell division is completed in 20 minutes to two hours. In some bacteria, multiplication also takes place by budding.

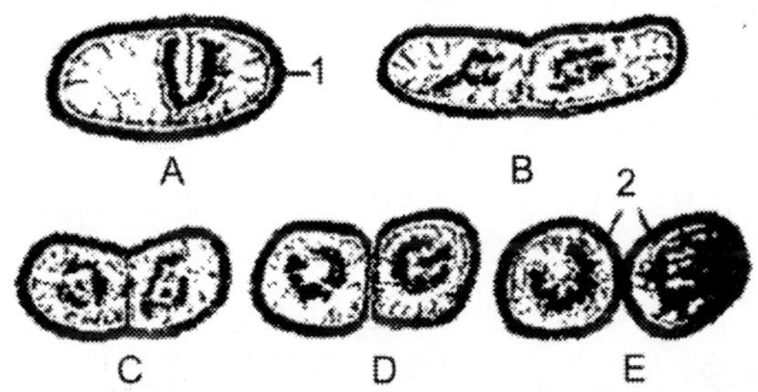


**Fig. 7.6.** Stages in reproduction of bacteria (binary fission). 1: Parent cell; 2: Daughter cell.

## 7.5 GENETIC RECOMBINATION (SEXUAL REPRODUCTION)

Genetic recombination in bacteria is of three types:

1. Transformation
2. Conjugation
3. Transduction

### 7.5.1 Transformation

Griffith (1928), while working on pneumococci, indicated that non-pathogenic strain produces rough colonies while a pathogenic strain produces smooth colonies. When the heat killed bacterial cells of the smooth colony were mixed with the bacteria having rough colonies, some of the resultant products were rough while some were smooth. The latter were also proved to be pathogenic. It was concluded that some substance from heat killed cells had transformed the rough colony producing cells to smooth colony producing cells. This process is called transformation and the transforming principle is a nucleic acid (Figs. 7.7 and 7.8).

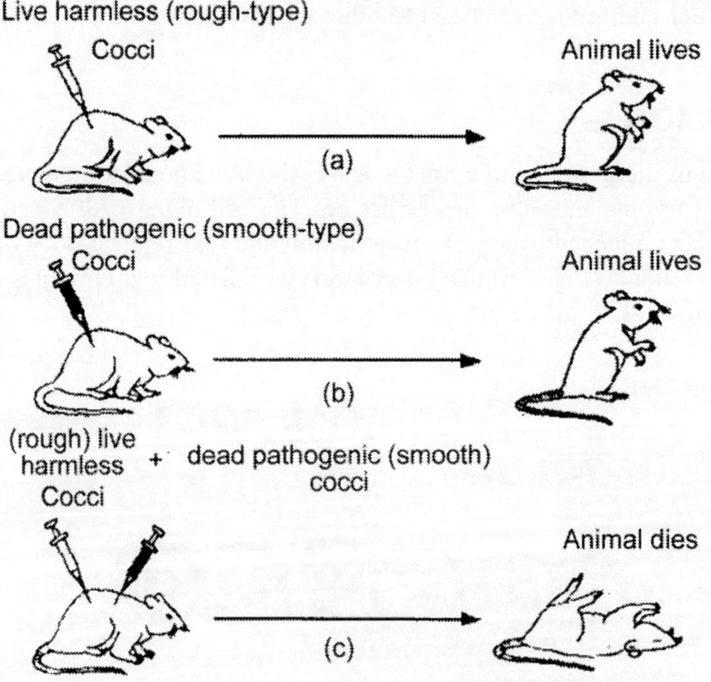


**Fig. 7.7.** Griffith's experiments on mice showing bacterial transformation.

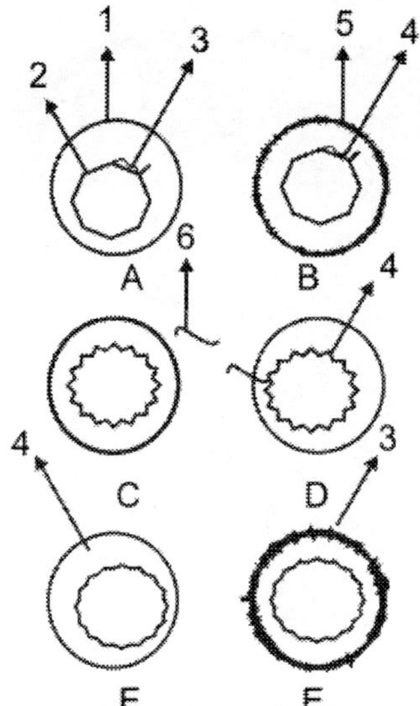


**Fig. 7.8.** Transformation of avirulent strain of bacterium into virulent strain. 1. A: Virulent cell; B: Avirulent cell; C: Host killed virulent cell; D: Entry of donor DNA; E: Virulent cell; F: Transformation cell; 1: Capsule; 2: DNA; 3: S gene; 4: R gene; 5: Rough wall; 6: DNA fragment.

### 7.5.2 Conjugation

Lederberg and Tatum (1946) worked on the strains of *Escherichia coli.* They observed that when auxotrophic strains of different strains were mixed and grown for several generations, some phototrophs were produced. This is due to genetic recombination called conjugation.

When a donor (male) and a recipient (female) bacteria are brought together, the F-factor, which is a plasmid, gets transferred from donor to recipient quickly through a cytoplasmic bridge between the two bacteria (Fig. 7.9). The donor forms a sex pilus with which it attaches itself to the recipient, establishing a physical contact. The donor is called F+ while the recipient is F-. Numerous recombinants are produced in high frequency.

### 7.5.3 Transduction

In this process the bacterial DNA is transferred from one cell to another through a virus. This process was discovered in 1952 by Zinder and Lederberg. The transfer of genetic material between two cells of *Salmonella* does not require cell to cell contact.

There are two types of transduction, namely generalized and specialized (Figs. 7.10 and 7.11). Specialized transduction occurs only in some temperate viruses, a specific group of host genes is integrated directly into the virus genome, usually replacing some of the virus genes and is transferred to the recipient during lysogenation. In general transduction, host genes become a part of mature virus particles in place or in addition to the virus genomes.

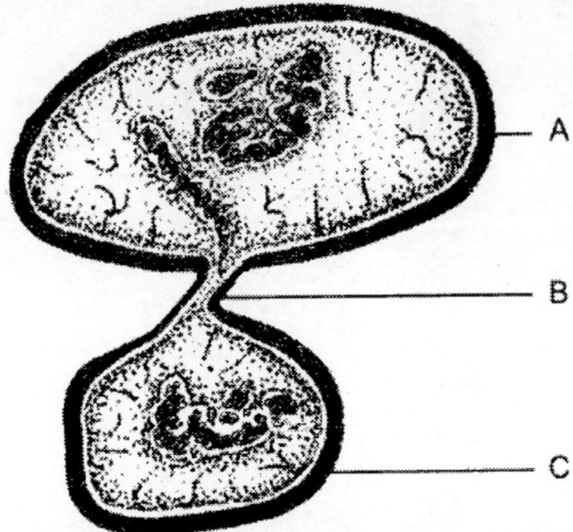


**Fig. 7.9.** Conjugation in bacteria. A: Donor cell; B: Conjugation bridge; C: Recipient cell.

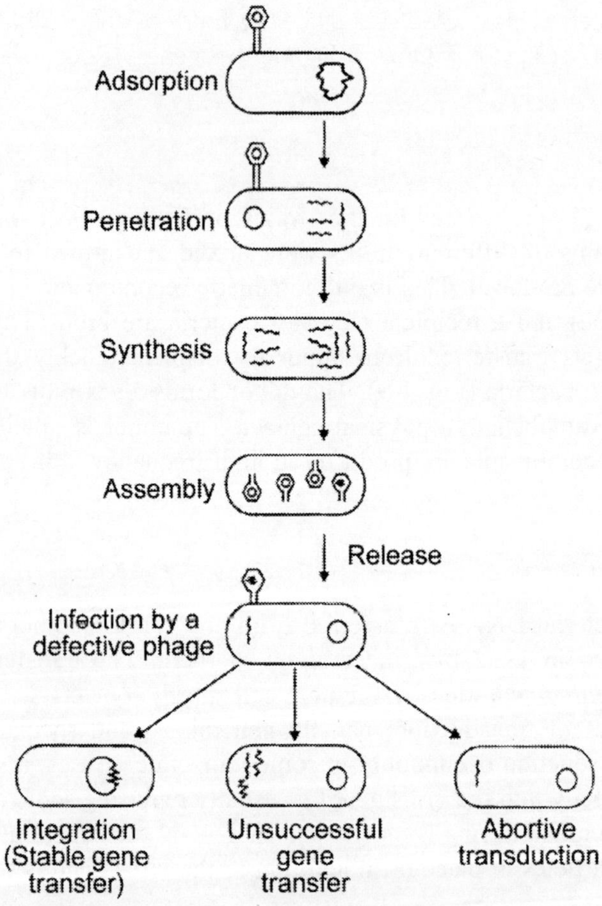


**Fig. 7.10.** Generalised transduction by bacteriophage.

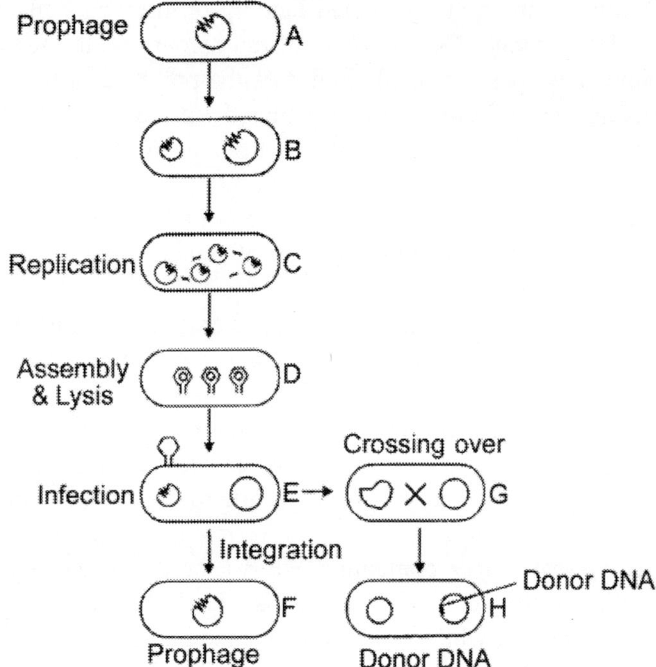


**Fig. 7.11.** Specialised transduction by temperate bacteriophage.

## 7.6 CHEMICAL AND BIOCHEMICAL CHARACTERIZATION OF BACTERIA

Following tests as described in 'Manual of Microbiological Methods', 'Microbiology Lab Manual' and 'Microbiological Methods' are performed for the identification of bacterial isolates:

1. Cell shape and Gram Reaction: For Gram reaction, see the staining method on annexure page 337.
2. Growth and Colony Characteristics: A loop full of the young growth is suspended in the dispersion fluid and streaked over nutrient agar to obtain discrete colonies. After required incubation, colony characters are recorded.
3. Oxidase Test: Cultures are streaked on nutrient agar and incubated for 2 days. Following incubation, test reagent-tetramethyl-p-phenylenediamine dihydrochloride is added. Development of black colouration on the surface of the colonies indicates cytochrome oxidase production, which represents a positive test.
4. Catalase Test: Cultures are streaked onto nutrient agar plates and incubated for 2 days. Few drops of 3% hydrogen peroxide (v/v) are allowed to flow over the surface of each plate culture. Evolution of bubbles of free oxygen was indicative of the presence of catalase.

5. Urease Test: Cultures are spot inoculated heavily on urea agar plates and observed for four days for colour change. Development of red colour due to the production of ammoniacal compounds and increase of pH indicates the presence of urease enzyme.
6. Starch Hydrolysis Test: Cultures are inoculated on starch agar plates and examined after 2 days for loss of turbidity due to hydrolysis of starch by the action of starch hydrolases.
7. Casein Hydrolysis Test: Cultures are inoculated on milk agar plates. The plates are observed after 2 days of incubation for loss of opacity of the medium to translucent, which indicates positive casein production.
8. Citrate Utilization Test: Citrate agar inoculant slants are inoculated with bacterial cultures and observed for change in pH. The change of colour from green (neutral pH) to blue (alkaline pH) was scored as positive for citrate utilization.
9. Indole Production Test: Tubes containing tryptone solution are inoculated, and after 2 days of incubation to the turbid growth, Kovac's reagent is added and mixed by rotating the tubes. The development of red colour due to the presence of indole is taken as a positive test.
10. Methyl Red (MR) Test: Tubes containing MR-VP broth are inoculated, and after 2 days of heavy growth in broth, methyl red solution is added. Red colouration (pH 4.5 or less) is recorded as a positive test, while yellow colour (alkaline pH) is taken as a negative test.
11. Voges Proskauer's (VP) Test: Tubes containing MR-VP broth are inoculated, and after 2 days of incubation, 0.5ml of 6% alpha-naphthol solution and 16% KOH and a pinch of arginine (alternatively creatine/ creatinine) are added and shaken vigourously. Development of red colour is taken as a positive test.
12. Decarboxylation of Lysine: The culture is inoculated on lysine decarboxylation medium and incubated for 2 days. Change of pH to alkaline (purple) was scored as positive for lysine decarboxylase activity.
13. Liquid Hydrolysis Test: Bacterial culture was streaked onto tributyrin agar plates and incubated for 2–4 days. Clear area surrounding the bacterial growth, i.e., zone of lyposis, indicates positive test.
14. Fermentation of Carbohydrates: Test tubes containing carbohydrate fermentation medium and Durham's tubes are inoculated with bacterial culture(s). After 2-4 days of incubation, production of acid and gas is recorded.
15. Gelatin Hydrolysis Test: Deep tubes containing nutrient gelatin medium are inoculated and incubated for 2 days. Following incubation, the cultures are placed in a refrigerator for 30 minutes. Cultures that remain liquefied due to gelatin hydrolysis indicate positive for gelatinase activity.
16. Nitrate Reduction Test: Test tubes containing tripticase nitrate broth are inoculated and incubated for 2 days. Few drops of sulfanilic acid solution followed by a few drops of dimethyl-alpha-naphtalamine solution are added to broth cultures. The development of red colour indicates nitrate reduction.
17. Sodium Malonate Utilization: Test tubes containing malonate broth are inoculated with bacterial cultures and incubated for 2 days. Blue colouration (alkaline pH) is recorded as positive test while yellow colour (acidic pH) is considered a negative test.

18. $H_2S$ Production Test: Test tubes containing TSI agar medium are inoculated with bacterial cultures and observed for change in colour of butt after incubation. Extensive blackening in the butt due to the precipitation of ferrous sulfide indicates $H_2S$ production.
19. Growth on NaCl Supplemented Agar: Bacterial cultures are streaked onto nutrient agar plates containing different levels of NaCl and incubated for 2 days. These are then recorded for the growth of the organism(s).
20. Motility and Lysine Deamination Test: Deep tubes containing MIL medium are stab-inoculated and incubated for 24 hours. Diffused growth indicates motility, and red-brown colouration at the top of the medium indicates lysine deamination.
21. Growth at Different Temperatures: Cultures are streaked on nutrient agar and incubated for 2 or more days at different temperatures ranging from 5°C to 65°C and observed for the presence of growth.

## 7.7 OUTLINES OF BACTERIAL CLASSIFICATION

Bacteria are small, unicellular prokaryotic organisms. Since the discovery of bacteria under microscope, several bacterial organisms have been isolated and characterized. As the information on newer organisms becomes more, it becomes necessary to arrange them in systematic order for the purpose of practical understanding. The first classification of bacteria was published in 1773, and several schemes of classification have appeared subsequently. Earlier classification schemes were mainly based on morphological characteristics, since microscopic observation was an important method for identification. Even today this is the first identification method for a given microorganism in laboratory. Information on microorganisms increased due to various studies on them. Therefore, other kinds of characteristics became increasingly important for classification. Each successive classification scheme reflected the level of knowledge available at that time, and this continues to be true even to the latest classification followed. Though many classification schemes exist for bacteria, the scheme followed in "Bergey's Manual of Determinative Bacteriology" is important because of its broad scope and wide acceptance. It provides descriptions of all established Genera and species of bacteria and also gives a practical arrangement of these organisms (taxa) for their identification.

Bergey's Manual of Determinative Bacteriology was first published in 1923, and subsequently it has undergone several revisions. The ninth edition of this manual is published in four volumes (since 1984–1989) with a changed title as "Bergey's Manual of Systematic Bacteriology", and it is called the first edition of 'Systematic Bacteriology'. As an abridged version of these four volumes is published in 1994 as 9th edition of 'Bergey's Manual of Determinative Bacteriology'. This abridged edition is most useful as a practical reference guide for every microbiology laboratory in identification of bacterial organisms. The second edition of 'Bergey's Manual of Systematic Bacteriology' is being published in five volumes by the 'Bergey's manual Trust'. Each volume gives extensive information on bacterial organisms with group of related organisms.

In the process of identification of bacterial organisms, the bacteria are subjected to various characters (ecological, pathogenic, staining, cultural/nutritional, biochemical/metabolic, immunology/antigenic, chemical, genetic/genomic (molecular identification). Every characteristic feature is important in identifying a given organism, its nomenclature and taxonomic positioning. The identification of bacteria, their classification and taxonomy are important parts of bacteriology. It is not possible to deal with extensive details of this subject in this small textbook. However, it is necessary to make the student understand the basic terminology and the concepts of bacterial classification.

### 7.7.1 Classification

Taxa are initially constructed from strains of a given organism in microbiology. A strain is made up of all the descendents of a pure culture. It is usually a succession of cultures derived from an initial colony. Each strain has a specific history and designation for itself. The basic taxonomic group (taxon) is the species, which is a collection of strains having similar characteristics. Bacterial species consist of a special strain called a type strain together with all other strains that are considered sufficiently similar to the type strain as to warrant inclusion in the species. The type strain is the strain that is designated to be the permanent reference specimen for the species. It is the strain that is most typical of all the strains included in the species, but to which all other strains must be compared to see if they resemble it closely enough to belong to the species. Therefore, type strains are particularly important and special attention is given to their maintenance and preservation by national reference collections such as the American Type Culture Collection (ATCC). In India, such preservation and maintenance centres are established, which are Microbial Type Culture Collection (MTCC) established at the Institute of Microbial Technology (IMTECH), Chandigarh and National Collection of Industrial Microorganisms (NCIM) at National Chemical Laboratory (NCL), Pune. Many other culture collections are maintained in India and throughout the world.

The meaning given to the bacterial species as explained above is similar to the type strain. Though at present there are no specific criteria for a bacterial species that are universally accepted, but certain criteria based on DNA homology experiments (i.e., genome analysis, rRNA gene sequence homology etc.) are probably more widely accepted today than any other. This may eventually lead to a unifying concept for defining a species.

A bacterial species is composed of a collection of similar strains. Similarly, a bacterial genus is composed of a collection of similar species. One of the species is designated as the type species, and this serves as the permanent example of the species to be included in the genus. Taxonomic groups of higher rank than genus are listed below.

**Family:** consists of a group of similar genera.
**Order:** consists of a group of similar families.
**Class:** consists of a group of similar orders
**Division:** consists of a group of similar classes
**Kingdom:** consists of a group of similar divisions

Generally, taxonomists strive to make classification that have the following two qualities.

(a) *Stability*: Classification that is subject to frequent, radical changes leads to confusion. Every attempt should be made to devise classification that need only minor changes as new information becomes available.
(b) *Predictability*: By knowing the characteristics of one member of a taxonomic group, it should be possible to assume that the order members of the same group probably have similar characteristics. If this cannot be done, the classification has little value.

Unfortunately, in the publication of Bergey's Manual of Determinative Bacteriology, every new edition has major variations from its previous one. This is due to the quantity of new information available regarding already known organisms and also discovery of new species of Genera. This is because of technology developments in identifying the bacterial organisms. Particularly, the molecular analysis of bacterial organisms for their genome and rRNA gene sequence data, which were not available in earlier editions of the manual. In recent times sequencing of DNA, DNA fingerprinting, polymer chain reaction, restriction fragment length polymorphism, ELISA and many other molecular tools have come into existence. These techniques not only help in identification of bacteria but also help in finding their phylogeny.

## 7.7.2 General Method of Classifying Bacteria

There are three important methods for arranging bacteria into taxa and their taxonomic grouping:

(a) ***The intuition method***: This is based on the experience of microbiologist with a given organism. A person who is thoroughly familiar with the properties of given organism, which he or she has been studying for several years, may decide that the organism represents one or more species or Genera. The disadvantage with this method is, that the characteristic of a given organism, which "A" person thinks important, may not be important to others. Therefore, different taxonomists arrive at a very different grouping of the same organism. This method does not have much significance in today's taxonomy.
(b) ***Numerical taxonomy***: In this method, taxonomists working with a given organism determine many characteristics (usually about 100–200) pertaining to that organism. Each characteristic is given equal weight. Using this data, the percentage similarity (%S) of each strain to every other known strain is calculated. For a given two strains, this is

$$\% \text{ S} = \text{NS}$$

$$\text{NS} + \text{ND}$$

where NS = the number of characteristics that are the same (positive or negative) for two strains.

ND = the number of characteristics that are different.

The strains having high %S to each other are placed into groups; those groups having a high %S to each other are in turn placed into larger groups. The degree of similarity needed to rank group as a species, genus, or other taxon is a matter of judgement on the part of the taxonomist. This method of classification has great practical usefulness as well as being relatively unbiased in its approach. It also yields classifications that have a high degree of stability and predictability.

(c) ***Genetic relatedness*:** This is the most reliable method of classification based on the degree of genetic relatedness between organisms. It is the most objective of all and is based on the most fundamental aspect of organisms, their hereditary material (DNA). Molecular biology provided techniques by which the DNA of one organism could be compared with other organisms. Initially, crude comparisons are made based on the G + C values. Generally, organisms of the same species are very closely related and will have very similar molecular percentage G + C values. Now it is known that organisms that are not related may also have similar molecular percentage G + C values. Therefore, much precise molecular methods are used, which include the DNA homology studies of various organisms and their comparison. This sequence is the most fundamental characteristic of an organism. Modern techniques have made possible such a comparison now. Another important comparison of molecular analysis is the 16S rRNA gene sequence analysis and its homology comparison. Now such studies are done on many organisms and the data is stored on international data banks so as to be available for comparison to others. This type of studies are done with use of bioinformatics methods for genetic analysis of bacterial organisms. The latest classifications and phylogenetic trees of bacterial organisms are arranged based on 16S rRNA gene sequence analysis due to fact that this gene sequence of an organism is the most conserved genetic structure. The DNA homology and 16S RNA homology studies are the most dependent information used for bacterial taxonomy and identification. The latest edition of Bergey's Manual of Systematic Bacteriology is mostly based on such data of bacterial organisms.

### 7.7.3 Nomenclature

It is necessary to name a given organism for its identity based on certain criteria and to be known to every scientific person throughout the world with the same name. This is generally known as scientific naming or nomenclature. In this, each species of organisms has only one officially accepted name by international agreement. This system provides for precise communication. If an organism were to be called *Staphylococcus aureus* in one country and with other name in another country, it becomes difficult to know that the same organism was being studied. Therefore, universally accepted names should be given to organisms so that they can be called with the same scientific names throughout the world and known similarly to everybody. Though there are several ways of naming a given organism, the universally accepted naming is by binomial (two words) nomenclature.

Bacteria are named according to the rules set down in the International Code of Nomenclature of Bacteria. The rule in bacteriological nomenclature is that a name must be written as a Latin or latinized binomial and must follow certain rules of Latin grammar. The first word in the binomial is the genus name and is always capitalized. The second word is the specific epithet and is never capitalized. Both the genus name and specific epithet are given in italics (or underlined if written in English, which means 'italics' to the printer). The bacterial names that have official standing in microbiology were published in the *Approved Lists of Bacterial Names* in 1880. Any new or additional names must be published in the *International Journal of Systematic Bacteriology* in order to achieve official recognition. The International Code of Nomenclature of Bacteria was developed with reference to the much established International Code of Zoological and Botanical Nomenclature. All these codes incorporate certain common principles, which are given below:

(i) Each distinct kind of organism is designated as a species.
(ii) The species is designated by a Latin binomial to provide a characteristic international label (binomial system of nomenclature).
(iii) Regulation is established for the application of names.
(iv) A law of priority ensures the use of the oldest available legitimate name.
(v) Designation of categories is required for classification of organisms.
(vi) Requirements are given for effective publication of new specification names, as well as guidance in coming new names.

### 7.7.4 Identification of Bacteria

A given bacterial organism is generally classified before it can be identified based on some preliminary characteristics. It is to recognize that the organism is different from any other known organisms. Once the organism is classified, a few of its characteristics are selected by which it can be identified by other microbiologists. In order to be useful for identification, the combination of characteristics chosen must occur only in that particular kind of organism and in no other. The characteristics chosen should also be ones that are easy to determine, such as shape, staining reactions, sugar fermentations etc., which can be easily performed in a simple microbiology laboratory. For example, DNA homology and 16S rRNA studies, though quite useful for identification and classification, it is not possible to do these studies in every laboratory that has no facility and expertise. Many identification schemes are in the form of keys, which give identifying characteristics arranged in a logical fashion. Identification tables are also useful and generally contain more characteristics than do keys, with the information arranged in an easy-to-read, summarized form. Bergey's Manual of Determinative Bacteriology is an important laboratory manual, which gives all such identification keys for easy approach in a laboratory. As of now, the 9th abridged 1994 edition of the above manual is available and used as a standard reference for every bacterial identification. The second edition of Bergey's Manual of Systematic Bacteriology is being published with extensive information on bacterial classification and identification. An outline of this classification is given below.

## 7.8 BERGEY'S MANUAL OF SYSTEMATIC BACTERIOLOGY

This manual is published in 5 volumes. These volumes give detailed characteristics of microorganisms (Archaea and Bacteria) with emphasis on identification and improvement of various groups as given below:

Volume 1: The Archaea, Cyanobacteria, Phototrophs and deeply branching genera
Volume 2: The Proteobacteria
Volume 3: The Low G + C Gram positive bacteria } (Firmicutes)
Volume 4: The High G + C Gram positive bacteria } (Firmicutes)
Volume 5: The Planctomycetes, Spirochaetes, Fibrobacteres, Bacteroids and Fusobacteria

### 7.8.1 Primary Domains

The important primary domains of living organisms as per Carl Woese proposal are Bacteria, Archaea and Eukarya. Note that each word of these domains is ending with 'a' unlike the earlier naming as Eubacteria. These are characterized based on the phylogenetic relationship derived from comparative sequencing of 16S rRNA or 18S rRNA. The identification characteristics of primary domains, Bacteria, Archaea and Eukarya, are mainly on the basis of:

(i) comparative ribosomal RNA sequencing;
(ii) cell walls;
(iii) lipids;
(iv) RNA polymerase; and
(v) features of protein synthesis.

Although the primary domains are defined on the basis of comparative ribosomal RNA sequencing, other studies have shown that each domain is characterized by a number of phenotypic properties. Some of these characteristics are unique to one domain, whereas others are found in two out of the three domains.

#### Cell Walls

All bacteria have cell walls containing peptidoglycan except the members *Planctomyces-Pirella* group, whose cells are composed of proteins, and the Chlamydia-Mycoplasma groups, which lack cell wall altogether. Therefore, peptidoglycan is considered a 'signature' molecule (for differentiation from other two phylogenetic domains) for species of bacteria. It is the muramic acid molecule, unique to peptidoglycan, which is analyzed in primary identification. Archaea and Eukarya lack pseudopeptidoglycan. Various cell types exist in archaea. These are made up of pseudopeptidoglycan, polysaccharides, protein or glycoprotein. Thus, great diversity is observed in the chemistry of microbes in the absence of peptidoglycan. Phylogenetically, it is the presence or absence of peptidoglycan, which is the most important characteristic for bacteria. Cell walls of eukaryotes are made up of cellulose and/or chitin.

## Lipids

Chemical nature of membranes is the most important characteristic for differentiating from archaea. Bacteria and eukaryotes have (synthesize) membrane lipids with fatty acids attached to glycerol with ester linkage. The nature of fatty acids may be highly variable in different groups of these organisms, the important point being their linkage to glycerol with ester bond. However, in archaea, the membranes have (synthesize) ether linkages consisting of long-chain (branched) molecules. These could be either phytanyl or biphytanyl type molecules in place of linear chain fatty acids present in bacteria or eukaryotes. The chirality of the glycerol moiety also differs in the lipid of representative of the three domains. The central carbon atom of glycerol molecule in bacteria and eukaryotes membrane lipids is of R form, and in archaea it is the L form. Such chemical structures and their chiral nature is an important criterion for domain differentiation.

## RNA Polymerase

DNA dependent RNA polymerase is the key enzyme in transcription of RNA from DNA template in all organisms. Bacterial cells have a simple RNA polymerase enzyme consisting of four sub-unit quaternary structures. This consists of four polypeptides named alpha, ß, ß' and o- in the ratio of 2:1:1:1, respectively, in the active polymerase molecule. The RNA polymerases of methanogens and halophiles contain eight polypeptides, five large ones and three smaller ones. The hyperthermophilic archaea contain an even more complex RNA polymerase consisting of at least ten distinct polypeptides. The major RNA polymerase of eukaryotes contains 10–12 polypeptides, and the relative size of the peptides coincide most closely with those from the hyperthermophilic archaea. Two other polymerases are known in eukaryotes, and each specializes in transcribing certain regions of the genome, one for rRNA. In terms of phylogenetic or molecular signatures, the RNA polymerase of bacteria is highly diagnostic and the other polymerases are too complex to be phylogenetically definitive.

## Features of Protein Synthesis

The ribosomes of bacteria and archaea are same in size (70S, as compared with 80S ribosomes in cytoplasm of eukaryotes), but several steps in archaeal protein synthesis resemble those in eukaryotes than in bacteria. The initiation codon (AUG) in bacteria codes for incorporation of modified (formyl) methionine, whereas in archaea and eukaryotes it is unmodified methionine coded by the same codon. Exotoxin of *Corynebacterium diphtheriae* is a potent inhibitor of protein synthesis in archaea and eukaryotes and not in bacteria. Most antibiotics that specifically inhibit protein synthesis in bacteria do not inhibit in archaea and eukaryotes. The ribosomal sub-unit cross reconstitution from archaea and yeast were found to be functional, whereas such cross reconstitution from bacteria and yeast were not functional. This shows functional similarity of ribosomal sub-units in archaea and eukaryotes and therefore their sensitivity to certain antibiotics.

The Bergey's Manual of Systematic Bacteriology gives extensive information on classification and identification of bacteria in its five volumes. It is not possible to give all such information in this limited textbook. The schematic content of this manual gives an overall idea of its contents, the major groups described in it, and their taxonomic position with representative bacterial examples wherever necessary. The readers are advised to consult the original manual for detailed information for use in identification of bacterial organisms and strategies used for identification, classification and taxonomical groupings.

Outline of Bergey's Manual of Systematic Bacteriology (Second Edition) (A brief scheme of classification up to sections with one example) is given below:

Volume 1: The Archaea, Cyanobacteria, Phototrophs and Deeply Branching Genera.

THE ARCHAEA

Kingdom I: Archaeota

SECTION I: Thermoprotei, sulfolobi and barophiles
- Phylum 1: Crenarchaeota
- Class I: Thermoprotei
- Order I: Thermoproteales
- Family I: Thermoproteaceae
- Genus: *Thermoproteus*
- Family II: Thermofilaceae
- Genus: *Thermofilum*
- Order II: Pyrodictiales
- Family I: Desulfurococcaceae
- Genus: *Desulfurococcus*
- Family II: Pyrodictiaceae
- Genus: *Pyrodictim*
- Class II: Sulfolobi
- Order I: Sulfolobales
- Family I: Sulfolobaceae
- Genus: *Sulfolobus*

SECTION II: The Methanogens
- Phylum II: *Euryarchaeota*
- Class I: Methanobacteria
- Class II: Methanobacteria
- Family I: Methanobacteriaceae
- Genus: *Methanobacterium, Methanobrevibacter, Methanotherms*
- Class II: Methanococci
- Order I: Methanococcales
- Family I: Methanococcaceae
- Genus: *Methanococcus*

| | |
|---|---|
| Order II: | Methanomicrobiales |
| Family I: | Methanomicrobiaceae |
| Genus: | *Methanomicrobium* |
| Family II: | Methanosarcinaceae |
| Genus: | *Methanosarcina* |
| Family II: | Uncertain Placement |
| Genus: | *Halomethanococcus* |

| | |
|---|---|
| SECTION III: | The Halobacteria |
| Class I: | Halobacteriae |
| Order I: | Halobacteriales |
| Family I: | Halobacteriaceae |
| Genus: | *Halobacterium, Halococcus, Natronobacterium, Natronorubrum* |

| | |
|---|---|
| SECTION IV: | The Thermoplasmas |
| Class I: | Thermoplasma |
| Order I: | Thermoplasmatales |
| Family I: | Thermoplasmataceae |
| Genus: | *Thermoplasma* |
| Family II: | Picrophilaceae |
| Genus: | *Picrophilus* |

SECTION V:The Thermococci

| | |
|---|---|
| Class I: | Thermococci |
| Order I: | Archaeoglobales |
| Order II: | Thermococcales |
| Family I: | Thermococcaceae |
| Genus: | *Thermococcus* |
| Family I: | Archaeoglobaceae |
| Genus: | Archaeoglobus |
| Class II: | Methanopyri |
| Order I: | Methanopyrales |
| Family I: | Methanopyraceae |
| Genus: | Methanopyrus |

DEEPLY BRANCHING GENERA OF BACTERIA

Kingdom II: Bacteria

| | |
|---|---|
| SECTION VI: | Bacteria |
| Phylum I: | Aquificae |
| Class I: | Aquificae |
| Order I: | Aquificales |
| Family I: | Aquificaceae |

| | |
|---|---|
| Genus | *Aquifex, Hydrogenobacter* |
| | |
| SECTION VII: | Thermotogas and Geotgas |
| Class I: | Thermotogae |
| Order I: | Thermotogales |
| Family I: | Thermotogaceae |
| Genus: | *Fervidobacterium, Thermotoga* |
| Class II: | Thermodesulfobacteria |
| Order I: | Thermodesulfobacteria |
| Family I: | Thermodesulfobacteriacae |
| Genus: | *Carboxydothermus, Thermodesulfobacterium* |
| | |
| SECTION VIII: | The Deinococci |
| Phylum II: | Xenobacteria |
| Class I: | Deinococci |
| Order II: | Deinococcales |
| Family I: | Dinococcaceae |
| Genus: | *Deinococcus* |
| | |
| SECTION IX: | Thermi and genera of uncertain affiliation |
| Class I: | Thermi |
| Family I: | Thermaceae |
| Genus: | *Thermus* |
| Family II: | The paraphyletic assemblage |
| Genus: | *Nitrospira, Thermodesulfovibrio* |
| | |
| SECTION X: | Chrysiogenes |
| Phylum III: | Chrysiogenetes |
| Class I: | Chrysiogenete |
| Order II: | Chrysiogenetes |
| Family I: | Chrysiogenaceae |
| Genus: | *Chrysioegenes* |
| | |
| SECTION XI: | The Chloroflexi and Herpetosiphons |
| Phylum IV: | Thermomicrobia |
| Class I: | Chloroflexi |
| Order I: | Chloroflexales |
| Family I: | Chloroflexaceae |
| Genus: | *Chloroflexus, Heliohrix* |
| Order II: | Herpetosiphonales |
| Family I: | Herpetosiphonaceae |
| Genus: | *Herpetosiphon* |

| | |
|---|---|
| SECTION XII: | Thermomicrobia |
| Class I: | Thermomicrobiae |
| Order II: | Thermomicrobiales |
| Genus: | *Thermomicrobium* |
| | |
| SECTION XIII: | The Cyanobacteria |
| Class I: | Prochlorophyta |
| Order I: | Chroococcales |
| Family I: | Chroococcaceae |
| Genus: | *Chroococcus, Cyanobacterium, Microcystis, Gleocapsa, etc.* |
| Order II: | Pleurocapsales |
| Family I: | Pleurocapsaceae |
| Genus: | *Cyanocystis, Microcoleus, Oscillatoria,*<br>*Spirulina, Lyngbya, Pseudoanabaena etc.* |
| Order IV: | Nostocales |
| Family I: | Nostocaceae |
| Genus: | *Anabaena, Nostoc, Cylindrospermum etc.* |
| Family II: | Scytonemataceae |
| Genus: | *Scytonema, Tolypothrix* |
| Family III: | Rivulariaceae |
| Genus: | Rivularia, Calothrix |
| Order V: | Stignematales |
| Family IV: | Stignemataceae |
| Genus: | *Stigonema, Chlorogloeopsis, Mastigocladus etc.* |
| | |
| SECTION XIV: | Chlorobia |
| Phylum VII: | Chlorobia |
| Class I: | Chlorobia |
| Order I: | Chlorobiales |
| Family I: | Chlorobiaceae |
| Genus: | *Chlorobium* |

Volume 2: The Proteobacteria

Kingdom II: Bacteria

| | |
|---|---|
| SECTION XV: | |
| Phylum VII: | α-Proteobacteria |
| Class I: | Rhodospirilli |
| Order I: | Rhodospirillales |
| Family I: | Rhodospirillaceae |
| Genus: | *Azospirillum, Rhodospirillum, Rhodospira, Rhodovibrio etc.* |
| Family II: | Acetobacteriaceae |
| Genus: | *Acetobacter, Acidiphilium* |

| | |
|---|---|
| Order II: | Rickettsiales |
| Family I: | Rickettsiaceae |
| Genus: | Rickettsia, Orientia |
| Family II: | Ehrlichiaceae |
| Genus: | Ehrlichia, Anaplasma |
| Family III: | Holosporaceae |
| Genus: | Holospora |
| Order III: | Rhodobacteriales |
| Family I: | Rhodobacteriaceae |
| Genus: | *Rhodobacter, Rhodovulum, Hypomonas etc.* |
| Order IV: | Sphingomonadales |
| Family I: | Sphingomonadaceae |
| Genus: | *Sphingomonas, Erythrobacter, Erythromicrobium, Zymomonas* etc. |
| Order V: | Caulobacteriales |
| Family I: | Asticcacaulis, Caulobacter |
| Order VI: | Rhizobiales |
| Family I: | Rhizobiaceae |
| Genus: | *Rhizobium, Agrobacterium* |
| Family II: | Phyllobacteriaceae |
| Genus: | *Phyllobacterium* |
| Family III: | Bartonellacae |
| Genus: | *Bartonella* |
| Family IV: | Brucellaceae |
| Genus: | *Brucella, Mycoplana* |
| Family V: | Methylobacteriaceae |
| Genus: | *Methylorhabdus, Protomonas* |
| Family VI: | Ethylocystaceae |
| Genus: | Methylocytis, *Methylosinus* |
| Family VII: | Beijerinkiacae |
| Genus: | *Beijerinckia, Derxia, Azorhizobium,* |
| Family VIII: | Hyphomicrobiaceae |
| Genus: | *Hyphomicrobium, Xanthobacter* |
| Family IX: | Bradyrhizobiaceae |
| Genus: | *Bradyrhizobium, Rhodospeudomonas Nitrococcus, Nitrospira* |

| | |
|---|---|
| SECTION XVI: | The ß-Proteobacteria |
| Class I: | Neisseriae |
| Order I: | Neisseriales |
| Family I: | Neisseriaceae |
| Genus: | *Neisseria, Aquaspirillium, Chromabacterium*, etc. |
| Order II: | Burkholderiales |

| | |
|---|---|
| Family I: | Burkholderiaceae |
| Genus: | *Burkholderia, Thermothrix* |
| Family II: | Oxalobacteriaceae |
| Genus: | *Oxalobacter, Telluria, Thermothrix* |
| Family III: | Alcaligenaceae |
| Genus: | *Achromobacter, Bordetella, Leptothrix* |
| Family IV: | Comamonadaceae |
| Genus: | *Herbaspirillum* |
| Order III: | Rhodocyclales |
| Family I: | Rhodocyclaceae |
| Genus: | *Rhodocyclus, Azoarcus* |
| Order IV: | Nitrosomonadales |
| Family I: | Nitrosomonadaceae |
| Genus: | *Nitrosomonas, Nitrospira, Nitrosococcus* |
| Family II: | Spirillaceae |
| Genus: | *Spirillum, Thiobacillus* |
| Family III: | Gallionellaceae |
| Genus: | *Gallionella* |

| | |
|---|---|
| SECTION XVII: | The γ-Proteobacteria |
| Class I: | Zymobacteria |
| Order I: | Chromatiales |
| Family I: | Chromatiaceae |
| Genus: | *Chromatium* |
| Family II: | Ectothiorhodospiraceae |
| Genus: | *Arhodomonas, Rhodochromatium, Thiocapsa Ectothiorhodospira* |
| Order II: | Xanthomonadales |
| Family I: | Xanthomonadaceae |
| Genus: | *Xanthomonas, Stenotrophomonas* |
| Order III: | Cardiobacteriales |
| Family I: | Cardiobacteriaceae |
| Genus: | *Cardiobacterium, Suttonella* |
| Order IV: | Thiotrichales |
| Family I: | Thiotrichaceae |
| Genus: | *Achromatium, Thiobacterium, Thiomicrospira Thiothrix, Vitreoscilla* |
| Family II: | Piscirickettsiaceae |
| Genus: | *Piscirickettsi* |
| Family III: | *Francisellaceae* |
| Order V: | Legionellales |
| Family I: | Legionellaceae |
| Genus: | *Legionella* |

| | |
|---|---|
| Family II: | Coxiellaceae |
| Genus: | *Coxiella, Rickettiella* |
| Order VI: | Methylococcales |
| Family I: | Methylococcaceae |
| Genus: | *Methylobacter, Methylobacterium, Methylococcus, Methylomonas* etc. |
| Order VII: | Oceanospirillales |
| Family I: | Oceanospirillaceae |
| Genus: | *Marionomonas, Oceanospirillum* |
| Family II: | Alomonadaceae |
| Genus: | *Deleya, Halomonas* |
| Order VIII: | Pseudodomonadales |
| Family I: | Pseudomonadaceae |
| Genus: | *Pseudomonas, Azotobacter* |
| Family II: | Moraxellaceae |
| Genus: | *Acinetobacter, Moraxella* |
| Order IX: | Alteromonadales |
| Family I: | Alteromonadaceae |
| Genus: | *Alteromonas, Pseudoaltermonas* |
| Order X: | Vibrionales |
| Family I: | Virbrionaceae |
| Genus: | *Photobacterium, Vibrio* |
| Order XI: | Aeromonadales |
| Family I: | Aeromonadaceae |
| Genus: | *Aeromonas* |
| Order XII: | Enterobacteriales |
| Family I: | Enterobacteriaceae |
| Genus: | *Citrobacter, Enterobacter, Erwinia, Escherichia, Hafnia, Klebsiella, Proteus, Salmonella, Serratia, Shigella, Yersinia, Pectobacterium* |
| Order XIII: | Pasteurellales |
| Family I: | Pasteurellaceae |
| Genus: | *Actinobacillus, Haemophilus, Pasteurella* |

| | |
|---|---|
| SECTION XVIII: | The δ-Proteobacteria |
| Class I: | Predibacteria |
| Order I: | Desulfovibrionales |
| Family I: | Desulfovibrionaceae |
| Genus: | *Desulfomicrobium* |
| Family II: | Desulfurellaceae |
| Genus: | *Desulfurella, Desulfovibrio* |
| Family III: | Desulfobacteraceae |
| Genus: | *Desulfohalobium, Geobacter* |

| | |
|---|---|
| Family IV: | Desulfobulbaceae |
| Genus: | *Bdellovibrio, Micavibrio* |
| Order II: | Myxococcales |
| Family I: | Myxococcaceae |
| Genus: | *Aiococcus, Myxococcus* |
| Family II: | Archangiaceae |
| Genus: | *Archangium* |
| Family III: | Cystobacteraceae |
| Genus: | *Cystobacter* |
| Family IV: | Polyangiaceae |
| Genus: | *Chondromyces, Stigmatella, Polyangium* |

| | |
|---|---|
| SECTION XIX: | The ε-Proteobacteria |
| Class I: | Campylobacters |
| Order I: | Campylobacterales |
| Family I: | Campylobacteraceae |
| Genus: | *Aerobacter, Campylobacter* |
| Family II: | Helicobacteriaceae |
| Genus: | *Helicobacter*, *Wolinella* |

Volume 3: The Low G + C Gram Positive

Kingdom II: Bacteria

Phylum VIII: Firmicutes

| | |
|---|---|
| SECTION XX: | The Clostridia and relatives |
| Class I: | Clostridia |
| Order I : | Clostridiales |
| Family I: | Clostridiaceae |
| Genus: | *Clostridium, Sarcina* |
| Family II: | Peptostreptococcaceae |
| Genus: | *Peptostreptococcus* |
| Family III: | Eubacteriaceae |
| Genus: | *Eubacterium, Helicococcus* |
| Family IV: | Lachnospiraceae |
| Genus: | *Butyrivibrio, Coprococcus* |
| Family V: | Peptococcaceae |
| Genus: | *Desulfotomaculum, Veukkibekka* |
| Family VI: | Syntrophomonadaceae |
| Genus: | *Syntrophomonas, Thermoanaerobacter* |
| Order II: | Haloanaerobiales |
| Family I: | Haloanaerobiaceae |
| Genus: | *Haloanaerobium* |

| | |
|---|---|
| Family II: | Halobacteroidaceae |
| Genus: | *Haloanaerobacter, Halocella* |
| | |
| SECTION XXI: | The Mollicutes |
| Class I: | Mollicutes |
| Order I: | Mycoplasmatales |
| Family I: | Mycoplasmataceae |
| Genus: | *Mycoplasma, Ureaplasma* |
| Family II: | Spiroplasma group |
| Genus: | *Mesoplasma, Spiroplasma* |
| Family III: | Acholeplasmataceae |
| Genus: | *Acholeplasma,Anaeroplasma* |
| Family IV: | Asteroleplasmataceae |
| Genus: | *Asteroleplasma* |
| Family V: | Erysipelothrichaeceae |
| Genus: | *Erysipelothrix, Holdemania* |
| | |
| SECTION XXII: | The Bacilli and Lactobacilli |
| Class I: | Bacilli |
| Order I: | Bacillales |
| Family I: | Bacillaceae |
| Genus: | *Bacillus* |
| Family II: | Planococcaceae |
| Genus: | *Kurthia* |
| Family III: | Caryophanaceae |
| Genus: | *Caryophanon* |
| Family IV: | Sporolactobacillaceae |
| Genus: | *Marinococcus, Sporolactobacillus* |
| Family V: | Paenibacillaceae |
| Genus: | *Paenibacillus* |
| Family VI: | Alicyclobacillaceae |
| Genus: | *Alicyclobacillus, Sulfobacillus* |
| Family VII: | Thermoactinomycetaceae |
| Genus: | *Thermoactinomyces* |
| Order II: | Lactobacillales |
| Family I: | Lactobacillaceae |
| Genus: | *Lactobacillus* |
| Family II: | Leuconostocaceae |
| Genus: | *Leuconostoc, Weissella* |
| Family III: | Streptococcaceae |
| Genus: | *Lactococcus, Pediococcus, Streptococcus* |
| Family IV: | Enterococcaceae |
| Genus: | *Enterococcus, Gagococcus* |

Family V: Carnobacteriaceae
Genus: *Carnobacterium, Lactosphaera*
Family VI: Aerococcaceae
Genus: *Abiotrophia,Tetragenococcus*
Family VII: Listeriaceae
Genus: *Brochothrix, Listeria*
Family VIII: Staphylococcaceae
Genus: *Staphylococcus, Gomella*
Family IX: Genera increate sedis
Genus: *Acetoanaerobium, Acetobacterium, Oscillospira*

Volume 4: The High G + C Gram Positives
Phylum VIII: Firmicutes

SECTION XXIII: The Actinobacteria
Class I: Actinobacteria
Subclass I: Acidimicrobidae
Order I: Acidimicrobiales
Suborder I: Acidimicrobiales
Family I: Acidimicrobiaceae
Genus: Acidimicrobium
Subclass II: Rubrobacteridae
Order II: Rubrobacterales
Suborder I: Rubrobacteridae
Family I: Rubrobacteriaceae
Genus: *Rubrobacter*
Subclass III: Cariobacteridae
Order III: Cariobacteriales
Suborder I: Cariobacteridae
Family I: Cariobacteriaceae
Genus: *Autoporium, Carbiobacterium*
Subclass IV: Sphaerobacterales
Order IV: Sphaerobacterales
Suborder I: Sphaerobacteridae
Family I: Sphaerobacteraceae
Genus: *Sphaerobacter*
Subclass V: Actinobacteridae
Order V: Actinomycedales
Suborder I: Actinomycineae
Family I: Actinomycetaceae
Genus: *Actinobaculum, Actinomyces*
*Arcanobacterium, Mobiluncus*
Suborder II: Micrococcinae

| | |
|---|---|
| Family II: | Micrococcaceae |
| Genus: | *Arthrobacter, Micrococcus* |
| Family III: | Brevobacteraceae |
| Genus: | *Brevibacterium* |
| Family IV: | Cellulonomanadaceae |
| Genus: | *Cellulomonas, Oerskovia, Rarobacter* |
| Family V: | Dermabacteraceae |
| Genus: | *Brachebaterium, Dermabacter* |
| Family VI: | Dermatophilaceae |
| Genus: | *Dermacoccus, Dermatophilus* |
| Family VII: | Intrasporangiaceae |
| Genus: | *Intrasporangium* |
| Family VIII: | Jonesiaceae |
| Genus: | *Jonesis* |
| Family IX: | Microbacteriaceae |
| Genus: | *Microbacterium, Aurobacterium* |
| Family X: | Promicromonosporaceae |
| Genus: | *Promicromonospora* |
| Suborder III: | Corynebacterineae |
| Family I: | Corynebacteriaceae |
| Genus: | *Corynebacterium* |
| Family II: | Dietziaceae |
| Genus: | *Dietzia* |
| Family III: | Gordoniaceae |
| Genus: | *Gordonia, Skermania* |
| Family IV: | Mycobacteriaceae |
| Genus | *Mycobacterium* |
| Family V: | Nocardiaceae |
| Genus: | *Micropolyspora, Nocordia, Rhodococcus* |
| Family VI: | Tsukamurellaceae |
| Genus: | *Tsukamurella* |
| Suborder IV: | Micromonospora |
| Family I: | Micromonosporaceae |
| Genus: | *Actinoplanes, Micromonospora* |
| Suborder VI: | Propionibacterineae |
| Family I: | Propionibacteriaceae |
| Genus: | *Leucoccus, Propionibacterium* |
| Family II: | Nocardioidaceae |
| Genus: | *Aeromicrobium, Nocardioides* |
| Suborder VII: | Pseudonocardineae |
| Family I: | Pseudonocardiaceae |
| Genus: | *Actinopolyspora, Amycolatopsis, Pseudonocardia, Saccharopolyspora, Saccharothrix* |

| | |
|---|---|
| Suborder VII: | Streptomycineae |
| Family I: | Streptomycetaceae |
| Genus: | *Streptomyces* |
| Suborder IX: | Streptosporangineae |
| Family I: | Streptosporangiaceae |
| Genus: | *Microbispora, Microtetraspora, Streptosporangium* |
| Family II: | Nocardiospaceae |
| Genus: | *Nocardiospsis, Thermobifida* |
| Family III:: | Thermomonosporaceae |
| Genus: | *Thermomonospora* |
| Suborder X: | Frankineae |
| Family I | Frankiaceae |
| Genus: | *Frankia* |
| Family II: | Geodermatophilaceae |
| Genus: | *Blastococcus, Geodermatophilus* |
| Family III: | Microsphaeraceae |
| Genus: | *Microsphaera* |
| Family IV: | Porichthyaceae |
| Genus: | *Sporichthya* |
| Family V: | Acidothermaceae |
| Genus: | *Acidothermus, Kineosporia* |
| Family VI: | Incertae sedis |
| Genus: | *Cryptosporangium* |
| Suborder XI: | Glycomycineae |
| Family I: | Glycomycetaceae |
| Genus: | *Glycomyces* |
| Order VI: | Bifidobacteriales |
| Family I: | Bifidobacteriaceae |
| Genus: | *Bifidobacterium, Falcivibrio* |
| Family II: | Unknown affiliation |
| Genus: | *Actinokineospora* |

Volume 5: The Planctomycetes, Spirochaetes, Fibrobacteres, Bacteroides and Fusobacteria

Kingdom: Bacteria

| | |
|---|---|
| Phylum IX: | Wall-less forms |

| | |
|---|---|
| SECTION XXIV: | The Planctomycetes and Chlamydia |
| Class I: | Planctomycetacia |
| Order I: | Planctomycetales |
| Family I: | Planctomycetaceae |
| Genus: | *Planctomyces* |

| | |
|---|---|
| Order II: | Chlamydiales |
| Family I: | Chalmydiaceae |
| Genus: | *Chlamydia* |
| | |
| SECTION XXV: | The Spirochetes |
| Phylum X: | Spirochetes |
| Class I: | Spirochaetes |
| Order I: | Spirochaetales |
| Family I: | Spirochaetaceae |
| Genus: | *Borrelia, Spirochaeta, Treponema* |
| Family II: | Serpulinaceae |
| Genus: | *Brachyspira, Serpulina* |
| Family III: | Leptospiraceae |
| Genus: | *Leptonema, Leptospira* |
| | |
| SECTION XXVI: | The Fibrobacters |
| Phylum XI: | Fibrobacter |
| Class I: | Fibrobacteres |
| Order II: | Fibrobacterales |
| Family I: | Fibrobacteraceae |
| Genus: | *Fibrobacter, Halophaga* |
| Family II: | Acidobacteriaceae |
| Genus: | *Acidobacterium* |
| | |
| SECTION XXVII: | The Bacteroides |
| Phylum XII: | Bacteroids |
| Class I: | Bacteroids |
| Order I: | Bacteroidales |
| Family I: | Bacteroidaceae |
| Genus: | *Activibrio, Bacteroides* |
| Family II: | Rikenellaceae |
| Genus: | *Marinilabilia* |
| Family III: | Prophyromonadaceae |
| Genus: | *Prophyromonas* |
| Family IV: | Prevotellaceae |
| Genus: | *Prevotella* |
| | |
| SECTION XXVIII: | The Flavobacteria |
| Phylum XIII: | Flavobateria |
| Class I: | Flavobacteria |
| Order I: | Flavobacteriales |
| Family I: | Flavobacteriaceae |
| Genus: | *Chryseobacterium, Flavobacterium* |

| | |
|---|---|
| Family II: | Myroideaceae |
| Genus: | *Myroides* |
| Family III: | Blattabacteriaceae |
| Genus: | *Blattabacterium* |

| | |
|---|---|
| SECTION XXIX: | The Sphingobacteria |
| Phylum XIV: | Spingobacteria |
| Class I: | Sphingobacteria |
| Order I: | Sphingobacteriales |
| Family I: | Sphingobacteriaceae |
| Genus: | *Pedobacter, Sphingobacterium* |
| Family II: | Saprospiraceae |
| Genus: | *Lewinella, Saprospria* |
| Family III: | Flexibacteraceae |
| Genus: | *Cytophaga, Flexibacter* |
| Family IV: | Flammeovirgaceae |
| Genus: | *Flammeovirga,* |
| Family V: | *Crenothrichaeceae* |
| Genus: | *Crenothrix, Rhodothermus* |

| | |
|---|---|
| SECTION XXX: | The Fusiforms |
| Phylum XV: | Fusibacteria |
| Class I: | Fusibacteria |
| Order II: | Fusobacteriales |
| Family I: | Fusobacteriaceae |
| Genus: | *Fusobacterium,* |
| Family II: | Genera incertae sedis |
| Genus: | *Cetobacterium, Streptobacillus* |

| | |
|---|---|
| SECTION XXXI: | Verrucomicrobium and relatives |
| Phylum XVI: | Verrucomicrobia |
| Class I: | Verrucomicrobiae |
| Order I: | Verrucomicrobiales |
| Family I: | Verrucomicrobiaceae |
| Genus: | *Prosthecobacter, Verrucomicrobium* |

# Viruses

## 8.1 INTRODUCTION

Virus diseases of plants were known long before the discovery of bacteria. Breaking of flower colour of tulips was mentioned as early as 1576. Transmission of leaf variegation was recorded in 1700. Anderson (1778) showed that seed tuber loss was due to virus. Adloph Mayer (1886), a Dutch scientist, was the first to point out that tobacco mosaic was readily transmissible and infected through sap.

Iwanowski Russian scientist, demonstrated that filtered sap from diseased tobacco plants was transmissible—filtrable bacterium. Beijernick, a Dutch microbiologist, provided evidence that the transmissible agent was not a bacterium. Stanley (1935) isolated and crystallized the infectious agent and called it Tobacco Mosaic Virus (virus = Latin for poison). Electron microscope was invented and this enabled viruses and bacteria to be studied in detail. Bawden (1964) defined viruses as submicroscopic, infective entities that multiply only intracellularly and are pathogenic. Hahons (1964) defined viruses as bits of infectious heredity in search of chromosome.

Bawden and Pirie (1936) isolated TMV and stated that it is a crystalline nucleo-protein containing nucleic acid of the pentose type. Hershey and Chase (1952) demonstrated the transfer of nucleic acid in viral replication when *Escherichia coli* was infected by a bacteriophage. Fraenkel-Conrat (1956) demonstrated the infectivity of naked TMV RNA. Electron microscopy and X-Ray crystallography have been the major techniques used to explore the structure of viruses. Diesner (1971) coined the term "viroid" to denote sub-viral pathogens, which have short strand of RNA with a molecular weight of about 75000–125000 daltons.

## 8.2 ORIGIN OF VIRUSES

Viruses are the living ultramicroscopic organisms, which are dead outside the living host and alive when live inside the host. If they are not cells and require living cells for replication, then viruses must not have been present prior to cellular evolution and could therefore have coevolved with cells. They were once cells but retained only information to replicate themselves using host machinery. Viruses have evolved from plasmids (plasmids are self-replicating independent DNA) or RNA viroids. These "early" viruses did not contain genes for capsids. As viroids moved from one cell to another, it picked up such genes.

## 8.3 GENERAL CHARACTERISTICS

Viruses (poisons) are obligate parasites, filtrable and infectious particles as demonstrated with plant sap. (20 nm to 14000 nm in size). Most viruses are smaller than bacteria (e.g., Adenovirus). Vaccinia virus is of the same size as small bacteria like Mycoplasma, Rickettsia and Chlamydia. These contain either DNA or RNA (never both) surrounded by a protein coat (capsid) and sometimes covered by an envelope. These are obligate intracellular parasites, which multiply inside living cells using cellular synthesis machinery. Viruses contain one type of nucleic acid, either DNA or RNA. Virion is the infectious agent and is produced from sole nucleic acid. These are incapable of growing or undergoing binary fission. Lipman system is absent in viruses (system for producing energy); viruses make use of ribosomes of their host cells.

## 8.4 HOST RANGE

Most viruses infect specific host cells i.e., are host specific. Host specificity is due to specific attachment sites on the host cells called receptors. Receptor sites for bacteriophage are found in bacterial cell walls, fimbriae, or flagella. Animal cell membranes contain receptors for animal viruses and cellular factors required for viral multiplication in the host cells.

## 8.5 STRUCTURE

Viruses contain only one nucleic acid, which may be linear or circular, single or double-stranded. It exists as a single piece or multi-segmented pieces. The protein coat surrounding the nucleic acid is called capsid. Capsid accounts for most of the mass of the virion. Its structure is determined by the viral nucleic acid and is made up of protein sub-units called capsomers (Fig. 8.1). Capsomers have single or several types of proteins. Arrangement of capsomers is characteristic of a particular type of virus and capsomers determine the antigenicity of the virion. In animal viruses, capsid is surrounded by an envelope. The envelope contains lipids, proteins and carbohydrates and is derived from host cells. They may also contain some proteins synthesized from virion genes or some material derived from normal host cell components which are covered by spikes. Spikes are made up of carbohydrate and protein complexes and project out into space from the surfaces of the envelope used for attachment to the host cells. Influenza viruses bind to RBC to from hemagglutination.

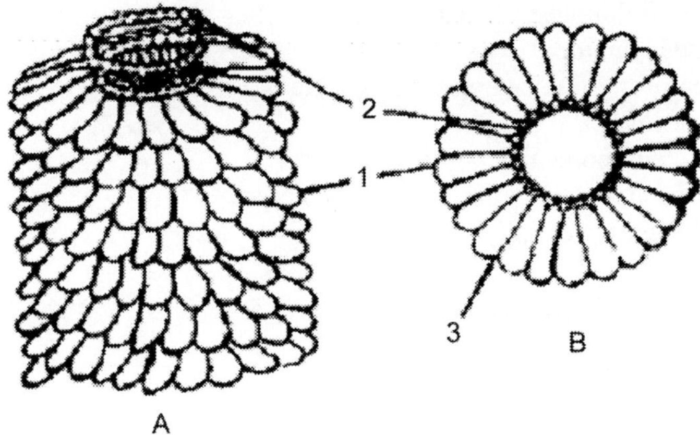


**Fig. 8.1.** Ultrastructure of Tobacco Mosaic Virus. A: Portion of virus in surface view showing structure; B: Section of the rod; 1: Capsomers; 2: RNA; 3: Core.

Four morphological types of viruses are recognized on the basis of structure of the capsid and envelope geometry:

1. **Helical viruses**—nucleic acid is surrounded by capsids, which form long hollow rods (TMV, bacteriophage M13).
2. **Polyhedral virsues**—capsids forming multiple (many) sides. Icosahedron is a regular polyhedron with 20 triangular faces and 12 corners (Adenovirus).
3. **Enveloped**—Pleomorphic to spherical shapes due to the presence of an envelope. Enveloped helical virus—Influenza virus, enveloped polyhedral virus—Herpes Simplex.
4. **Complex**—Complex structures in case of Lambda, T-odd and T-even, polyhedral capsid + nucleic acid forms the head, helical tail (sheath, tail fibres + pin, and baseplate).

## 8.6 GENERAL PROPERTIES OF VIRUSES

Viruses are very common and every living organism seems to be the host for some viruses. Viruses can multiply only within a host living cell. A given virus can usually grow only within one species of plant or animal or only within a few related species or genera. Viruses are of two types—temperate (latent or lysogenic) and virulent. Plant viruses do not attack animals and animal viruses do not attack plants. Temperate viruses are embedded within the host's nucleic acid. Bacteria, algae, fungi, plants, animals, and humans are attacked by viruses. Often viruses reside in their host without causing any disease or symptom.

### 8.6.1. Structure

1. **Nucleic Acid**—Contains 3-400 genes.

**Deoxyribonucleic Acid (DNA)—Unique Features**

- Single and/or double-stranded
- Glycosylated and/or
- Gaps present in double-stranded molecule
- Circular or linear
- Bound protein molecules
- Unique purine and/or pyrimidine bases present
- Ribonucleotides present

### Ribonucleic Acid (RNA)

- Single or double-stranded
- Segmented or unsegmented
- Bound protein molecules
- Unique purine and/or pyrimidine bases present

2. **Capsid**—The capsid accounts for most of the virus mass and is the protein coat of the virus. It is a complex and highly organized entity, which gives shape to the virus. Subunits called protomeres aggregate to form capsomeres, which in turn aggregate to form the capsid.
3. **Envelope**—This is an amorphous structure composed of lipid, protein and carbohydrate, which lies outside the capsid. It contains a mosaic of antigens from the host and the virus. A naked virus is one without an envelope.
4. **Spikes**—These are glycoprotein projections, which have enzymatic and/or adsorption and/or hemagglutinating activity. They arise from the envelope and are highly antigenic. Spikes are present on envelope of animal viruses.

## 8.6.2. Morphology (Symmetry)

**Icosahedral**—The protomeres aggregate in groups of five or six to form the capsomere. Capsomeres are recognized as regularly spaced rings with a central hole. The shape and dimensions of the icosahedron depends on the characteristics of its protomeres. All icosahedral capsids have 12 corners each occupied by a penton capsomere and 20 triangular faces, each containing the same number of hexon capsomeres. Icosahedral symmetry is identical to cubic symmetry (Fig. 8.2).

**Iridoviridae** **Herpesviridae**
**Adenoviridae** **Papovaviridae** **Parvoviridae**

2. **Helical**—The protomeres are not grouped in capsomeres, but are bound to each other so as to form a ribbon-like structure. This structure folds into a helix because the protomeres are thicker at one end than at the other. The diameter of the helical capsid is determined by characteristics of its protomeres, while its length is determined by the length of the nucleic acid it encloses.

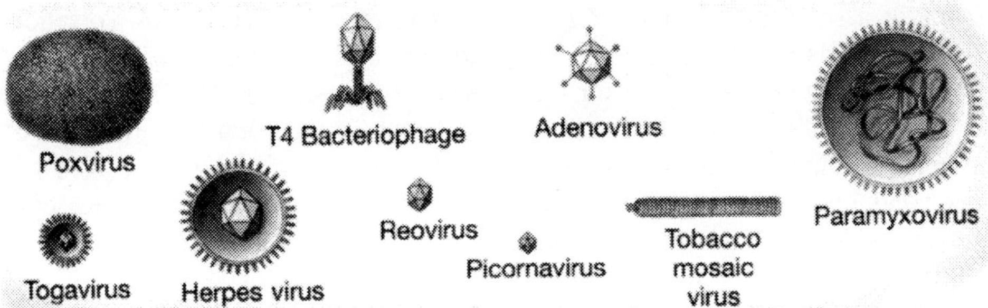


**Fig. 8.2.** Some types of viruses.

3. **Complex**—This group comprises all those viruses which do not fit into either of the above two groups e.g., that exhibited by poxvirus and rhabdovirus.

## 8.7 REPLICATION CYCLE (MULTIPLICATION)

Virus multiplication takes place in living cells and is divided into five phases.

1. **Adsorption**—Viruses can enter cells via phagocytosis, viropexis or adsorption. Adsorption is the most common and highly specific process. It requires the interaction of a unique protein on the surface of the virus with a highly specific receptor site on the surface of the cell.
2. **Penetration**—This occurs by one or more processes.
    - Enveloped viruses fuse their envelope with the membrane of the host cell. This involves local digestion of the viral and cellular membranes, fusion of the membranes and concomitant release of the nucleocapsid into the cytoplasm.
    - Naked viruses bind to receptor sites on the cellular membrane, digest the membrane and enter into the cytoplasm intact.
    - Both naked and enveloped viruses can be ingested by phagocytic cells. However, in this process they enter the cytoplasm enclosed in a cytoplasmic membrane derived from the phagocytic cell.

3. **Uncoating**—During this stage, cellular proteolytic enzymes digest the capsid away from the nucleic acid. This always occurs in the cytoplasm of the host cell. The period of the replication cycle between the end of the uncoating stage and maturation of new viral particles is termed the eclipse. Thus, during the eclipse stage, no complete viral particles can be viewed within the cell.
4. **Replication of Nucleic Acid**—Replication of viral nucleic acid is a complex and variable process. The specific process depends on the type of nucleic acid.

   **DNA virus replication:** With the exception of the poxviruses, all DNA viruses replicate in the nucleus. In some cases, one of the DNA strands is transcribed (in others both strands of a small part of the DNA may be transcribed) (step 4) into specific mRNA, which in turn is translated (step 5) to synthesize virus-specific proteins such as tumour antigen and enzymes necessary for biosynthesis of virus DNA. This period encompasses the early virus functions. Host cell DNA synthesis is temporarily elevated and is then suppressed as the cell shifts over to the manufacture of viral DNA (step 6). As the viral DNA continues to be transcribed, late virus functions become apparent. Messenger RNA transcribed during the later phase of infection (step 6) migrates to the cytoplasm and is translated (step 7). Proteins for virus capsids are synthesized and transported to the nucleus to be incorporated into the complete virion (step 8).

Assembly of the protein sub-units around the viral DNA results in the formation of complete virions (step 9), which are released after cell lysis.

The single-stranded DNA viruses first form a double-stranded DNA, utilizing a host DNA-dependent DNA polymerase. They then undergo a typical replication cycle.

**RNA virus replication:** With the exception of the orthomyxoviruses and retroviruses, all RNA viruses replicate in the cytoplasm of the host cell. The exact process varies with the species of virus. The single-stranded RNA that is released after uncoating will act as either: (a) the mRNA to synthesize viral-coded proteins; or (b) a template to synthesize mRNA; or (c) a template to synthesize double-stranded RNA, which is then used as a template to synthesize mRNA; or (d) a template to synthesize double-stranded DNA, which is then utilized as a template to synthesize mRNA. This latter process occurs only with the retroviruses (oncornaviruses).

The replication of poliovirus, which contains a single-stranded RNA as its genome, provides a useful example. All of the steps are independent of host DNA and occur in the cell cytoplasm. Polioviruses absorb to cells at specific cell receptor sites (step 1), losing in the process one virus polypeptide. The sites are specific for virus coat-cell interactions. After attachment, the virus particles are taken into the cell by viropexis (similar to pinocytosis) (step 2), and the viral RNA is uncoated (step 3). The single-stranded RNA then serves as its own messenger RNA. This messenger RNA is translated (step 4), resulting in the formation of an RNA-dependent RNA polymerase that catalyzes the production of a replication intermediate (RI), a partially double-stranded molecule consisting of a complete RNA strand and numerous partially completed strands (step 5). At the same time, inhibitors of cellular RNA and protein synthesis are produced. Synthesis of (+) and (–) strands of RNA occurs by similar mechanisms. The RI consists of one complete (–) strand and many small pieces of newly synthesized (+) strand RNA (step 6). The replicative form (RF) consists of two complete RNA strands, one (+) and one (–).

The single (+) strand RNA is made in large amounts and may perform any one of three functions: (a) serve as messenger RNA for synthesis of structural proteins; b) serve as template for continued RNA replication; or (c) become encapsulated, resulting in mature progeny virions. The synthesis of viral capsid proteins (step 7) is initiated at about the same time as RNA synthesis.

The entire poliovirus genome acts as its own mRNA, forming a polysome of approximately 350S, and is translated to form a single large polypeptide that is subsequently cleaved to produce the various viral capsid polypeptides. Thus, the poliovirus genome serves as a polycistronic messenger molecule. Poliovirus contains four polypeptides.

5. **Maturation and Release**

   *Naked viruses*—Maturation consists of two main processes: the assembly of the capsid, and its association with the nucleic acid. Maturation occurs at the site of nucleic acid replication. After they are assembled into mature viruses, naked virions may become concentrated in large numbers at the site of maturation, forming inclusion bodies. Naked virions are released in different ways, which depends on the virus and the cell type. Generally, RNA-containing naked viruses are released rapidly after maturation and there is little intracellular accumulation, therefore, these viruses do not form predominant inclusion bodies. On the other hand, DNA-containing naked icosahedral viruses that mature in the nucleus do not reach the cell surface as rapidly, and are released when the cells undergo autolysis or in some cases are extruded without lysis. In either case they tend to accumulate within the infected cells over a long period of time. Thus, they generally produce highly visible inclusion bodies.

   *Enveloped viruses*—In the maturation of enveloped viruses, a capsid must first be assembled around the nucleic acid to form the nucleocapsid, which is then surrounded by the envelope. During the assembly of the nucleocapsid, virus-coded envelope proteins are also synthesized. These migrate to the plasma membrane (if assembly occurs in the cytoplasm) or nuclear membrane (if assembly occurs in the nucleus) and become incorporated into that membrane. Envelopes are formed around the nucleocapsids by budding of cellular membranes.

   *Complex viruses*—These viruses, of which the poxvirus is a good example, begin the maturation process by forming multilayered membranes around the DNA. These layers differentiate into two membranes: The inner one contains the characteristic nucleoid, while the external one acquires the characteristic pattern of the surface of the virion. These form very characteristic cytoplasmic inclusion bodies. The viruses are generally released from the cell via cell lysis.

## 8.8 COMPOSITION AND STRUCTURE OF BACTERIOPHAGE

### 8.8.1 Composition

Although different bacteriophages may contain different materials, they all contain nucleic acid and protein. Depending upon the phage, the nucleic acid can be either DNA or RNA but not both, and it can exist in various forms. The nucleic acids of phages often contain unusual or modified bases. These modified bases protect phage nucleic acid from nucleases that break down host nucleic acids during phage infection. The size of the nucleic acid varies depending upon the phage. The simplest phages have enough nucleic acid to code for 3–5 average size gene products, while the more complex phages may code for over 100 gene products.

The number of different kinds of protein and the amount of each kind of protein in the phage particle will vary depending upon the phage. The simplest phage may have many copies of only one or two different proteins, while more complex phages may have many different kinds. The proteins function in infection and to protect the nucleic acid from nucleases in the environment.

### 8.8.2 Structure

Bacteriophage comes in many different sizes and shapes. The basic structural features of bacteriophages are illustrated in Fig. 8.3, which depicts the phage T4.

1. *Size*—T4 is among the largest phages; it is approximately 200 nm long and 80–100 nm wide. Other phages are smaller. Most phages range in size from 24–200 nm in length.
2. *Head or capsid*—All phages contain a head structure, which can vary in size and shape. Some are icosahedral (20 sides), others are filamentous. The head or capsid is composed of many copies of one or more different proteins. Inside the head is found the nucleic acid. The head acts as the protective covering for the nucleic acid.
3. *Tail*—Many but not all phages have tails attached to the phage head. The tail is a hollow tube through which the nucleic acid passes during infection. The size of the tail can vary and some phages do not even have a tail structure. In the more complex phages like T4, the tail is surrounded by a contractile sheath, which contracts during infection of the bacterium. At the end of the tail the more complex phages like T4 have a base plate and one or more tail fibres attached to it. The base plate and tail fibres are involved in the binding

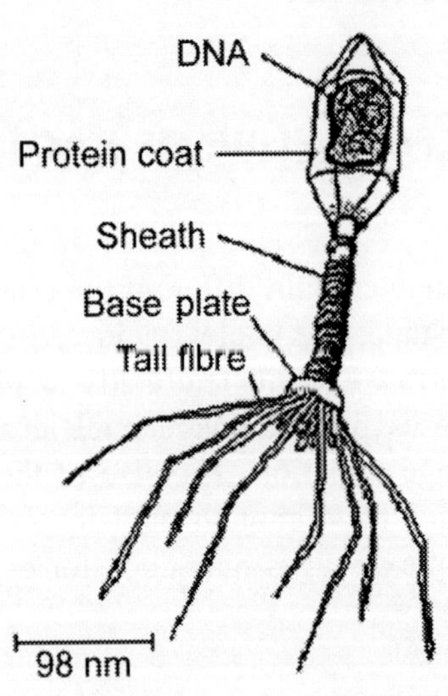


**Fig. 8.3.** Bacteriophage.

of the phage to the bacterial cell. Not all phages have base plates and tail fibres. In these instances, other structures are involved in binding of the phage particle to the bacterium.

### 8.8.3 *Phage Lambda* (λ)

This belongs to a group of phages, which exhibit the property of lysogeny and are known as lysogenic phages. In the lysogenic state, the viral genome gets integrated with the host chromosome and replicates along with the host during cell division. In this condition, the phage is called a temperate phage or prophage. Under certain conditions, the phage nucleic acid is separated from the host chromosome, becoming a plasmid. It then replicates as other phages do and causes lysis of the cell, releasing progeny virus particles. This state is known as the lytic condition and the phage is then said to be a virulent phage (Fig. 8.4).

Phage "l" is the most extensively studied lysogenic phage. The virion has an icosahedral head, 65 nm in diameter and a helical tail (150 nm) with a tail fibre of 23 nm. The nucleic acid is a linear dsDNA, with a molecular weight of 31 × 106 daltons. It assumes a circular form when occurring freely in a host cell. In this form it has been sequenced completely. There are 48503 base pairs. Genes responsible for both lysogenic and lytic actions are present in the genomes, and which action prevails is dependent on the cell environment.

Phage "l" has been a very useful vehicle as a cloning vector, as it is possible to form hybrid DNA by replacing parts of its DNA with foreign DNA. This is possible because the entire central part of its genome is not necessary for its replication but is required only to induce lysogeny (integration with host chromosome). It has been possible to introduce upto 35 to 40 eukaryotic genes into its central portion. Such a hybrid is known as a cosmid. When such a DNA-Cos hybrid is introduced into *E. coli,* it behaves as a plasmid. Thus, eukaryotic genes can be introduced into bacteria. This is an important aspect of biotechnology. Human health products such as insulin and interferon are now produced this way.

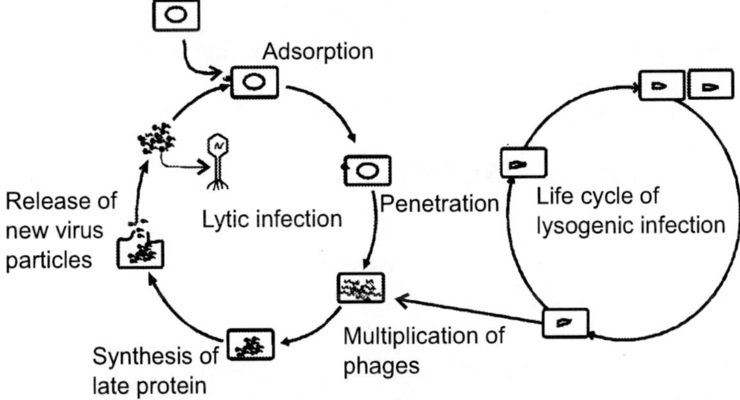


**Fig. 8.4.** Lytic and lysogenic life cycle of bacteriophage.

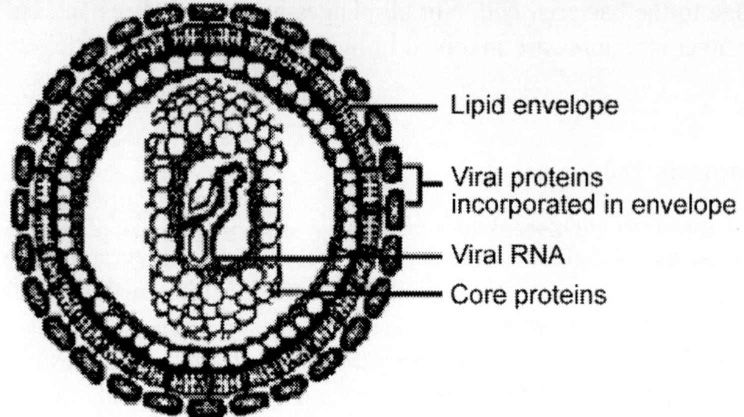


**Fig. 8.5.** Model structure of HIV virus.

### 8.8.4 *HIV (Retroviridae)*

The human immunodeficiency virus (HIV) is the causative agent of AIDS (Acquired Immunodeficiency Syndrome), which has become a worldwide problem in recent years (Fig. 8.5). It attacks the immune system, especially T-cells and macrophages in the blood, as a result of which the patient gradually loses resistance and succumbs to other diseases. It is transmitted through direct contact of blood. Hence, homosexuals and drug users are more vulnerable. Transfusion of blood collected from HIV positive people is also an important source.

The retroviruses are also oncogenic (cancer producing), implicated in some cancers. They are all unique in that their genome is RNA, but they replicate through a DNA intermediate. This is coded by the enzyme RNA-dependent DNA polymerase (reverse transcriptase), and hence the name is enveloped and complex. It is about 100 nm in diameter. The lipoprotein envelope has spikes made up of two proteins. There is an inner protein shell enclosing an icosahedral capsid. The core consists of several enzymes including reverse transcriptase, besides the two RNA strands. The DNA intermediate gets integrated with the host chromosome, hence control of the HIV with vaccines has been difficult. Much current research is engaged on this aspect. One approach is to develop a ribozyme, an enzyme that breaks up viral RNA so that the DNA intermediate is not formed.

## 8.9 CLASSIFICATION OF VIRUSES

A universal system for classifying viruses, and a unified taxonomy has been established by the International Committee on Taxonomy of Viruses (ICTV) since 1966. The system makes use of a series of ranked taxons, with the:

- Order (-virales) being the highest currently recognised
- Family (-viridae)

- Subfamily (-virinae)
- Genus (-virus)
- Species (e.g., tobacco mosaic virus)

For example, the Ebola virus from Kikwit is classified as:

- Order Mononegavirales
- Family Filoviridae
- Genus Filovirus
- Species *Ebola virus Zaire*

*Taxonomic criteria:* The most important taxonomic criteria are—Host Organism(s), Particle Morphology and Genome Type.

Although a number of other criteria such as disease symptoms, antigenicity, protein profile, host range, etc. are important in precise identification, consideration of the above three criteria are sufficient in most cases to allow identification of a virus down to familial if not generic level.

Hosts of viruses include all classes of cellular organisms described to date. Broadly they are:

1. Prokaryotes—Archaea, Bacteria, Mycoplasma and Spiroplasma
2. [illegible]—Algae, Plants, Protozoa, Fungi, Invertebrates and Vertebrates

Certain virus families/groupings cross "kingdom" or phylum boundaries. For example, virus families infecting two kingdoms of organisms are:

- Bunyaviridae (animals and plants)
- Partitiviridae (plants and fungi)
- Reoviridae (animals and plants)
- Rhabdoviridae (animals and plants)
- Phycodnaviridae (protozoa and plants)
- Picornaviridae (plants and animals—tentative)
- Totiviridae (protozoa / fungi and insects—tentative)

Virus families infecting across different phyla (all infecting insects and vertebrates) are:

- Flaviridae
- Iridoviridae
- Parvoviridae
- Poxviridae
- Togaviridae

# Growth and Kinetics of Microorganisms

## 9.1 INTRODUCTION

Growth is defined as an increase in the quantity of cellular constituents and structures of an organism. When growth occurs in the absence of cell division, it usually results in an increase in the size and weight of the cell. In most microorganisms, growth is followed by cell division. This results in an increase in cell number, the new cells formed eventually attaining the same size as the original cell.

It is important to distinguish between the growth of individual cell and the growth of populations of cells. The growth of a cell results in an increase in its size and weight and is usually a prelude to cell division. Population growth, on the other hand, results in an increase in the number of cells as a consequence of cell growth and division. Because of their small size, the study of cell growth in microbes is difficult; most growth studies in microbiology deal with populations of cells.

The growth of a population of bacterial cells generally occurs at an exponential rate. Exponential growth is a consequence of the fact that after a cell divides to form two daughter cells, each daughter cell in turn divides and produces two new cells, so that at each division period, the population doubles. If the number of cells is plotted versus time on an arithmetic graph, the line obtained curves upward at a progressively increasing rate, whereas when the logarithm of the number of cells is plotted, a straight line results (Fig. 9.1). Population growth of this type is called exponential growth and is characteristic of unicellular organisms.

The rate of exponential growth is usually expressed as generation time, or doubling time, which is the time taken for the population to double. The growth rate can also be expressed as the number of generations per hour.

The simplest way to determine generation time is from a graph in which the logarithm number of cells is plotted against time; the time for the population to double is read directly from the graph.

One of the characteristics of exponential growth is that the rate of increase in cell number is slow initially but increases at a faster rate, resulting in an increase in cell number.

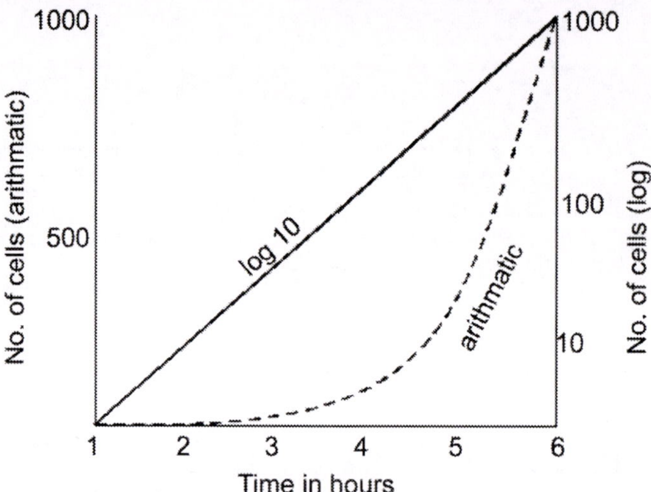


**Fig. 9.1.** Arithmetic and exponential growth.

The number of cells in a population is measured directly by counting under the microscope, a method called the direct microscopic count. The small size of most microbes and their large population densities make it necessary to use special chambers such as the Petroff-Hausser or haemocytometer to count the number of cells in a sample. Direct microscopic counting cannot distinguish living from dead cells, and it is not very sensitive.

## 9.2 MEASUREMENT OF GROWTH

Growth of microorganisms is an orderly increase of all chemical components. Orderly increase means that during a period of balanced growth, the doubling of biomass is accompanied by doubling of all other measurable properties of the population e.g., protein, RNA, DNA and intracellular water. Thus different properties are used for measurement of bacterial growth by estimation of cell number, cell mass or density and protein content.

### 9.2.1 Bacteria

#### (a) Viable count

Viable cell counting is usually done by determining the number of cells in a population capable of dividing and forming colonies. This is the most commonly used method for determination of viable cell count. It is based on the principle that a single viable cell gives rise to a single colony. The colony forming units (cfu) per ml or per g of the culture can be calculated on the basis of dilution factor. The plate count method is most widely used. Different techniques viz. streak method, spread method and pour plate method for counting bacterial population has been described in Chapter IX.

The plate count is very sensitive because in principle any viable cell, when placed on an appropriate medium, will give rise to a colony. In addition to sensitivity, the colony that is formed can serve as the inoculum for a pure culture, which can be identified taxonomically. Furthermore, since different organisms often produce colonies of different shape, size, texture, and colour. Several kinds of organisms can be counted in a mixture. For dense suspensions, viable counting requires considerable dilution of the sample, a condition that introduces the possibility of error.

### (b) Turbidometric method

This technique is based on the fact that small particles scatter light proportionally within certain limits, to their concentration. When a beam of light is passed through a suspension of bacteria, the reduction in the amount of light transmitted as a consequence of scattering is thus a measure of the bacterial mass present.

### (c) Estimation of protein

Different methods such as Lowry's procedure can be used for the estimation of protein (see annexure).

### (d) Estimation of bacterial population by most probable number (MPN) method

If for some reason agar cannot be used, viable counts can still be performed with liquid media using the most-probable-number (MPN) technique.

The most probable number (MPN) technique is based on a determination of the presence/absence of microorganisms in several individual aliquots of each of several consecutive dilutions of soil or other materials.

A prerequisite of the method is that the microorganism whose population is to be determined must be able to bring about some characteristics and readily recognisable transformation in the medium into which it is inoculated, or else the microorganism itself, after it has undergone multiplication, must be easily recognisable in the substrate.

A tenfold dilution series from $10^1$ to $10^8$ are prepared and 1 ml of aliquot from each of the dilutions is disposed into tubes containing the appropriate medium for the growth of microorganism concerned. Sizeable number of replications are maintained for each dilution and the tubes are incubated at 30–32°C for the required period of growth.

To calculate the MPN count of microorganisms in the original sample, three parameters, $P_1$, $P_2$ and $P_3$ are considered. Where $P_1$ is the number of positive tubes in the least concentrated dilution in which all tubes are positive, or in which the greatest number of tubes are positive, and $P_2$ and $P_3$ are the number of positive tubes in the next two higher dilutions.

From the MPN table in which $P_1$ and $P_2$ correspond to the values observed experimentally are noticed. In the next step, row number across the table corresponding to the $P_3$ value is found. The figure at the point of interaction is the most probable number organisms in the quantity of the original sample represented in the inoculum added in the second dilution. This figure is multiplied by the dilution factor to obtain MPN of the original sample.

Table of most probable numbers for use with 10-fold dilutions and 5 tubes per dilution (Cochran, 1950).

| | | Most probable number for indicated values of $p_3$ | | | | | |
|---|---|---|---|---|---|---|---|
| $p_1$ | $p_2$ | 0 | 1 | 2 | 3 | 4 | 5 |
| 0 | 0 | - | 0.018 | 0.036 | 0.054 | 0.072 | 0.090 |
| 0 | 1 | 0.018 | 0.036 | 0.055 | 0.073 | 0.091 | 0.11 |
| 0 | 2 | 0.037 | 0.055 | 0.074 | 0.092 | 0.11 | 0.13 |
| 0 | 3 | 0.056 | 0.074 | 0.093 | 0.11 | 0.13 | 0.15 |
| 0 | 4 | 0.075 | 0.094 | 0.11 | 0.13 | 0.15 | 0.17 |
| 0 | 5 | 0.094 | 0.11 | 0.13 | 0.15 | 0.17 | 0.19 |
| 1 | 0 | 0.020 | 0.040 | 0.060 | 0.080 | 0.10 | 0.12 |
| 1 | 1 | 0.040 | 0.061 | 0.081 | 0.10 | 0.12 | 0.14 |
| 1 | 2 | 0.061 | 0.082 | 0.10 | 0.12 | 0.15 | 0.17 |
| 1 | 3 | 0.083 | 0.10 | 0.13 | 0.15 | 0.17 | 0.19 |
| 1 | 4 | 0.11 | 0.13 | 0.15 | 0.17 | 0.19 | 0.22 |
| 1 | 5 | 0.13 | 0.15 | 0.17 | 0.19 | 0.22 | 0.24 |
| 2 | 0 | 0.045 | 0.068 | 0.091 | 0.12 | 0.14 | 0.16 |
| 2 | 1 | 0.063 | 0.092 | 0.12 | 0.14 | 0.17 | 0.19 |
| 2 | 2 | 0.093 | 0.12 | 0.14 | 0.17 | 0.19 | 0.22 |
| 2 | 3 | 0.12 | 0.14 | 0.17 | 0.20 | 0.22 | 0.25 |
| 2 | 4 | 0.15 | 0.17 | 0.20 | 0.23 | 0.25 | 0.28 |
| 2 | 5 | 0.17 | 0.20 | 0.23 | 0.26 | 0.29 | 0.32 |
| 3 | 0 | 0.078 | 0.11 | 0.13 | 0.16 | 0.20 | 0.23 |
| 3 | 1 | 0.11 | 0.14 | 0.17 | 0.20 | 0.23 | 0.27 |
| 3 | 2 | 0.14 | 0.17 | 0.20 | 0.24 | 0.27 | 0.31 |
| 3 | 3 | 0.17 | 0.21 | 0.24 | 0.28 | 0.31 | 0.35 |
| 3 | 4 | 0.21 | 0.24 | 0.28 | 0.32 | 0.36 | 0.40 |
| 3 | 5 | 0.25 | 0.29 | 0.32 | 0.37 | 0.41 | 0.45 |
| 4 | 0 | 0.13 | 0.17 | 0.21 | 0.25 | 0.30 | 0.36 |
| 4 | 1 | 0.17 | 0.21 | 0.26 | 0.31 | 0.36 | 0.42 |
| 4 | 2 | 0.22 | 0.26 | 0.32 | 0.38 | 0 44 | 0.50 |
| 4 | 3 | 0.27 | 0.33 | 0.39 | 0.45 | 0.52 | 0.59 |
| 4 | 4 | 0.34 | 0.40 | 0.47 | 0.54 | 0.62 | 0.69 |
| 4 | 5 | 0.41 | 0.48 | 0.56 | 0.64 | 0.72 | 0.81 |
| 5 | 0 | 0.23 | 0.31 | 0.43 | 0.58 | 0.76 | 0.95 |
| 5 | 1 | 0.33 | 0.46 | 0.64 | 0.84 | 1.1 | 1.3 |
| 5 | 2 | 0.49 | 0.70 | 0.95 | 1.2 | 1.5 | 1.8 |
| 5 | 3 | 0.79 | 1.1 | 1.4 | 1.8 | 2.1 | 2.5 |
| 5 | 4 | 1.3 | 1.7 | 2.2 | 2.8 | 3.5 | 4.3 |
| 5 | 5 | 2.4 | 3.5 | 5.4 | 9.2 | 16 | - |

The 95% confidence limits for MPN values can be calculated from the prepared table. To find the upper limit, the MPN value is to be multiplied by the appropriate factor from the table. To get the lower limit, the MPN value is to be divided by the same factor.

Factors for calculating the confidence limits for the most probable number count (Cochran, 1950).

| Number of tubes per dilution (*n*) | Factor for 95% confidence limits with indicated dilution ratios | | | |
|---|---|---|---|---|
| | **2** | **4** | **5** | **10** |
| 1 | 4.00 | 7.14 | 8.32 | 14.45 |
| 2 | 2.67 | 4.00 | 4.47 | 6.61 |
| 3 | 2.23 | 3.10 | 3.39 | 4.68 |
| 4 | 2.00 | 2.68 | 2.88 | 3.80 |
| 5 | 1.86 | 2.41 | 2.58 | 3.30 |
| 6 | 1.76 | 2.23 | 2.38 | 2.98 |
| 7 | 1.69 | 2.10 | 2.23 | 2.74 |
| 8 | 1.64 | 2.00 | 2.12 | 2.57 |
| 9 | 1.58 | 1.92 | 2.02 | 2.43 |
| 10 | 1.55 | 1.86 | 1.95 | 2.32 |

The tables apply only to the particular number of tubes and dilutions ratios specified. Examination of the table factors for 95% confidence limits reveals that the confidence intervals can be narrowed by increasing the number of replications in each dilution or by narrowing the dilution ratios.

**Example:**

| Concentration | Number of positive tubes |
|---|---|
| $10^4$ | 5 |
| $10^5$ | 5 |
| $10^6$ | 4 |
| $10^7$ | 3 |
| $10^8$ | 0 |
| MPN $= 2.8 \times 10^6$ | |

From the table with 5 replications in each dilution and a dilution ratio of 10, the factor is 3.3.

Therefore, the upper confidence limit $= 2.8 \times 3.3 \times 10^6$

$= 9.24 \times 10^6$

$$\text{Lower confidence limit} = 2.8 / 3.3 \times 10^6 = 0.848 \times 10^6 = 8.48 \times 10^5$$

$$\text{Number of cells / ml} = 8.48 \times 10^5 - 9.24 \times 10^6$$

### Measurement of Microbial Mass

For many studies, especially those dealing with the biochemistry of growth processes, we are interested in determining the biomass, which can be measured directly by determining either the dry or the wet weight of a sample taken from the population. Dry weight is usually about 20 to 25 percent of the wet weight.

The average weight of a single cell can be calculated by measuring the weight of a large number of cells and dividing by the total number of cells present, as measured by direct microscopic count. Prokaryotic cells range in dry weight from less than $10^{-15}$ g to greater than $10^{-11}$ g, and the cells of eukaryotic microorganisms range from about $10^{-11}$ g to about $10^{-7}$ g.

In dense suspensions, turbidity measurements can be used to quantify growth, and a variety of commercial instruments are available for this purpose. Turbidity is more closely related to cell mass than cell number. Although turbidity measurements can be made quickly and accurately without disturbing a culture, they are not very sensitive, as a suspension of bacteria does not show visible turbidity until the density is about $10^7$ bacteria per millimeter.

## 9.2.2 Growth Curve

A typical growth curve of bacterial cells is illustrated in Fig. 9.2. This growth curve can be divided into several distinct phases, namely lag phase, exponential phase, stationary phase, and death phase.

***Lag phase:*** When a microbial population is inoculated into a fresh medium, growth usually does not begin immediately, but only after a period of time called the lag phase, which may be brief or extended, depending on the conditions. If an exponentially growing culture is inoculated into the same medium under the same conditions of growth, a lag is not seen and exponential growth continues at the same rate. However, if the inoculum is taken from an old (stationary phase) culture and inoculated into the same medium, a lag usually occurs even if all the cells in the inoculum are alive. This is because the cells are usually depleted of various essential coenzymes or other cell constituents, and time is required for resynthesis. A lag also ensures when the inoculum consists of cells that have been damaged (but not killed) by treatment with heat, radiation, or toxic chemicals due to the time required for the cells to repair the damage.

A lag is also observed when a population is transferred from a rich culture medium to a poorer one. This happens because for growth to occur in a particular culture medium, the cells must have a complete complement of enzymes for the synthesis of essential metabolites not present in the medium. On transfer to a new medium, time is required for the synthesis of new enzymes.

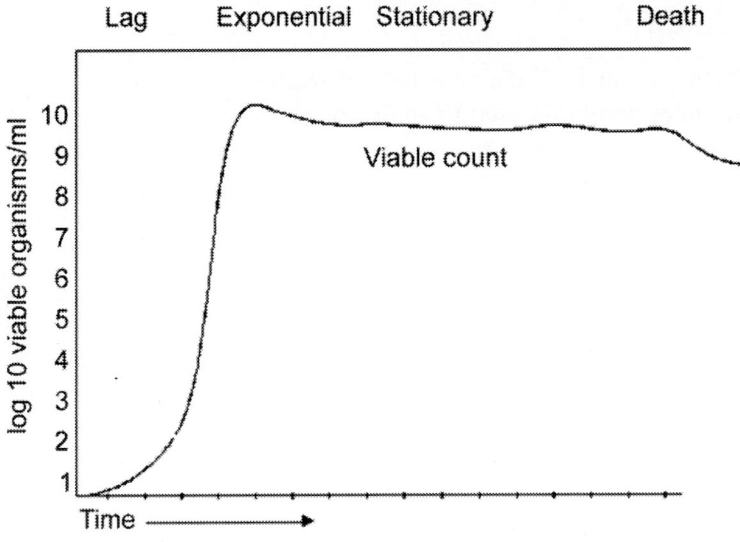


**Fig. 9.2.** Growth curve of bacteria.

***Exponential phase:*** The exponential phase was shown to be a consequence of the fact that each cell in a population divides into two cells. Under optimal conditions, growth rates of different organisms vary widely, apparently, owing to inherent genetic limitations. In general, prokaryotes grow faster than eukaryotes, although there are some notably slow-growing prokaryotes. For a given organism, growth rate varies with environmental conditions and composition of the culture medium. In a rich medium with many performed cell constituents, growth rate is usually faster than in a poor medium in which the cell must synthesize these things for itself.

***Stationary phase:*** Cells cannot grow indefinitely at an exponential rate. Growth limitation occurs either because of the exhaustion of essential nutrient(s) or accumulation of some toxic metabolic product. The period during which the growth of cells remains static is called the *stationary phase* (Fig. 9.2). Thus, it is the activity of the organism itself that induces the changes in the medium leading to the stationary phase.

***Death phase:*** After the stationary phase, the cells may remain alive and continue to metabolize, but often they die. If the latter occurs, the population is said to be in the *death phase.* During death phase, the direct microscopic count may remain constant but the viable count decreases. In some cases, death is accompanied by cell lysis, leading to a decrease in the direct microscopic count concurrent with the drop in viable count.

## Batch Culture

It is a form of culture in which a given volume of liquid medium is inoculated with bacterial cells capable of growth in that medium. The inoculated medium is incubated for an appropriate period of time. The growth curve is normally obtained by monitoring a batch culture commonly exhibiting four main phases of growth.

## Synchronous Culture

It is a form of culture in which all the cells pass through the same stage of the cell cycle at the same time. Synchronous culture is used to study specific stages of the cell cycle.

Synchrony may be achieved in various ways. In certain algae, it can be obtained by controlled cycles of light and dark. In bacterial cultures, synchrony may be achieved by subjecting the cells for a given period of time to an inhibitor of protein synthesis like chloramphenicol. Yeast cultures have been made synchronous by the use of EDTA or the ionophore. Synchronous cultures of endospore forming bacteria have been obtained by giving heat shock of a non-synchronous culture to an elevated temperature.

A common method of applying synchronous culture is termed selection synchrony where cells at a given stage of the cell cycle are separated out from a non-synchronous culture by physical means such as density gradient centrifugation. The growth occurring in a fixed volume of a culture medium is eventually altered by the actions of the growing organisms so that it is no longer suitable for growth. In the early stages of exponential growth in batch cultures, conditions may remain relatively constant.

## Continuous Culture

In the early stages of exponential growth in batch cultures, conditions may remain relatively constant, but in later stages, drastic changes usually occur. For many studies, it is desirable to keep the culture in constant environments for long periods, and this is done by employing *continuous cultures*. A continuous culture is essentially a flow system of constant volume into which the medium is added continuously and from which a device allows continuous removal of any overflow. Once such a system is in equilibrium, cell number and nutrient status remain constant, and the system is said to be in a steady state.

The most common type of continuous culture device used is called a chemostat, which permits control of both the population density and the growth rate of the culture (Fig. 9.3). Two elements are used in the control of chemostat—the flow rate, and the concentration of a limiting nutrient, such as a carbon (energy) or nitrogen source (growth factor). At low concentrations of an essential nutrient, growth rate is proportional to the nutrient concentration. However, at these low nutrient concentrations, the nutrient is quickly used up by growth. In batch cultures, growth ceases at the time the nutrient is used up, but in the chemostat, continuous addition of fresh medium containing the limiting nutrient permits continued growth.

The cell density in the chemostat is controlled by the level of limiting nutrient. If the concentration of this nutrient in the incoming medium is raised, with dilution rate remaining constant, cell density will increase although growth rate will remain the same. Thus, by adjusting dilution rate and nutrient level, one can obtain at will a variety of population densities at a variety of growth rates.

Chemostats have been quite useful in genetic, physiological, and ecological studies of microorganisms. For genetic studies, chemostats have been especially useful in studying the mutation process, since they permit the maintenance of growing populations of constant size for long periods of time. In physiological studies, chemostats have been useful in studying the relationships between growth and protein and nucleic acid synthesis, and the control of

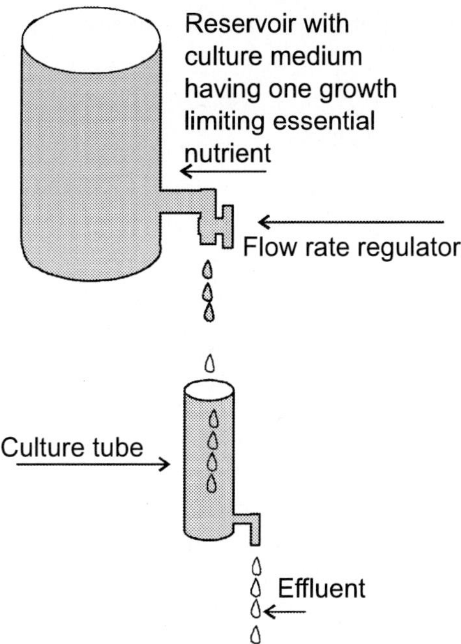


**Fig. 9.3.** Chemostat assembly for continuous culturing of bacteria.

enzyme synthesis. Another physiological use is that an organism can be cultured under a variety of different environmental conditions but at the same rate in each, so that the effect of the environment on the properties of the organism can be studied without concern that the environment may merely be affecting growth rate. In this way, the effects of temperature variation on the properties of an organism have been studied, with identical growth rates at the temperatures. In ecological studies, chemostats have served as useful models for microbial growth in nature under conditions of low nutrient concentration, and have found special use in studying bacterial growth in lakes, rivers, and oceans, where concentrations of organic substances are often low. Finally, chemostats are useful in the selection, from either natural populations or laboratory cultures, and strains able to carry out specific processes of interest. The selective properties of a chemostat should be clear, since the strain best adapted to the conditions used should be able to outgrow other strains and dominate the culture.

## 9.3 CONDITIONS AFFECTING THE GROWTH OF MICROORGANISMS

A number of microorganisms also require one or other of the trace elements, vitamins, or other additions. When a nutrient solution can be made up entirely from defined chemical compounds, it is called a synthetic medium. It is desirable to find for each microorganism the minimal nutrient requirements and to develop a minimal medium that contains only the components actually required for growth, e.g., a minimal medium developed for *Leuconostoc mesenteroides* contains 40 components.

### 9.3.1 Hydrogen Ion Concentration

The $H^+$ and $OH^-$ ions are the most mobile of all ions. Therefore, even small changes in their concentrations have large effects. The attainment of the optimal initial pH and the maintenance of the pH during the growth of a culture are therefore of great importance. Most organisms grow best when $H^+$ and $OH^-$ ions are present in equal concentration (pH 7.0). Bacteria also prefer higher pH values, i.e., a slightly alkaline medium. Some examples of these are the nitrifiers, rhizobia, actinomycetes, and urea digesting bacteria. Only a few are acid tolerant (*Lactobacilli, Acetobacter, Sarcina ventriculi*). Fungi prefer lower pH ranges. Inoculation of soil samples into complex media of different pH values will give predominantly fungi on the media at pH 5–6, and bacteria on the media at pH 7–8.

The maintenance of a constant pH during the growth of a culture is especially important for those organisms that produce acid but are not acid tolerant (lactobacilli, enterobacteriaceae, many pseudomonads). To avoid autolysis (self-killing) of long-term cultures by acid production, as in *Zymosarcina*, the media must be buffered or free from fermentable substrates. A certain buffering capacity, though limited to values above pH 7.2, is given by inorganic phosphates. In the case of more pronounced acid production, addition of calcium carbonate to the medium is recommended.

Many bacteria, however, are fairly insensitive to small pH variations in the range 6–9. If the pH of the medium changes rapidly, there may be a transient change in the intracellular pH, but this is usually readjusted to the original pH in a short period.

The damage produced by unsuitable pH is not actually due to the hydrogen and/or hydroxyl ions; these only increase the undissociated proportion of weak acids or bases, which penetrate more readily into the cell than their dissociation products. The undissociated acids are the physiologically more active forms.

### 9.3.2 Carbon Dioxide

Nutrient media for $CO_2$-fixing, autotrophic bacteria usually contain sodium bicarbonate and are incubated in a $CO_2$-containing atmosphere in a closed system. Alternately, they can be bubbled with air or $CO_2$-enriched air. In all cases, the relationship between pH, bicarbonate concentration, and $CO_2$ partial pressure of the gaseous atmosphere must be taken into consideration.

Heterotrophic organisms that assimilate organic carbon also need $CO_2$. Many parasitic bacteria that occur in blood, tissues, or gastrointestinal tracts are adapted to a higher $CO_2$ content in their atmosphere than that in air. Such bacteria must be incubated in a gaseous atmosphere containing 10% $CO_2$ by volume.

### 9.3.3 Water Content and Osmotic Pressure

Microorganisms vary widely in their requirements with regard to water content. To compare solid materials and aqueous solutions with respect to available water, the parameters water activity ($a_W$) or relative humidity are used. These parameters are related to the vapour phase, which is in equilibrium with a solid material or an aqueous solution. They give the ratio of the water concentration in the vapour phase over the material in question, and the water concentration in the gaseous phase over pure water at a given temperature.

Microorganisms can grow over a range of water activities of 0.998–0.6 ($a_W$). The tolerance of lowest water activity for growth has been recorded for the osmotolerant yeast *Saccharomyces rouxii*, which grows at $a_W$ of 0.6. *Aspergillus glaucus* and other moulds can grow at $a_W$ 0.8, but most bacteria need water activities of more than 0.98. The only exceptions are halophilic bacteria with $a_W$ requirements of 0.75.

### 9.3.4 Temperature

Bacteria also differ with regard to the temperatures they require for growth. Most of the soil and aquatic bacteria are mesophilic, i.e., they have their maximal growth rates at temperatures in the range of 20–42°C. Thermotolerant organisms are those that can grow in temperatures of upto 50°C (*Methylococcus capsulates*). Thermophilic bacteria grow at maximal rates at temperatures above 40°C and their temperature limit is in the region of 70°C (*Bacillus stearothermophilus, Thermoactinomyces vulgaris*). Extreme thermophiles are organisms whose growth optimum is above 65°C (*Thermus aquaticus, Sulfolobus*); some of these can grow above 70°C (several species of the genera *Bacillus* and *Clostridium*) or 80°C (*Sulfolobus acidocaldarius*) and even at 105°C (a strictly anaerobic, sulphur-reducing bacterium). On the other hand, the temperature range of growth for psychrophilic (or cryophilic) organisms varies from 15–20°C. Among these are predominantly some marine bacteria (photobacteria) and the iron bacteria (*Gallionella*); they reach their maximum growth rates at temperatures below 20°C.

### 9.3.5 Aeration

All microorganisms that are obligate aerobes need oxygen as the final electron acceptor. When these bacteria grow on the surface of agar plates or in thin layers of liquid, in contact with air, the oxygen supply is usually sufficient. However, in liquid media of greater depth, aerobic bacteria can grow only in the surface layers because oxygen is continuously consumed and the deeper layers become anaerobic. When aerobic bacteria are required to grow throughout a liquid medium, it is necessary to ensure a continuous oxygen supply by aeration. Microorganisms can utilize only dissolved oxygen, whereas the mineral salts and organic nutrients of a medium can be supplied at concentrations sufficient for their growth. Most microorganisms are adapted to very low concentration of dissolved oxygen, but a certain minimal value, the 'critical oxygen concentration', must be maintained for adequate respiration of the cells.

### 9.3.6 Anaerobic Culture

Exclusion of oxygen from the culture medium is essential for the growth of obligate anaerobic bacteria. Anaerobic techniques include the use of de-aerated boiled nutrient media, free of air bubbles in sealed bottles; oxygen-free gaseous atmosphere in desiccators or anaerobic jars; the use of oxygen-absorbing substances (alkaline pyrogallol, dithionite) and other means. Addition of reducing agents (ascorbic acid, thioglycollate, cysteine, or sulphide) to nutrient media to a certain extent can reduce or counteract the toxic effects of aerobic oxygen. Even very oxygen-sensitive bacteria can be transferred in the open atmosphere, provided a continuous stream of oxygen-free nitrogen is passed through the culture vessel that prevents the contact of the medium with air (Hungate technique). Alternatively, one can use inoculation chambers that are filled with oxygen-free nitrogen, argon, or hydrogen. As an indicator for anaerobic conditions, resauzurin can be added to media. In the presence of oxygen, this is blue, and if the conditions are anaerobic, it is colourless, and after reoxidation, it is red. Also, a beaker with alkaline glucose-methylene blue solution can be added to the incubation jar, which gets decolourised under anaerobic conditions.

## 9.4 NUTRITION AND METABOLISM

### 9.4.1 Nutrient Requirements of Microorganisms

Microbial cell is made up of a number of chemical elements. All such elements together make a biologically functional unit, the cell, capable of performing the activities of living cells. All living organisms of life, from microorganisms to human beings, need a variety of chemical elements, called nutrients, for their survival, growth and reproduction. These elements exist in a variety of compounds in nature, which may be either inorganic or organic. Basic information about such nutritional requirements of microorganisms is necessary when these are to be grown in the laboratory by providing nutrients under controlled conditions than from their natural environments. Microbiologists use media that stimulate or even improve on natural conditions of microbial growth. The major elements for cell growth include carbon, nitrogen, hydrogen, oxygen, sulphur, phosphorus, magnesium, potassium, sodium, calcium, iron and others, which are the minor nutrients. A general list of nutrients required for the growth of microorganisms and their available forms in nature is given in Table 9.1. The nature of nutrients availability, quantity and quality of such elemental nutrients for different microorganisms or other living organisms may vary. A brief description of some of the important major elemental nutrients is given below.

***Carbon:*** Carbon is the most important chemical element required for the growth of microorganisms. In fact every living cell requires carbon in some form or other. Carbon is present in biological system and function in its relatively reduced carbon compounds. In general, the carbon is available in nature in organic (generally reduced form of carbon as variety of organic carbon compounds) or inorganic (oxidized form of carbon as $CO_2$ compounds) form. Depending on the nature of carbon compounds and their availability, living organisms may assimilate it as inorganic or organic compounds. The major classes of organic carbon nutrients are carbohydrates, proteins, lipids. These nutrients provide the energy and macromolecular building blocks to the cell and cellular material. However, inorganic carbon compounds do not provide energy but they serve as carbon nutrient source and on reduction become part of building blocks.

**Table 9.1.** Nutrient requirements of microorganisms

| Elemental Nutrient | Stable form found in nature | Chemical form supplied in microbiological media |
|---|---|---|
| Carbon (C) | $CO_2$, organic compounds | Glucose, malate, acetate, pyruvate and other organic compounds |
| Hydrogen (H) | $H_2O$, organic compounds | $H_2O$, organic compounds |
| Oxygen (O) | $H_2O$, $O_2$, organic compounds | $H_2O$, $O_2$, organic compounds |
| Nitrogen (N) | $NH_3$, $NO_3^-$, $N_2$, organic compounds | $NH_4Cl$, $(NH_4)_2SO_4$, $KNO_3$, $N_2$, amino acids and many other organic compounds |
| Phosphorus (P) Sulphur (S) | $PO_4^{-3}$, $H_2S$, $SO_4^{-2}$, organic S, metal sulfide compounds | $KH_2PO_4$, $Na_2HPO_4$ etc. |
| Sodium (Na) | $Na_2S_2O_3$, $Na_2S$, $Na_2SO_4$, cystein, other organic compounds | NaCl |
| Potassium (K) | $K^+$ ions in solution or salts | KCl, $KH_2PO_4$ |
| Magnesium (Mg) | $Mg^{+2}$ in solution | $MgCl_2$, $MgSO_4$ and other salts |
| Calcium (Ca) | $CaSO_4$ and other salts | $CaCl_2$ |
| Iron (Fe) | $Fe^{+3}/Fe^{+2}$ in solutions, FeS, $Fe(OH)_3$ and other salts | $FeCl_3$, $FeSO_4$, other chelates like EDTA etc. |

***Nitrogen:*** All living organisms require nitrogen in some form or other. This element is an essential part of amino acids, proteins, nucleic acids, and vitamins of living cells. In biological structure and function, nitrogen is always present in its reduced form in a variety of organic molecules. In nature, nitrogen is available as organic or inorganic compounds. The organic forms of nitrogen are as mentioned above and inorganic forms are nitrites, nitrates and ammonium salts. Molecular nitrogen, $N_2$, in atmosphere also serves as nutrient to some of the microorganisms called nitrogen fixing organisms. Not all microorganisms are capable of utilizing the atmospheric nitrogen for growth. In fact, the phenomenon of biological nitrogen fixation (or biological nitrogen reduction) is a property limited exclusively to prokaryotic microorganisms. Bacteria are particularly versatile in the utilization of molecular nitrogen. Unlike eukaryotic cells, some nitrogen-fixing bacteria (e.g., *Azotobacter, Rhizobium,* cyanobacteria, *Azospirillium* etc.) can use atmospheric nitrogen for cell synthesis through a process called biological nitrogen fixation at their normal growth conditions. Others use inorganic nitrogen compounds or organic nitrogen compounds for cell synthesis.

***Hydrogen and oxygen:*** Hydrogen and oxygen are the structural constituents of many organic compounds of cells. The water is the universal solvent for all living systems. These elemental nutrients are present in nature as free atmospheric gases, water and a variety of inorganic (oxides and hydrides) and organic compounds. Some chemolithotrophs, aerobic and anaerobic $H_2$-oxidizing bacteria (e.g., *Alkaligenes*), and archeae are capable of assimilating $H_2$ gas. Majority of organisms assimilate hydrogen and oxygen in the form of water, inorganic compounds and $O_2$ gas itself.

***Sulphur and phosphorus:*** Sulphur is essential for cellular structures and functions by its presence in amino acids, cystein, cystine, methionine, some vitamins and co-enzymes. Phosphorus is also essential for cellular structures and functions like nucleic acids, high energy compounds (e.g., adenosine tri-phosphate, ATP), which are important for energy storage and transfer. Some of these elements are found in water, as components of various nutrients. In organic ions such $SO_4^{-2}$ and $PO_4^{-3}$ can also supply the main elements needed by microorganisms. Some bacteria require organic sulphur compounds and some can even use elemental sulphur. Phosphorus is usually used in the form of phosphate. However, it must be available in the form of soluble phosphate. Apart from high energy compounds, it is an essential component of nucleic acids, phospholipids, teichoic acids and other cellular components.

***Other elements:*** Many other essential elements are required, though in smaller amounts than the elements already listed above. The metal ions such as Na, K, Ca, Mg and Fe are also required for normal growth. These essential elements are often required as co-factors for enzymes and some as structural part of biomolecules (e.g., Mg in chlorophyll, Fe in haemoglobin). Microbes differ in the concentration of ions they require. Some salt loving bacteria (halophiles) cannot grow with less than 15 per cent NaCl in the growth environment. Some other metal ions are also needed but usually in very small amounts, such as Zn, Cu, Mn, Mo, Ni, B and Co, and these are called 'trace elements'. These may be added as salts to microbiological media, but usually occur as impurities of other components of culture media in amounts sufficient to support microbial growth. Media and even water used need to be purified to ensure the absence of contaminating trace elements when the nutritional requirements for organisms(s) are studied.

All living organisms contain vitamins and vitamin-like compounds for their cell function. These function either as co-enzymes for enzymes or become part of co-enzyme. Some bacteria are capable of synthesizing their entire requirements of vitamins from other simpler nutrients in culture medium, but others cannot do so, and therefore, required vitamins are supplied in the culture medium.

Water is an universal solvent for all biological systems. All nutrients must be in aqueous solution before they can enter the living cells. Certain special properties of water make it an unequalled solvent for living systems. It is a highly polar compound having the ability to dissolve or disperse cellular components and to provide a suitable milieu for the various metabolic reactions of the cell. High specific heat of water provides resistance to sudden transient temperature changes in the environment. It is also a chemical reactant required for many hydrolytic reactions carried out by a living cell.

According to the mechanism of energy conversion into the biochemically useful form of ATP, organisms can be divided into two principal metabolic types: phototrophic and chemotrophic.

The organisms that can use light as the energy source for growth are called phototrophs (photosynthetic). The phototrophic organisms comprise two large groups: the anaerobic phototropic bacteria, which do not evolve oxygen, and the aerobic phototrophs, i.e., cyanobacteria, algae, and green plants, which produce oxygen in light. The term chemotroph (or chemosynthetic) indicates energy gain by oxidation–reduction reactions of nutrient substrates, irrespective of the way in which the biochemical energy is obtained either by respiration or fermentation.

***Hydrogen donors and carbon sources:*** All organisms that use organic compounds as hydrogen donors are called organotrophs in contrast to lithotroph, which is used to designate the ability to use inorganic hydrogen donors such as $NH_3$, $H_2S$, S, CO, $Fe^{2+}$, and others. The concepts 'autotroph' and 'heterotroph' have been narrowed down in meaning and refer only to the source of cellular carbon. Microorganisms are autotrophs if they can obtain the major part of their cellular carbon by carbon dioxide fixation. They are heterotrophs if they obtain their cell carbon by assimilation of organic compounds.

In general, classification by the basic processes of energy conversion and hydrogen donor is sufficient. Green plants, cyanobacteria, and purple sulphur bacteria are photolithotrophs, nitrifying bacteria are chemolithotrophs, and animals and the vast majority of microorganisms are chemoorganotrophs (the source of cellular carbon is included in the designation only in special cases). Since in most microorganisms lithotrophy is linked to autotrophy, one may also speak of photoautotrophs and chemoautotrophs.

The viruses have been omitted from the discussions of nutritional types because their nutrition is of a different type, unlike that of any other kind organism. Viruses have been called paratrophic since their energy, and some materials, being extorted from the host cells due to their obligate nature of growth.

## 9.5 CLASSIFICATION OF ORGANISMS ACCORDING TO SOURCES OF CARBON AND ENERGY

### 9.5.1 Autotrophs and Heterotrophs

Two definite types of organisms were differentiated on the basis of whether (a) they were restricted to an entirely inorganic diet, including carbon dioxide (or carbonates) as a sole source of carbon or whether (b) in addition to a mineral diet they also required one or more organic substances, i.e., carbon in organic combination. The former (a) were designated true or strict autotrophs (Gr. Autos = self; trophe = nourishing); the latter (b) were called heterotrophs since it was originally supposed that carbon in organic form could only have been prepared by other living forms (Gr. Hetero = other).

#### Autotrophs

These typically possess the enzymic equipment to catalyze the direct combination of two carbon atoms or the combination of carbon from carbon dioxide with an organic complex already existing in the cell. Typical autotrophs are represented by familiar green plants and a number of species of bacteria important in agriculture and industry.

### Heterotrophs

Heterotrophic organisms (i.e., those that cannot live without organically combined carbon) are the commonest and most widely distributed types. They are represented by Homo sapiens and all other animals, and by most species of eumycetes and bacteria.

With advances in knowledge, the differentiation between autotrophs and heterotrophs lost much of its usefulness because various overlappings, exceptions, and borderline cases have come to light that blur the lines of demarcation. For example, some soil inhabiting bacteria (Genus *Hydrogenomonas*) and others of the Family Athiorhodaceae are capable of living either autotrophically or heterotrophically. Some species living on an otherwise completely inorganic diet must have certain vitamin-like (i.e., organic) substances. Certain common intestinal bacilli can grow well in a simple and completely inorganic solution if provided with sodium citrate as a source of carbon.

Another question arose when it was found that many supposedly typical heterotrophs require carbon dioxide in addition to organic carbon. Although it is not surprising that in the course of billions of years of organic evolution some metabolic aberrations have occurred, it makes classification difficult.

## 9.5.2 Chemotrophs and Phototrophs

The primeval earth is believed (by some) to have been completely dark because of the heavy clouds of water vapour in the skies. It may (or may not) be a result of their origin and evolution in the dark, but most species of bacteria and higher fungi do not depend on sunlight as their source of energy. On the contrary, they thrive in complete darkness, and many species are injured and killed by direct sunlight. These primitive organisms obtain the energy to live and synthesize themselves from chemical reactions (oxidations) that are independent of the sun. They are therefore, said to be chemosynthetic or chemotrophs. Organisms that obtain their energy for life and self-synthesis from the sun (or artificial radiant sources) are said to be photosynthetic or phototrophs.

## 9.5.3 Lithotrophs and Organotrophs

Among the chemotrophs, differentiations are made on the basis of the kinds of substrates that are used as electron donors (oxidized) as sources of energy. Among the chemotrophs, some species oxidize only inorganic substances as sources of energy.

For example:

$$NaNO_2 + \tfrac{1}{2} O_2 \rightarrow NaNO_3 + \text{energy}$$

$$H_2S + 2O_2 \rightarrow H_2SO_4 + \text{energy}$$

$$CO + \tfrac{1}{2} O_2 \rightarrow CO_2 + \text{energy}$$

These are mainly strictly autotrophic bacteria of the soil. When the oxidized substrates (electron donors) are entirely inorganic or "mineral," the chemotrophic organisms using them are called chemolithotrophs (Gr. *Lithos* = stone or mineral).

Chemotrophs that obtain energy by oxidizing organic substrates like glucose are said to be chemoorganotrophs.

Typical phototrophic organisms use only radiant energy, but they differ in respect to the hydrogen donors, which they can use in their photosynthetic processes. Green plants use water as a hydrogen donor, for the reduction of carbon dioxide, in the synthesis of their cellular organic compounds (see photosynthesis). They do not require organic carbon. Such organisms are said to be photolithotrophic or photolithotrophs. Some species of bacteria are photolithotrophic, but, unlike all other photosynthetic organisms, they use hydrogen sulfide as hydrogen donor, not water. Some phototrophic bacteria utilize as hydrogen donors neither water nor hydrogen sulfide, but certain specific organic compounds. Such bacteria are said to be photoorganotrophs.

The relationships of the various nutritional types may be tabulated as follows:

**A. As carbon source:**

1. Autotrophs: can use carbon dioxide as sole source of carbon.
2. Heterotrophs: require organic source of carbon.

**B. As energy source:**

3. Chemotrophs: obtain energy from oxidation; independent of sunlight for energy.
4. Phototrophs: obtain energy from sunlight.
5. Lithotrophs: obtain energy from oxidation of inorganic substrates.
6. Organotrophs: obtain energy from oxidation of organic substrates.

Terms 3 through 6 are commonly combined as:

I. *Chemolithotrophs* obtain energy only from oxidation of inorganic substrates; most are autotrophic soil bacteria.
II. *Chemoorganotrophs* obtain energy only from oxidation of organic substrates. All are heterotrophs: all animals and eumycetes and most bacteria.
III. *Photolithotrophs* obtain energy from radiant sources; use inorganic hydrogen donors photosynthesis; all green plants, blue-green algae and a small group of bacteria.
IV. *Photoorganotrophs* obtain energy from radiant sources; use organic hydrogen donors in chemosynthesis; heterotrophs; bacteria of families Thiorhodaceae and Athiorhodaceae.

## 9.6 ENZYMES

An enzyme is a protein that changes the rate of a chemical reaction but does not affect the nature of the final products; in other words, it acts as a catalyst.

Enzymes are denoted by the suffix—ase. For example, an enzyme that catalyzes the breakdown of proteins is called proteinase; one that catalyzes an oxidation is called an oxidase, and so on.

The combination of a substrate with an enzyme is a very specific process, i.e., each catalyst has a unique surface containing a precise spot where the substrate molecule can join it. Because of the unusual folded surface of proteins, only very specifically shaped molecules can gain access to the particular chemical groups. For this reason, the enzyme-substrate complex is often compared to a lock and key mechanism. We can write the reaction in the following way:

$$E + S \underset{k_2}{\overset{k_1}{\rightleftharpoons}} ES \xrightarrow{k_3} E + \text{products}$$

where

E = Enzyme
S = Substrate
ES = Enzyme-substrate complex
$k_1$ = Rate of combination of E and S to form ES
$k_2$ = Rate for the dissociation of ES to E and S
$k_3$ = The rate of conversion of ES to the products is indicated by the constant

Here $k_1$ refers to the rate of combination of E and S to form ES, and $k_2$ is the rate for the dissociation of ES and E and S, after the substrate combines with the enzyme to form ES. One can assume that enough activation energy is acquired so that ES is converted into the product of the reaction, which dissociates from the enzyme surface to allow the enzyme to combine with another substrate molecule. The rate of conversion of ES to the products is indicated by the constant $k_3$. At a low substrate concentration, some of the enzyme is free, and the maximum rate is not observed. With excess substrate, all the enzyme is converted into ES, and the reaction takes place at its maximum rate—i.e., maximum velocity (Vm) = $k_3$ (ES) = $k_3$ (E).

An enzyme-substrate combination is specific in which a particular enzyme and its substrate must possess shapes that complement one another if they are to fit together. In other words, the general configuration of the protein molecule must be such that the substrate molecule can fit into a specific site on the enzyme. The specificity and geometry of enzyme reactions are illustrated by the phenomenon of competitive inhibition (Fig. 9.4).

Competitive inhibitors abound, and some of them are effective antibacterial agents. One of these is sulfanilamide. This substance inhibits the growth of certain bacteria and it was observed that the B vitamin, p-aminobenzoic acid (or PABA), removed inhibition. Subsequent studies indicated that this was an example of competitive inhibition.

The success of sulfanilamide and other bacterial killers opened the exciting possibility that all disease producing organisms, and cancer as well, might be susceptible to the antimetabolite approach, but, unfortunately, the host and disease producing organisms have very similar

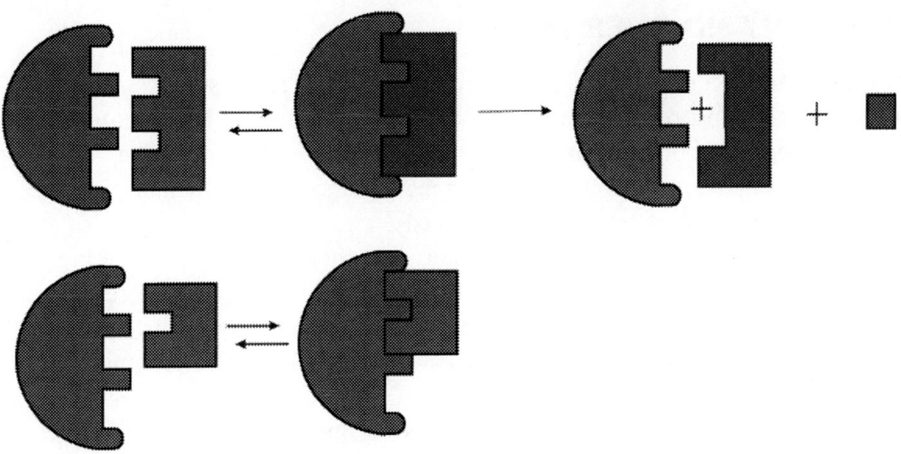

**Fig. 9.4.** Competitive inhibition.

**Table 9.2.** Enzyme Inhibitors

| Inhibitor | Group on enzyme affected | Example of enzyme |
|---|---|---|
| p-Chloromercuribenzoic acid | –SH | Urease |
| Iodoacetic acid | –SH and others | Triose phosphate dehydrogenase |
| Mercury and silver cyanide | SH and anions | Glutamic hydrogenase |
| and azide | Metalloporphyrins and metals | Catalase |
| Carbon monoxide | Metalloporphyrin | Cytochrome oxidase |
| Citrate, Oxalate | Removes metals | Hexokinase Chymotrysin |
| Diisopropylfluorophosphate | Combines with OH group of serine | |

enzyme machinery, which means that the altered vitamin or substrate is also quite toxic to the host cells. Since there are some differences in rates, affinities, etc. between the host and its enemy's enzyme-catalyzed reactions, by careful screening one may be able to find the appropriate chemical that will kill the disease producing organism before harming the host cells. PABA, for example, is one essential part of folic acid, a key substance for the transfer of methyl ($-CH_3$) groups in both animals and plants.

Compounds such as cyanide and carbon monoxide, by combining with the cytochromes (iron containing catalysts), completely block the electron transport to oxygen. Diisopropylphospho-fluoridate (DPF) is used as an enzyme inhibitor. For example, when the proteinase chymotrypsin combines with DPF, one molecule of inhibitor combines with one molecule of enzyme, causing a complete inhibition of catalytic activity.

In some instances, for an enzyme to function it must be associated with a small molecule called a coenzyme. For example, iron is necessary for transport; magnesium is essential for the transfer of phosphate groups, and so on.

## 9.7 OXIDATION AND ENERGY

The oxidation of carbohydrates or related compounds is the main source of energy for many organisms. Energy yielding reactions must be coupled with energy consuming reactions in an intimate way, if energy is not to be lost as heat. Under these circumstances, the energy is not "liberated" as such but is redistributed in the reacting molecules. In many of these coupled reactions, the overall process proceeds spontaneously.

The oxidative reactions of the cells are the ones that most actively redistribute or "liberate" energy. Oxidation is any process in which hydrogen ions or electrons are removed. The simplest type of oxidation is called dehydrogenation. Hydrogen is removed, as occurs in the oxidation of alcohol, $RCH_2OH \rightarrow RCHO + 2H$. However, oxidations and reductions cannot proceed, of course, by these half reactions. Every substance undergoing oxidation must be accompanied by a substance undergoing reduction, and vice-versa. Thus, in the reaction below, $AH_2$ is the reductant and is being oxidized by B.

$$AH_2 + B \rightarrow A + BH_2$$

Since the reaction is reversible, one can start with the chemicals on the right; then $BH_2$ is the reductant and A is the oxidant. The substances that take part in these reactions have a tendency either to reject or attract electrons, and this tendency is directly proportional to their strength as reducing or oxidizing agents.

In the next step, the oxidation process occurs, and energy in the molecule is redistributed to form the "energy-rich bond".

The coupling of this bond energy to other systems is the key reaction in biosynthetic processes. Instead of hydrolyzing the oxidation product, one can split it with phosphoric acid ($H_3PO_4$). The phosphate group prevents the loss of energy of the link, resulting in the formation of an energy-rich link between the carbon and phosphorus.

Another important phosphorylated compound, adenosine diphosphate (ADP), will react this energy rich phosphate group. The product of the reaction is adenosine triphosphate (ATP). ATP is proved to be universally distributed in plants, microorganisms, and animals, and it serves as an initial storehouse for the energy generated in oxidative reactions. This bond energy of the phosphate groups directly or indirectly drives all the energy-requiring processes of life.

Thus, the energy of an oxidation reaction creates an energy-rich group, and this bond energy must couple many reactions that require energy before they can proceed to any significant degree. It turns out that even the second phosphate in ATP is energy-rich, but the third one, which is on the number 5 carbon of the pentose sugar, is not energy-rich. ATP, therefore, is capable of supplying two energy-rich units. The link between two energy-rich phosphates, as in ATP, is called a pyrophosphate bond, and has proven to be the primary grouping in the conservation of oxidation-reduction energy. This pyrophosphate bond energy of ATP is used in biological reactions.

When ATP is hydrolyzed to liberate the two terminal phosphates, a large amount of heat is liberated. The free energy of hydrolysis of the terminal pyrophosphate bond in ATP is approximately 8000 calories at pH 7.0, in contrast to the free energy of hydrolysis of an ordinary ester, which may not be more than 1000 or 2000 calories. The energy-rich phosphate is then transferred to ADP to form more ATP, leaving pyruvic acid as a final residue at this stage in the dissimilation of the glucose molecule. The formation of new energy-rich phosphate bonds in derived portions of the substrate itself is generally called substrate level phosphorylation, and is best exemplified in the Embden-Meyerhof–Parnas (EMP) pathway after the formation of the trioses (Fig. 9.5).

## 9.7.1 Oxidative-Chain Phosphorylation

In the presence of available oxygen, when the cytochrome system is evolved in substrate oxidations, at least three additional phosphorylations occur during the progress of the substrate electrons from FAD, down the "steps" of the oxidative chain, to lower and lower energy levels in the electron-transporting cytochrome system. At each "descent", the electrons give up some of their energy. In the presence of inorganic phosphate (iP), ADP and several enzymes, more energy-rich (e-r)ATP is formed. These oxidative-chains or oxidative phosphorylations, as they are called, yield much additional energy. As previously mentioned, the use of free oxygen as a final hydrogen ($e^-$) acceptor is often called respiration. If, in the absence of free oxygen, an alternative inorganic hydrogen ($e^-$)acceptor is used, e.g., $NaNO_3$, the biooxidation is sometimes called anaerobic respiration.

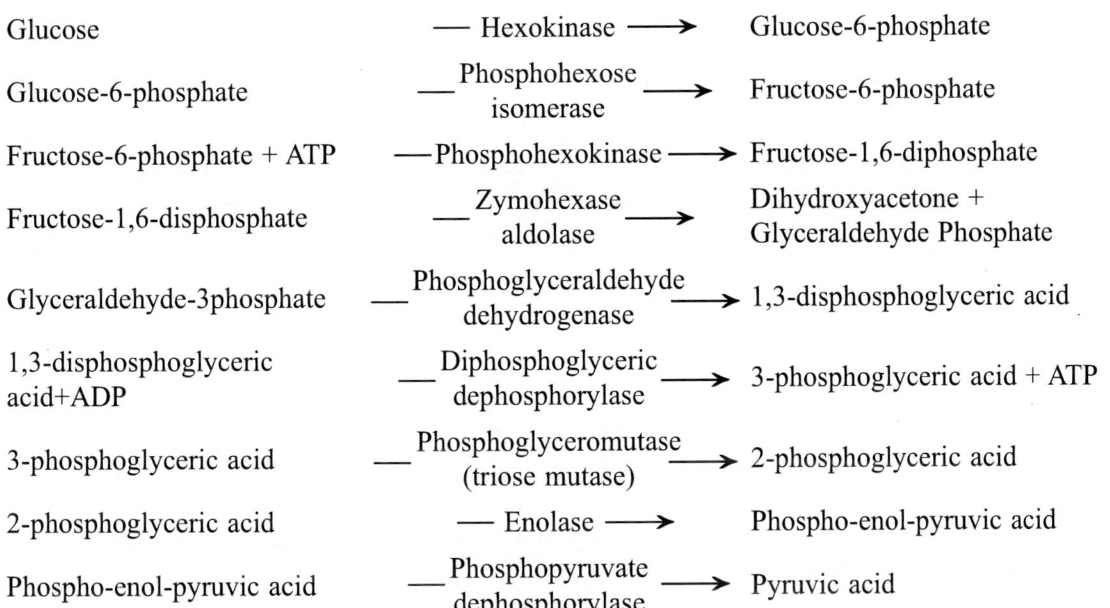

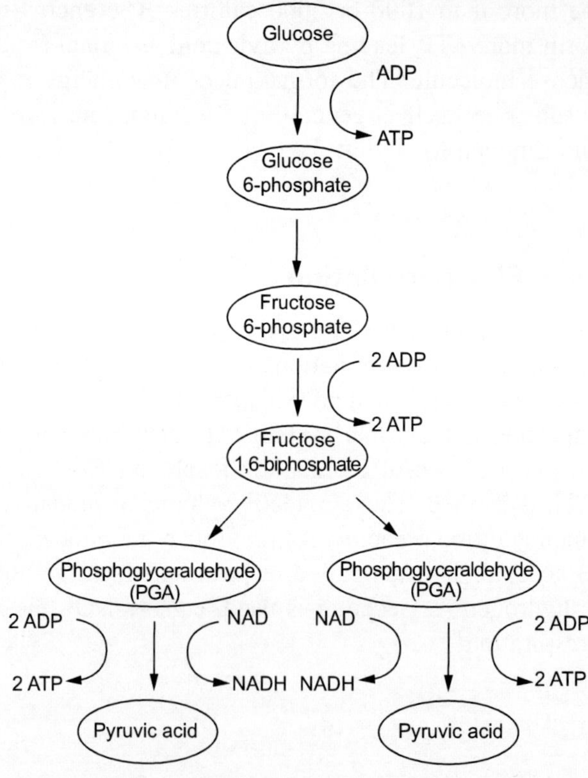


**Fig. 9.5.** Embden-Mayerhoff-Parnas pathway (Glycolysis pathway).

### 9.7.2 Photophosphorylation

In addition to the phosphorylations as described, a third type of phosphorylation, termed photophosphorylation occurs in photosynthetic species of cells. The process derives its energy from the effect of light on electrons in chlorophyll. The net result is the same as in substrate level or oxidative-chain phosphorylations, i.e., generation of ATP. However, the reaction is light dependent.

### 9.7.3 Phosphate-Bond Energy

The energy of the phosphate bonds in ATP formed during all biooxidations, including photosynthesis, is released when the phosphate bonds are hydrolyzed. ATP is thereby degraded to ADP; the energy released is used in synthesis and other vital functions of the cell. The ADP is then available to accept a new e-r phosphate group. Thus ADP and ATP act together as an energy transfer and storage system; this is one of the most important mechanisms in biooxidation.

The energy of the phosphate bonds may also be readily transferred into some other types of e-r bonds (e.g., thioesters) at about the same energy level, and vice versa. In some reactions, both the phosphate bonds of ATP are hydrolyzed, forming ADP and AMP successively. In other reactions, pyrophosphate is split off from ATP, leaving AMP. AMP is restored to ATP by successive phosphorylations or by combination of AMP with pyrophosphate or by combination of AMP with pyrophosphate. Similar nucleotides of other nucleosides, e.g., guanidine, and cytidine carry on similar functions in other energy-yielding reactions and energy transfer systems (see Kreb's cycle, protein synthesis, etc.) in various cells.

## 9.7.4 Complete and Incomplete Biooxidations

### Biooxidation in the Presence of Available Oxygen

In most aerobic and some facultative bacteria, biooxidation of a substrate is complete. If the substrate is an organic compound such as glucose, the end products are $CO_2$ and $H_2O$. This is illustrated by the aerobic utilization of glucose by Bakers' yeast (*Sachharomyces cerevisiae*).

$$C_6H_{12}O_6 + 6O_2 \rightarrow 6CO_2 + 6H_2O + 688{,}500 \text{ cal.}$$

If the substrate is inorganic, the oxidation may be equally complete, even though the end products differ as in the oxidation of sulphur to sulphuric acid. Such oxidations theoretically release all the energy available from the substance as seen in aerobic metabolism. Actually, parts of the substrate are usually utilized in cell synthesis.

However, even in the presence of free oxygen, biooxidation may be incomplete. Much depends on the species, the enzymic equipment of the cell and the availability of an alternate hydrogen ($e^-$) acceptor. An example of aerobic but incomplete oxidation is the formation of vinegar (acetic acid) from wine (alcohol) by *Acetobacter*. These aerobic bacteria grow on the surface of wine and convert it to acetic acid:

$$C_2H_5OH + O_2 \rightarrow CH_3COOH + H_2O + 118 \text{ kilocalories}$$

Much of the available energy of the alcohol is left in the acetic acid. Many aerobic fungi can utilize this energy by oxidizing the acetic acid to $H_2O$ and $CO_2$.

$$CH_3COOH + 2O_2 \rightarrow 2CO_2 + 2H_2O$$

### Biooxidation in the Absence of Available Oxygen

Biooxidations may occur in the total absence of available oxygen in several genera of strict anaerobes. Biooxidations can occur only in the total absence of free oxygen. Biooxidation in the total absence of free available oxygen, when the hydrogen ($e^-$) acceptor is an organic substance, is called fermentation.

### 9.7.5 Fermentation

Obligate aerobes are enzymically equipped to use only free oxygen; facultative aerobes are equipped to use either free oxygen or some alternative, reducible, inorganic substrate, commonly a nitrate as final hydrogen ($e^-$) acceptor. In biooxidations which occur in the absence of available inorganic oxygen and in which case both hydrogen donors and acceptors are organic substrates, the substrates are generally, but not necessarily, carbohydrates. In fermentations, usually only NAD or NADP functions as hydrogen ($e^-$) carrier. FAD and the cytochrome system are not required since the final hydrogen ($e^-$) acceptor is not oxygen but an organic substance, commonly pyruvic acid.

The term fermentation is sometimes erroneously used for certain strictly aerobic processes, especially industrial processes, e.g., the aerobic oxidation of alcohol to acetic acid by *Acetobacter* sp. in vinegar fermentation.

Fermentation may be caused by facultative organisms under anaerobic conditions, e.g., brewer's yeast (*Saccharomyces cerevisiae*); by obligate anaerobic organisms such as *Clostridium* or even in the presence of free oxygen by different organisms that do not ordinarily utilize free oxygen like species of *Lactobacillus.*

#### Products of Fermentation

Depending on the conditions of growth, the substrate and the organisms involved, the end products of fermentations vary greatly. *Clostridium* species commonly ferment glucose to yield butyl and other alcohols and certain acids; *Lactobacillus lactis* yields almost lactic acid entirely, while *L. brevis* yields lactic and acetic acids, ethyl alcohol and carbon dioxide. Other microorganisms produce numerous other valuable end products of fermentation.

***Pyruvic Acid (Pyruvate)***

Pyruvate is physiologically the most important first intermediate product in most biooxidations of glucose, aerobic or anaerobic. It is commonly produced from glucose in stages. However, several other analogous series of reactions occur in many bacteria and other microorganisms, e.g., the Entner-Doudoroff cycle in bacteria of the large and important genus *Pseudomonas*. In this cycle, the pathway to pyruvate progresses via glucose-6-phosphate, 6-phosphogluconic acid, 2-keto-3-deoxy-6-phosphogluconic acid, to pyruvate plus 3-phosphoglyceraldehyde.

Pyruvate may also be reached via the metabolism of sugars other than glucose, and by pathways other than those cited via the metabolism of fatty acids and amino acids. The monosaccharides, fatty acids and amino acids are, of course, derived mainly from the preliminary digestive hydrolysis of starch, glycogen, cellulose, fats, poly-beta-hydroxybutyric acid, chitin or proteins depending on the organism(s). Thus, many kinds of organic nutrient may contribute to the pyruvate content of the cell.

Therefore, pyruvate is the main source for a wide variety of metabolic substrates and end products. It serves as an electron acceptor in most fermentations, being reduced in many types of cell to lactic acid. It may also be decarboxylated and the product reduced to ethyl alcohol.

Conversely, it may serve as a source of amino acids, fatty acids, aldehydes or other important cell building materials.

For example, a series of reactions by which many amino acids are changed to pyruvate is summarized in the oxidative deamination of alanine:

Long-chain fatty acids undergo somewhat more complex changes, but the residues of the changes often become pyruvate, e.g., propionic acid may be oxidized to pyruvate. Lactic acid is a common source of pyruvate in fermentations.

If a substance, e.g., acetic acid, $CH_3COOH$, lacks a carbon atom to form pyruvate, it can be changed (via a series of enzymic reactions, with energy from CoA~S) to a three-carbon acid by addition of $CO^{--}$. If it has too many carbon atoms, it may be split enzymically. Thus pyruvate is the product in which many molecules are made to fit.

A wide variety of end products of incomplete oxidation of glucose may be derived under common conditions of growth For example, from a common saprophyte, the gram negative, facultative *Klebsiella aerogenes,* the following end products have been shown to occur: hydrogen, carbon dioxide, ethyl alcohol, acetic acid, acetyl-methyl-carbinol, 2-3-butylene glycol, lactic acid, glycerol, succinic acid. There are undoubtedly others in smaller amounts. Some of the end products of metabolism, e.g., acetyl-methyl-carbinol, $H_2$ and $CO_2$ are easily tested for in a culture tube and are often distinctive of certain species. Distinctive products of metabolism of various substances, glucose among others, are extremely useful in identifying certain organisms. Some of the products, e.g. lactic acid and butyl alcohol, are of great industrial value.

### 9.7.6 Alcoholic Fermentation

Although the alcoholic fermentation is known since prehistoric times, not much information was available till 1860 when the French chemist Gay-Lussac had succeeded in describing the production of alcohol from sugar by the equation:

$$\underset{\text{Glucose}}{C_6H_{12}O_6} \rightarrow \underset{\text{Carbon dioxide}}{2CO_2} + \underset{\text{Ethyl alcohol}}{2CH_3CH_2OH}$$

In the mid-nineteenth century, two schools of controversy developed concerning the mechanism of alcoholic fermentation. One group considered the process to be strictly a chemical one, while the other claimed that it was a process intimately associated with the activities of microorganisms. During this controversy, Pasteur was studying how lactic acid is produced in milk and how acetic acid is formed in certain wines. From his experiments he concluded that fermentation takes place in the absence of oxygen (anaerobically) in the presence of certain microorganisms. When these organisms were excluded, no fermentation occurred, thus indicating that fermentation is a physiological process closely associated with the life cells. This observation contradicted Lavoisier's earlier conclusion that all living cells need oxygen for respiration. Challenging Lavoisier's views, Pasteur suggested that there were substitutes for molecular oxygen, and that fermentation, indeed, is the result of life activities being carried out in the absence of air. The experiments and reports of Pasteur did much to establish the belief that fermentation is directly associated with the growth and metabolism of living organisms.

The next major breakthrough in the investigation of fermentation came after 20 years. In 1897, Buchner's discovery had not only opened the insight of fermentation, but also to those of modern enzyme chemistry as well. He ground yeast with sand, mixed it with certain other compounds and squeezed out the juice with a hydraulic press to obtain a cell-free extract. He observed fermentation of yeast juice in cell-free extract in the complete absence of living cells for the first time. This observation made it possible to study the processes of alcoholic fermentation independently of all other processes such as growth, multiplication, excretion, and the additional complications that accompany fermentation in the living yeast cell. Buchner's work was soon followed by intensive studies of the properties of yeast juice, which was found to be capable of fermenting a large number of sugars such as glucose, fructose, mannose, sucrose, and maltose. Glucose was converted by the cell-free extract of juice into ethyl alcohol and carbon dioxide, according to the Gay-Lussac equation.

The first important analysis of the activity of yeast juice was made by Harden and Young in 1905. When they added fresh yeast juice to a solution of glucose at pH 5, fermentation began almost at once. Although the rate of carbon dioxide production soon fell off, they restored it by introducing inorganic phosphate. The recovery was only temporary, however, because the phosphate was soon used up, and the rate of fermentation dropped as the phosphate concentration declined. By adding more phosphate, they sparked another burst of fermentation.

Later Robison isolated another sugar phosphate. When examined in detail, the sugar phosphate proved to consist of an equilibrium mixture between glucose-6-phosphate (phosphate attached to 6 carbon atoms of glucose) and fructose-6-phosphate. Since both of these sugar phosphates could be fermented, it seemed clear that these substances must arise from the coupling of inorganic phosphate with glucose on carbon atom number 6.

The first step in the series consists of transfer of one of the phosphate groups of ATP to glucose to form glucose-6-phosphate and ADP, in a reaction which is catalyzed by hexokinase enzyme (all enzymes concerned with reactions of this type that involve ATP are called kinases—fructokinase, glucokinase, etc.) Next, the glucose-6-phosphate molecule, after a series of transformations, reacts with ATP again to form the hexose diphosphate. The hexose diphosphate is now split into two triose phosphate molecules (three-carbon sugars, each of which contains a phosphate). Although the initial products of the splitting reaction are different, they can be converted into one another in a reversible reaction (Fig. 9.6). The crucial intermediate in the formation of alcohol is glyceraldehyde-3-phosphate.

### 9.7.7 The Oxidation of Glyceraldehyde-3-Phosphate

In the sugar fermentation, the enzyme, triosephosphate dehydrogenase catalyzes this oxidative step, in which hydrogen is removed. The glyceraldehyde-3-phosphate adds to the SH group of the enzyme and then oxidation occurs to form the energy-rich group on the enzyme. This group, in turn, reacts with inorganic phosphate to form 1,3-diphosphoglyceric acid. In the next stage, the energy-rich phosphate group is transferred from carbon 1 to a molecule of ADP to form 3-phosphoglyceric acid and ATP, a reaction that is catalyzed by an enzyme called phosphoglyceric acid kinase, i.e., the reaction is reversible.

The metabolism of 3-phosphoglyceric acid takes place in the following manner. The 3-phosphate ester is converted into the 2-phosphate ester (2-phosphoglyceric acid), which is

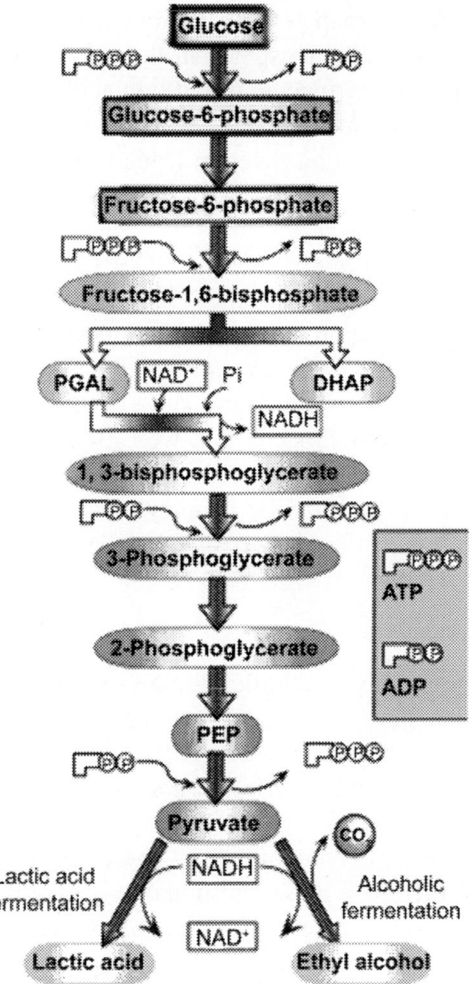


**Fig. 9.6.** Alcoholic fermentation.

then dehydrated, causing the energy in the molecule to be redistributed into an energy-rich phosphate group, phosphoenolpyruvic acid. The formation of phosphoenolpyruvic acid from 2-phosphoglyceric acid is catalyzed by enolase enzyme. The energy-rich phosphate is then transferred to ADP to form more ATP, leaving pyruvic acid as a final residue at this stage in the dissimilation of the glucose molecule. The formation of new energy-rich phosphate bonds in derived portions of the substrate itself is generally called substrate-level phosphorylation, and is best exemplified in the Embden-Meyerhof pathway after the formation of the trioses. In the presence of phosphate, enolase is strongly inhibited by fluoride, because of the removal of magnesium (as magnesium fluorophosphate), which is essential for the action of enolase. If fluoride is used in yeast juice fermentation, the metabolism of 2-phosphoglyceric acid can be prevented, a fact that enabled biochemists to accumulate 3-phosphoglyceric acid in quantities that could be isolated and identified.

This new energy-rich phosphate group is not different from the one formed in the oxidation of glyceraldehyde-3-phosphate. The phosphate in the presence of a specific kinase can be transferred to ADP to form ATP and pyruvic acid. Pyruvic acid is an example of an alpha keto acid. The first step in the metabolism of such acids is the loss of carbon dioxide (decarboxylation). In yeast extracts, the decarboxylation reaction leads to the production of carbon dioxide and acetaldehyde.

$$CH_3COCOOH \rightarrow CH_3CHO + CO_2$$

The enzyme that catalyzes this reaction is called carboxylase and requires a cofactor (cocarboxylase) and thiamine. For thiamine to be enzymatically functional, it must be phosphorylated to form thiamine pyrophosphate (TPP). In the final stage of the production of alcohol, acetaldehyde is reduced to alcohol by the hydrogens that were passed on to DPN in the oxidation of glyceraldehyde-3-phosphate. The reduction of the acetaldehyde to alcohol or the oxidation of alcohol to form acetaldehyde is catalyzed by an enzyme called alcohol dehydrogenase. The coenzyme necessary for the action of this enzyme is DPN.

### 9.7.8 Initial Steps in Glycolysis

Glycogen is a branched, complex polysaccharide with a molecular weight as high as four million. Enzymes attack it at either the 1–4 or the 1–6 linkages. Many hydrolytic enzymes (glycosidases-amylases) can break down glycogen to smaller polysaccharides or free-glucose units.

The enzyme that catalyzes this reaction is phosphorylase, and the product of the reaction is glucose-1-phosphate and the remainder of the glycogen molecule. Once the outer tier of 1–4 glycosidic linkages is broken, the reaction stops. There are hydrolytic enzymes, however, that split the 1–6 linkage and thereby expose additional 1–4 glycosidic linkages, which phosphorylase will attack. The new phosphate ester formed in the breakdown of glycogen is readily converted into glucose-6-phosphate by the enzyme phosphoglucomutase. This phosphate ester, is the compound formed when ATP reacts with glucose. All these reactions are reversible and the glycogen synthesis occurs if there is an excess of glucose and ATP.

After the formation of glucose-6-phosphate, glycolysis and alcoholic fermentation follow a common path until pyruvic acid is formed. Here the pathways again diverge. Pyruvic acid will, nevertheless, react with the reduced DPN and oxidize it in much the same way as the acetaldehyde did in the yeast extract. When pyruvic acid is reduced by DPNH, lactic acid is formed, and the enzyme that catalyzes this reversible oxidation-reduction reaction is called lactic dehydrogenase.

In the fermentation of free glucose, however, two of the new bonds are used in the preliminary phosphorylation of the glucose molecule, so there is a net gain of only two molecules of ATP. In the first stage in glycolysis, on the other hand, the glycogen is split by inorganic phosphate and not by ATP, so that when the product, glucose-1-phosphate, is converted into glucose-6-phosphate, only one molecule of ATP is needed to produce each molecule of hexose diphosphate.

The reoxidation of DPNH, the action of inorganic phosphate, and the utilization of ATP are the essential processes.

The importance of iron compounds in oxygen utilization was also suggested by the observation that low concentrations of cyanide and carbon monoxide inhibit respiration. Warburg found that cyanide blocked iron catalyzed oxidations and concluded that it affected cellular respiration by combining with an iron-containing "respiratory enzyme". In addition, he observed that the respiratory inhibition by carbon monoxide could be reversed by visible light.

Subsequent studies by Keilin clearly revealed the importance of a number of haemes, or iron-containing pigments, cytochromes for electron transport. He was able to demonstrate three different haeme proteins, which he named cytochromes a, b, and c.

Warburg and Christian also isolated from cells a second coenzyme that was needed to set off oxidative reactions. They were able to isolate from yeast a yellow protein capable of catalyzing the transfer of electrons from reduced DPN to some other acceptor. They removed the yellow component by dialysis and prepared it in pure form. It proved to be a riboflavin compound that could accept electrons from DPNH. The reduced flavins are colourless.

Two forms of riboflavin involved in electron transport occur in nature. One is flavin mononucleotide (FMN) or riboflavin phosphate, and the other is flavin adenine dinucleotide (FAD), which is analogous to DPN, a substance formed from another B vitamin, nicotinic acid amide. The flavins can exist in either the oxidized or the reduced form, and DPN is one of the better reducing agents in the presence of certain enzymes. Although reduced flavins are readily reoxidized by oxygen when they are free in solution, they are often poorly oxidized when bound to an enzyme.

The flavin operates as an intermediate carrier of hydrogen between reduced DPN and other electron acceptors before oxygen is reduced. A number of investigations have indicated that the cytochromes are the oxidizing agents for the reduced flavins. Cytochrome pigments are present in all aerobic cells and can be alternately reduced and oxidized by a number of agents. Extensive studies indicated that the cytochromes contain a protein and an iron porphyrin group. The iron porphyrin of the cytochromes is made up of four pyrrole rings and an atom of iron, and is very similar to the pigment in haemoglobin. Cyanide and carbon monoxide are inhibitors of these iron containing catalysts. In cytochromes, it is the iron in the porphyrin that undergoes oxidation–reduction: Fe+++ to Fe++. One of the cytochromes, cytochrome C, appears to be particularly important in the transfer of electrons to oxygen. The porphyrin of this cytochrome attaches itself to a protein by forming a linkage with two sulphur atoms in the cysteine residues of the polypeptide chain. An additional enzyme is necessary to oxidize the reduced cytochrome C by molecular oxygen and is called cytochrome C oxidase. The evidence indicates, therefore, that the mechanism of the electron transport to oxygen is from the flavin coenzymes to the cytochromes and then to molecular oxygen. In all aerobic organisms studied, the pathway of electron flow seemed to occur in the following sequence:

$$\text{DPN} \rightarrow \text{Flavins} \rightarrow \text{cytochromes} \rightarrow \text{molecular oxygen}$$

Kreb discovered that they are involved in a cyclic process of electron transport. The catalytic effect of these acids on respiration explained Krebs results from their participation in a cyclic series of reactions that is needed for the oxidation of pyruvic acid to $CO_2$ and water. In these reactions, the electrons are transported to oxygen over the DPN-flavin-cytochrome system. The net effect of this series of reactions is the complete combustion of pyruvic acid to $CO_2$ and water.

The citric acid cycle is initiated by the condensation of the active acetic acid with oxaloacetic acid to form citric acid, a six carbon compound containing three carboxyl groups. The energy of the acetyl-CoA link is used in the condensation, and, as a result, the coenzyme A molecule is readily removed. In the ensuing reactions, citric acid is rearranged by the removal and addition of water to form isocitric acid.

Isocitric acid is then oxidized to form a new keto acid, which rapidly loses $CO_2$ to form α-ketoglutaric acid. Since this keto acid is metabolized in a series of steps that are analogous to the initial steps in pyruvic acid metabolism, the oxidative decarboxylation of α-ketoglutaric acid results in the formation of succinic acid, $CO_2$, DPNH, and ATP. Succinic acid is then oxidized to fumaric acid, which, upon hydration, leads to the formation of malic acid. The oxidation of malic acid regenerates the oxaloacetic acid and thus closes the cycle by entering the initial reaction, the condensation of acetyl CoA to form citric acid. When all the necessary enzymes are present, this cycle can be initiated by the addition of any of the compounds within the cycle. To keep the cycle going, however, acetyl CoA must be supplied.

The acetyl Co-A ($CH_3 \cdot CO \cdot S{\sim}CoA$) then combines with oxalacetate to form citric acid (citrate) by means of a condensing enzyme, liberating HS and CoA. Citrate then undergoes a series of changes called citric acid, tricarboxylic, or Krebs cycle (Fig. 9.7). The overall reaction shows the complete oxidation of pyruvate to $CO_2$ and $H_2O$.

### 9.7.9 Photosynthesis

The process of photosynthesis is divided into two phases: (1) the light phase in which radiant energy is trapped by chlorophyll and transferred to ATP with concomitant formation of NADH and NADPH; and (2) the dark phase, in which the energy stored in ATP is used in the

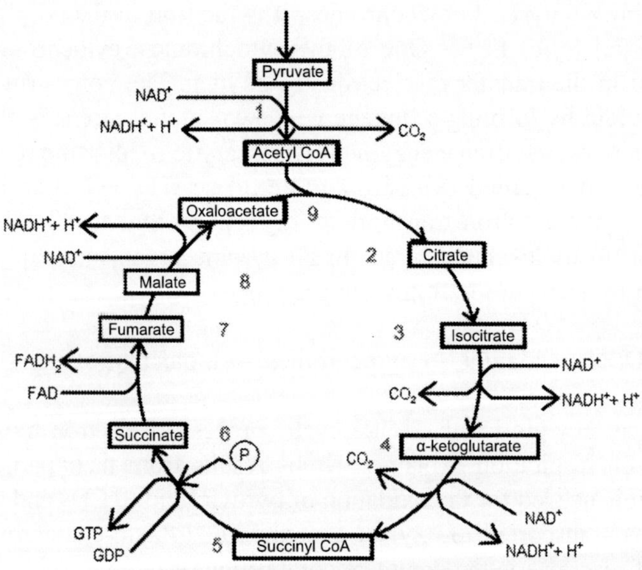


**Fig. 9.7.** Tricarboxylic acid cycle (Krebs cycle).

synthesis of organic compounds. The dark phase is independent of light and occurs in both photosynthetic and chemosynthetic cells, when energy is provided in the form of ATP.

## Light Phase

The first effect of radiation of chlorophyll is to raise the energy level of certain of its electrons in such a manner that one electron leaves each molecule of chlorophyll. Excited electrons are potent reducing agents. The extra energy of the excited electrons is used in two processes: cyclic phosphorylation and noncyclic phosphorylation. Each process generates ATP. Noncyclic phosphorylation also generates reducing potential in the form of NADPH.

## Cyclic Phosphorylation

In this process, the "high energy" electrons driven out of chlorophyll by light are passed along an electron transfer system consisting in part of flavin nucleotides and cytochromes, analogous to the electron transport systems in biooxidation. The electrons are transferred from coenzyme to coenzyme, yielding energy at each step. The energy of each electron generates two ATP from two ADP and two $H_3PO_4$ indicated as Pi (inorganic phosphoric acid). The electron then returns to the chlorophyll, completing the cycle.

## Noncyclic Phosphorylation

In this reaction the electrons are taken up by ferredoxin. Ferredoxin is an iron containing metalloprotein having enzyme-like properties and a uniquely high capacity for electrons, i.e., it thus becomes a potent reducing agent. The electrons acquired by ferredoxin from irradiated chlorophyll are immediately transferred along two paths. Most of the energy is used to generate ATP from ADP and Pi. The remaining energy is used to oxidize $H_2O$, yielding $2H^+$ and $O^{--}$. The hydrogen ions are then reduced by electrons from ferredoxin and combine with NADP to form NADPH. With energy from ATP, this carries the hydrogen, which is used to reduce $CO_2$ to organic compounds in the cell.

The electrons left attached to the oxygen after the water is split are immediately returned at a very low energy level to restore the chlorophyll that was robbed of its electrons by sunlight. The oxygen, minus its electrons, is given off as gaseous molecular oxygen. This terminates the light phase of photosynthesis.

## Dark Phase

The ATP, NADH and NADPH resulting from the light phase reaction cooperate to introduce $CO_2$ into organic combination in the cell. This process can proceed in the dark as well as in the light. At least two processes appear to be involved. In one, that has been most thoroughly studied by means of $^{14}CO_2$ by Nobel Prize winner, Calvin and others, $CO_2$ is added directly to 1,5-ribulose diphosphate, a 5-carbon ketose. The 6-carbon compound thus formed is immediately split by the enzyme carboxydismutase into two molecules of 3-phosphoglyceric acid (PGA).

## Bacterial Photosynthesis

Photosynthetic bacteria do not contain any of the chlorophylls found in eukaryotic green plants or the cyanophyceae. The last contain only chlorophyll a plus accessory phycocyanin. Instead, the photosynthetic bacteria contain the structurally similar pigments, bacteriochlorophyll or, depending on the species, *Chlorobium* chlorophyll. Unlike eukaryotic photosynthetic cells, bacterial chlorophylls and the chlorophyll a of cyanophyceae are not contained in complex, membrane enclosed, highly differentiated, independently replicating chloroplasts, but in vesicular particles or disks or tubular structures in the cytoplasm called chromatophores (Gr. *Chroma* = colour; *phoras* = carrier) or, in the case of *Chlorobium* species, ovoid or spindle-shaped bodies called "*Chlorobium* vesicles." In all groups, the chromatophores appear to be associated with the cytoplasmic membrane and contain oxidative and synthetic enzymes characteristic of cell membrane and photosynthesis. Prokaryotic chromatophores are now seen to be very suggestive of the eukaryotic chloroplast. The structures differ in different families of bacteria.

The chlorophylls of bacteria absorb most light energy in the red (720 to 750 mm; green *Chlorobium* chlorophyll) and invisible infrared (purple bacteriochlorophyll, 850 to 950 mm) zones. Light of these relatively long wavelengths may be the very first to reach the primeval earth and permit the development of primitive forms of photosynthesis. Since the energy of radiations is inversely proportional to their wavelength, it is evident that the chlorophylls of bacteria furnish less energy than the chlorophylls of green plants, which use light of much shorter wavelengths. The energy trapped by bacteriochlorophylls is, in fact, insufficient to split $H_2O$, a process that requires high level of energy. Consequently, photosynthetic bacteria cannot use water as a source of hydrogen to reduce $CO_2$ and therefore, do not give off free oxygen as a result of photosynthesis. However, they can do so as a result of photosynthesis. They split $H_2S$, a chemically analogous substance ($2H^+ + S^{--}$), or thiosulfates or certain organic compounds, as hydrogen donors in an equivalent manner; some can use molecular hydrogen directly. These relationships have been summarized by Van Niel as analogous to eukaryotic plant photosynthesis in a generalized formula:

**1. Plant photosynthesis:**

$$CO_2 + 2H_2O \rightarrow (CH_2O) + H_2O + O_2$$

**2. Bacterial photosynthesis:**

$$CO_2 + 2H_2A \rightarrow (CH_2O) + H_2A + A_2$$

In reaction 2, A may be sulphur or an organic residue, depending on the species. Some sulfide splitting species make themselves conspicuous by storing the sulphur as intracellular globules.

# Growth and Kinetics of Algae and Fungi

## 10.1 INTRODUCTION

The carbon fixation has been studied extensively in the unicellular algae. Exposure of *Chlorella* and *Scenedesmus* to $C^{14}$ labelled $CO_2$ for short time period showed that fixation by carboxylation of ribulose 1–5, biphosphate (RUBP) resulted in the formation of phosphoglyceric acid (PGA) as stable product. Ribulose biphosphate carboxylase was the enzyme that played an important role. $C_4$ pathway was important in seaweeds. PEP carboxylase activity has been low in Chlorophyceae, Phaeophyceae, Rhodophyceae and Bacillariophyceae. Most of the algae are $C_3$ plants biochemically, and physiologically resemble $C_4$ plants in the lower $CO_2$ compensation points. The uptake of inorganic carbon is predominant during photosynthesis. Heterotrophic $CO_2$ fixation also occurs and some algae are capable of utilizing organic substrates via photosynthesis. Heterotrophic nutrition is not that much uncommon in some algal classes.

Dark respiration in algae is identical with other aerobic plants. The glycolytic pathway or Embden–Meyerhoff–Paranas pathway runs in parallel with the hexose monophosphate shunt (HMP), while the tricarboxylic acid or Krebs cycle is in series with the processes. The system produces ATP for the cell and still larger quantities are produced under aerobic conditions by membrane associated oxidative phosphorylation using reductant generated by dehydrogenases acting on organic substrates. Reductant is mainly NADPH and it is also generated through hexose monophosphate shunt. The enzymes of the EMP and HMP pathways lie in the cytosol and also occur in the chloroplast. The TCA enzymes are located in the mitochondrial matrix, while oxidative phosphorylation processes are associated with the inner mitochondrial membrane.

In Cyanophyceae, the EMP, HMP and TCA enzymes are found in the cytosol and oxidative phosphorylation is associated with the membranes. Almost all ATP and NADPH produced during photosynthesis are utilized in the reduction of carbohydrate.

Photorespiration and glycolate metabolism are also present in charophytes. The oxygen uptake occurs and glycolate oxidase catalyses the conversion of glycolate to glyoxylate. This oxidation is due to the action of glycolate dehydrogenase. It is established that algae have glycolate pathway. Glycolate is an extracellular product in many algae. Algae may also release other carbohydrates, amino acids, lipids, phenolics and enzymes. The storage product in eukaryotic algae are carbohydrates; storage products being sugars and sugar alcohols. Chlorophyceae produces typical starch. The starch in Rhodophyceae and Cryptophyceae are with higher amylopectin component. Cyanophyceae starch closely resembles glycogen.

## 10.2 NUTRITION

The elements like carbon, hydrogen, oxygen, phosphorus, nitrogen, magnesium, zinc, copper, manganese, iron, molybdenum, sulphur, potassium and calcium are required by all algae. Sodium, iron, boron, iodine, bromine, vanadium, chlorine and silica are also required by some algae. Nitrogen and phosphorus are required in larger quantities. Others such as copper and molybdenum are required as cofactors in enzyme systems.

These are required in small amounts and are utilized in electron transport. Nitrogen fixation is done by Cyanophyceae and this process requires nitrogenase, oxygen sensitive iron-sulphur- and molybdenum containing enzyme complex. This brings about reduction of other substrates containing triple covalent bonds.

Acetylene reduction to ethylene and the simple separation of substrate and product by gas chromatography has been adopted as the standard assay for nitrogen fixation. Many algae utilize organically combined nitrogen such as urea etc. Nitrogen is stored as nitrate, ammonium or low molecular weight organic compound and used for growth at later stage.

Phosphorus is required to form nucleic acids, phospholipids and various ester phosphates such as phosphorylated sugars, ATP and NADP. Calcification in algae involves the precipitation of $CaCO_3$ around or within algal cells. Trace elements are required for the growth of algae in smaller quantities. Algae may also require vitamins and growth factors for their survival. Some algae are auxotrophic for thiamine.

## 10.3 GROWTH KINETICS

Oxygen and carbon dioxide uptake, except at high pH when active transport is necessary for the uptake of bicarbonate at rates sufficient to support net photosynthesis, is by passive diffusion. Other inorganic nutrients are generally taken up in ionic forms, and this uptake usually involves active transport linked to photosynthesis, which provides the energy necessary for the uptake.

The mechanisms and kinetics of active uptake in algae have been the subject of considerable research. The main stimulation for this work came from two papers (Dugdale, 1967; Eppley and Coatsworth, 1968), which examined the role of nutrient uptake kinetics in the growth of phytoplankton. The rate of nutrient uptake was shown to be related to the extracellular nutrient concentration.

$$V = V_{max}\ (S/K_s + S)$$

where $V$ = the nutrient uptake rate, $V_{max}$ = the maximum uptake rate, $S$ = the concentration of the nutrient, and $K$ = the half saturation constant or substrate concentration at which $V = V_{max}/2$.

This implies that at low nutrient concentrations the rate of uptake increases rapidly with increase in external nutrient concentration, but at higher concentration, increase in concentration adds progressively less to uptake rate until a high point is reached at which uptake is virtually constant and thus independent of the external concentration of the nutrients.

The growth rate of an alga will decline if the concentration of the nutrients in the external medium drops below that which will support an uptake rate sufficient to maintain the current growth rate. Under such conditions, the growth rate of the alga and the nutrient concentration show relationship analogous to the Michaelis-Menten equation for nutrient uptake. This is the Monod equation (Monod, 1942), which was originally applied to the growth of bacteria in chemostat culture, but also describes nutrient-limited, steady-state growth of algae (Droop, 1974).

$$\mu = \mu_{max}\ (S/K_s + S)$$

where $\mu$ is the specific growth rate, $\mu_{max}$ is the maximum specific growth rate, S is the substrate concentration, and $K_s$ is the half saturation constant of growth.

Other studies have shown that growth rate is also related to the cellular concentration of the limiting nutrient rather than its concentration in the external environment. This relationship is expressed in Droop equation:

$$\mu = \mu_{max}\ (1 - Q_0/Q)$$

where Q is the amount of nutrient in the algal cells, $Q_0$ is the lowest level of the Q at which the alga can grow and $\mu_{max}$ is the growth rate at infinite Q.

Under steady-state conditions, both Monod and Droop equations are applicable to the description of nutrient limited growth. The Droop equation is more practical, as the higher nutrient levels of the cell quota are more readily measured than the very low external nutrient concentrations.

The theoretical basis of nutrient uptake by macroalgae is essentially the same as that for microalgae. It is, however, complicated by the size of the plants and the increase in the amount of apoplastic free space, which must be considered in any uptake experiments, and in the measurements of the cell quota.

## 10.4 NUTRITION, METABOLISM AND GROWTH OF FUNGI

The fungi constitute one of the largest groups of plants with richest arrays of genera and species. Studies on physiology and biochemistry of fungi are quite fascinating, intriguing and also rewarding. These non-chlorophyllous plants agree with the green ones in having their protoplasm encased in cell walls. They, however differ essentially in their mode of nutrition. The fungi are remarkable in expressing a very wide range of diversity of forms and functions. There is so much variability that it is risky to pack up all examples within one scheme. We are still groping in dark for answers to many important questions. Possibly we may be able to stand on a more firm ground if the dimensions of physiological studies are expanded to cover a larger variety of fungi. It is rather disappointing that the initial attempts on fungal physiology were not only slow but somewhat misdirected and emphasis was mainly on the nutritional aspects with the main object of cultivating and maintaining the cultures in the mycological laboratories. The slow pace in this area can be gauged from the fact that the existence of many important pathways including the presence of functional TCA cycle in fungi was doubted till the middle of the present century. It is further depressing to record that most of the biochemical information about fungi has come from a limited group of organisms like yeasts, *Neurospora, Penicillium, Aspergillus* and *Fusarium* etc., while a rich forest of fungal species still awaits exploration by competent physiologists and biochemists. Some recent studies on fungal metabolism have been, however, of substantial help to the biologists as many elaborate and intricate biochemical processes have been discovered in selected group of fungi. After absorbing raw materials from the substrate, the young growing mycelium undertakes the enormous task of synthesizing nucleic acids, proteins, fats, glycogen and various cell wall substances. Simultaneously, it also accomplishes the task of breaking down the complex compounds, particularly the carbohydrates with the liberation of respiratory energy, some of which is rechannelized for synthetic reactions. Fungi are by and large the most suitable tools for physiological experiments because their forms as well as the phase of growth can be easily manipulated by careful adjustment of the environment. Practically in all the fungal organisms, the initial phase leads to the production of a homogenous mycelial growth due to rapid feeding. This results in rapid respiration and synthesis of a very wide range of complex organic substances.

Besides carbohydrates and fats, sometimes even structural components of cells are used during respiratory degradation. Initiation of morphogenesis reflects the advanced stage of mycelial growth, and this is also indicative of a major shift in metabolic speed/pathway or both. In most of the fungi, the production of various types of reproductive and protective structures is the apparent sign of differentiation and morphogenesis. Practically all such morphological changes in fungi can be controlled by suitable manipulation of the nutritional conditions. In nature, many fungi are able to grow quite extensively and efficiently as vegetative mycelium, but it is only during adverse conditions that they are provoked into producing reproductive bodies more as a means of survival rather than propagation. The variability with regard to production of different types of reproductive spores in fungi is so wide that it cannot be attributed to any one factor. This is, however, certain that the conditions required for production of spores in fungi are much more demanding than those needed for growth. Starvation or exhaustion of nutrients induces sporulation in many fungi. In general, the metabolic activities in spore bearing hyphae are higher than the vegetative mycelium. The nucleic acid contents in the spores are also comparatively richer. Factors favouring sporulation differ with the nature of the spore organism. Those favouring production of asexual spores are distinct from those inducing characteristic sexual or perfect stages. The advent of De Bary towards the middle of the nineteenth century gave a new direction to the study of fungi and he laid the foundation of experimental mycology and pathology. De Bary suspected the role of "chemicals" in accomplishment of sexual reproduction of some lower fungi. It was, however for Couch (1926) and Bishop (1940) to establish convincingly the operation of hormones in sexual reproduction of *Dictyuchus* and *Sapromyces* respectively. Implication of a multihormonal system in sexual reproduction of a variety of fungi is now well established. Diversity in forms has perhaps led to diversity in functions also, and in fact, fungi appear to be physiologically well adapted to the environment. As compared to bacteria, they can tolerate a wider range of pH (pH 2.0 to 9.0), osmotic pressure and desiccation etc. Their temperature range is also sufficiently wide. Some fungi can grow at freezing temperature and cause food spoilage under cold storage, while the thermophillic species can thrive well even at 62°C. Fungi possess the ability to utilize a wide spectrum of nutrients as source of energy. Being nutritional heterotrophs, they obtain their carbon requirements from organic carbon sources, including various carbohydrate and non-carbohydrate compounds. With an effective battery of enzymes, they are able to utilize a wide array of such substances with high degree of efficiency. A large number of enzymes have been located in fungal mitochondria, cytoplasm and spores. In nitrogen nutrition, they appear to be more versatile. They can fulfill their nitrogen requirements from inorganic, ammonium as well as organic sources. Even utilization of gaseous nitrogen has been claimed for some fungi, although this needs further confirmation. Most of the studies pertaining to nitrogen metabolism of fungi are based on *Neurospora crassa* alone, and there is a need for undertaking such studies with other groups of fungi. Similarly, they obtain their mineral requirements from their surrounding without much difficulty. Details about pathways of sulphur metabolism are fragmentary and most of the information in this area has been gathered during the last two decades on the basis of studies with some biochemical mutants of fungi and bacteria. Regarding vitamins, most of them are prolific producers of vitamins and they do not require vitamins from external sources, but some of the species are known to be auxoheterotrophic as they require one or the other vitamin from some external source. None of the fat-soluble vitamins like A, D, E and K have so far been found to be synthesized by fungi and probably they do not require these growth factors. Blending of all these characteristics has made the fungi a heterogenous but interesting group of organisms. Some recent investigations on their nutritional efficiency have yielded still more interesting results.

Minocha and Devay (1971) could maintain species of *Cephalosporium* and *Fusarium* on a medium devoid of any known complex organic carbon source. These fungi assimilated carbon from labelled $^{14}CO_2$. When grown as heterotrophs, they were able to revert to autotrophism, on subsequent transfer to autotrophic conditions. Degradation of glucose to pyruvate or lactate through glycolysis has more than one limitation as far as its functioning as an open biological process is concerned. The main shortcomings of this process are due to (i) the possible toxic effects of pyruvic acid accumulation, (ii) the limiting action of the coenzyme NAD+ due to its continued reduction to NADH, and (iii) inadequate free energy I gradient, with pyruvic acid still retaining more than four-fifths of the available energy of the carbohydrate source. These shortfalls may be overcome biologically in various ways, as for instance the pyruvic acid may itself act as an oxidant to reoxidize the NADH, and thus may get rid of its own toxicity; or it may get decarboxylated into acetaldehyde, which may then act as the hydrogen-acceptor to reoxidize the NADH, etc. However, the best recourse for the improvement of glycolysis as a respiratory process, both chemically and thermodynamically, is the full aerobic oxidation of pyruvic acid. It is perhaps on this account that great majority or most of the aerobes and facultative anaerobes utilize their glycolytic product pyruvate through an entirely aerobic sequence of reactions variously called the Krebs cycle, tricarboxylic acid cycle or citric acid cycle, at least under adequate aeration. The primary step in the aerobic oxidation of pyruvate is, however, an oxidative decarboxylation of pyruvic acid leading to the formation of acetyl coenzyme A through a series of closely knit reactions.

## 10.5 NUTRITION OF FUNGI

Monosaccharides are carbohydrates that are assimilated easily. Among these glucose has been reported to be the most efficient source of carbon and energy for most of the fungi. Sugar alcohols are obtained by reduction of monosaccharide. Sugar alcohols are less efficient carbon sources. These are widely distributed in various complex polysachharides. However, sugar acids are not utilized much by fungi. Sugars with two or more monosaccharide units linked together through glycoside bonds are known as the oligosaccharides. Trisaccharides like raffinose and disaccharides like sucrose, maltose, cellobiose, lactose, and melibiose found important carbon sources in fungal nutrition. Chitin forms major component of fungal cell wall and chitinase enzyme has been elaborated by many fungi. Starch is usually found as a storage compound in green plants and majority of fungi utilize this with efficiency. Organic acids like citrate, succinate and malate are acceptable to fungi. Nitrate has been reported as an excellent source for fungi than nitrite. The ammonium salts are generally inferior to ammonium nitrate. A large number of amino acids are recorded as good nutritional sources for fungi, which include glycine, L-aspartic acid, leucine, isoleucine etc.

Oxygen, hydrogen, phosphorus, sulphur, potassium, magnesium and microelements (Fe, Zn, Cu, Mo) are essential for the growth of fungi. Vitamins like thiamine, riboflavin, pyridoxine, niacin, biotin are required by fungi.

## 10.6 METABOLISM OF CARBON

Fungi are capable of utilizing a variety of carbon compounds. The non-carbohydrates are also metabolized, which include aliphatic and aromatic compounds.

Synthesis of organic acids by fungi was first confirmed by the classical studies of Wehmer during 1896–97. Most of the modern concepts regarding the mechanism of their synthesis presuppose the glycolytic conversion of the carbohydrate substrate into pyruvic acid from which acids are produced by varied transformations. Respiration involves three correlated phases:

(1) Dehydrogenation of the substrate with the help of a hydrogen acceptor, generally nicotinamide adenine dinucleotide, oxygen etc. serve as hydrogen acceptor.
(2) Substrate degradation by cleavage of the carbon to carbon bonds.
(3) Energy transfer that involves incorporation of energy liberated from the respiratory substrate into ATP. Glycolytic reactions in fungi are similar to other organisms. The only pathway operating through the utilization of a double phosphorylated hexose is the Embden-Meyerhoff (EM) pathway and is widespread in fungi. Pentose phosphate sequence, the alternative pathway of glucose metabolism, is chiefly aerobic. This

pathway has two distinct phases: (a) oxidative phase; (b) the non-oxidative phase. Entner doudoroff sequence belongs to hexose monophosphate pathway. This pathway includes (1) dehydration of the 5-phosphogluconate leading to the formation of 2-keto-3-deoxy-6-phosphogluconate. (2) cleavage of the product into glyceraldehyde-3-phosphate and pyruvate.

Degradation of glucose to pyruvate or lactate through glycolysis has more than one limitation as an open biological process is concerned.

## 10.7 NITROGEN METABOLISM

Nitrate reduction in fungi is to a great extent similar to that reported in bacteria and chlorophyllous plants. This is carried out by a group of enzymes in various steps, ultimately leading to formation of ammonia. Nitrite, nitric oxide, hyponitrite and hydroxylamine are the principal intermediate substances, which are produced during the course.

Nitrate reduction in fungi is carried out by nitrate reductase enzyme. This enzyme was formerly considered to be only assimilatory in nature, but subsequently their dissimilatory nature was also established (Walker and Nicholas, 1961) and in some fungi they are functional even under anaerobic condition when oxygen supply is limited or scanty.

In contrast to bacterial system, the dissimilatory nitrate reductase enzymes of fungi require iron as well as molybdenum for their activity.

## 10.8 METABOLISM OF AMINO ACIDS

Ammonia, which is the key intermediate in nitrate metabolism, combines with carbon compounds to form organic nitrogenous substances. The details of the mechanism by which this key process proceeds is not known in most of the fungi. Studies on biosynthetic processes in germinating spores of some fungi lead to conclusion that skeletal material metabolic activities or amounts of some cellular component provide only indirect methods for growth measurements. Mendels (1965) has discussed in detail the various methodologies of measuring fungal growth and needs no further elaboration. Whatever be the method employed, the quantitative study of fungal growth has not been as simple as that of bacteria, particularly in case of filamentous fungi. The unicellular fungi, albeit, follow generally the same pattern of growth as do the bacteria, and a typical exponential growth curve is obtained in their case. They also exhibit the same growth phases, including (i) Lag-phase, during which cells do not divide, although cell enlargement and enhancement in their metabolic activity may occur; (ii) acceleration phase, during which cell division starts and increases to a level, characteristic of the next phase; (iii) logarithmic or exponential phase, marked by cell division as a uniform and maximum rate; (iv) deceleration phase, during which rate of cell division shows a decline; (v) stationary phase, an apparent cessation of cell-division, but in fact this may represent only a balance between cell-formation and death of old cells; (vi) phase of decline or death, during which an actual decline in the number of cells takes place. The various phases of growth may be recognized in case of filamentous fungi also, but since their growth is not autocatalytic and is restricted to the hyphal tips only, they do not actually manifest any exponential phase of growth. This phase may at best be described as phase of rapid growth. However, Plomley (1959) demonstrated that an entire hypha of *Chaetomium globosum* along with its branches grows exponentially. He also recorded exponential increase in the radius of a young colony. Later, growth rate of central hyphae of the colony declined, but the marginal hyphae grew as earlier. Thus the phases recognizable in filamentous fungi may be designated as (1) lag-phase, (2) linear phase, with rapid and almost linear growth rate and (3) decline phase, marked with decrease in dry weight of the mycelium owing to autolysis.

# Biology and Nutrition of Protozoa

## 11.1 INTRODUCTION

The protozoa are considered to be a subkingdom of the kingdom Protista, although in the classical system they were placed in the kingdom Animalia. More than 50,000 species have been described, most of which are free-living organisms; protozoa are found in almost every possible habitat. The fossil record in the form of shells in sedimentary rocks shows that protozoa were present in the pre-Cambrian era. Antony van Leeuwenhoek was the first person to see protozoa using microscopes he constructed with simple lenses. Between 1674 and 1716, he described, in addition to free-living protozoa, several parasitic species from animals, and *Giardia lamblia* from his own stools. Virtually all humans have protozoa living in or on their body at some time, and many persons are infected with one or more species throughout their life. Some species are considered commensals, i.e., normally not harmful, whereas others are pathogens and usually produce disease. Protozoan diseases range from very mild to life-threatening. Individuals whose defences are able to control but do not eliminate a parasitic infection become carriers and constitute a source of infection for others. In geographic areas of high prevalence, well-tolerated infections are often not treated to eradicate the parasite because eradication would lower the individual's immunity to the parasite and result in a high likelihood of reinfection.

Many protozoan infections that are inapparent or mild in normal individuals can be life-threatening in immunosuppressed patients, particularly patients with acquired immune deficiency syndrome (AIDS). Evidence suggests that many healthy persons harbour low numbers of *Pneumocystis carinii* in their lungs. However, this parasite produces a frequently fatal pneumonia in immunosuppressed patients such as those with AIDS. *Toxoplasma gondii,* a very common protozoan parasite, usually causes a rather mild initial illness followed by a long-lasting latent infection to AIDS patients; however, can develop fatal toxoplasmic encephalitis. *Cryptosporidium* was described in the 19th century, but widespread human infection has only recently been recognized. *Cryptosporidium* is another protozoan that can produce serious complications in patients with AIDS. Microsporidiosis in humans was reported in only a few instances prior to the appearance of AIDS. It has now become a more common infection in AIDS patients. As more thorough studies of patients with AIDS are made, it is likely that other rare or unusual protozoan infections will be diagnosed.

*Acanthamoeba* species are free-living amoebas that inhabit soil and water. Cyst stages can be air-borne. Serious eye-threatening corneal ulcers are due to *Acanthamoeba* species infection. AIDS patients, however, can develop fatal toxoplasmic encephalitis.

The parasites presumably are transmitted in contaminated lens-cleaning solution. Amoebas of the genus *Naegleria*, which inhabit bodies of freshwater, are responsible for almost all cases of the usually fatal disease, primary amoebic meningoencephalitis. The amoebas are thought to enter the body from water that is splashed onto the upper nasal tract during swimming or diving. Human infections of this type were predicted before they were recognized and reported, based on laboratory studies of *Acanthamoeba* infections in cell cultures and animals.

The lack of effective vaccines, the paucity of reliable drugs, and other problems, including difficulties of vector control, prompted the World Health Organization to target six diseases for increased research and training. Three of these were protozoan infections—malaria, trypanosomiasis, and leishmaniasis. Although new information on these diseases has been gained, most of the problems with control persist.

## 11.2 STRUCTURE

Most parasitic protozoa in humans are less than 50 μm in size. The smallest (mainly intracellular forms) are 1 to 10 μm long, but *Balantidium coli* may measure 150 μm. Protozoa are unicellular eukaryotes. As in all eukaryotes, the nucleus is enclosed in a membrane. In protozoa, other than ciliates, the nucleus is vesicular, with scattered chromatin giving a diffused appearance to the nucleus; all nuclei in the individual organism appear alike. One type of vesicular nucleus contains a more or less central body, called an endosome or karyosome. The endosome lacks DNA in the parasitic amoebas and trypanosomes. In the phylum *Apicomplexa*, on the other hand, the vesicular nucleus has one or more nucleoli that contain DNA. The ciliates have both, a micronucleus and macronucleus, which appear quite homogeneous in composition.

The organelles of protozoa have functions similar to the organs of higher animals. The plasma membrane enclosing the cytoplasm also covers the projecting locomotory structures such as pseudopodia, cilia, and flagella. The outer surface layer of some protozoa, termed pellicle, is sufficiently rigid to maintain a distinctive shape, as in the trypanosomes and *Giardia.* However, these organisms can readily twist and bend when moving through their environment. In most protozoa, the cytoplasm is differentiated into ectoplasm (the outer, transparent layer) and endoplasm (the inner layer containing organelles); the structure of the cytoplasm is most easily seen in species with projecting pseudopodia, such as the amoebas. Some protozoa have a cytosome or cell "mouth" for ingesting fluids or solid particles. Contractile vacuoles for osmoregulation occur in some, such as *Naegleria* and *Balantidium.* Many protozoa have subpellicular microtubules; in the *Apicomplexa*, which have no external organelles for locomotion, these provide a means for slow movement. The trichomonads and trypanosomes have a distinctive undulating membrane between the body wall and a flagellum. Many other structures occur in parasitic protozoa, including the Golgi apparatus, mitochondria, lysosomes, food vacuoles, conoids in the *Apicomplexa,* and other specialized structures. Electron microscopy is essential to visualize the details of protozoal structure. From the point of view of functional and physiologic complexity, a protozoan is more like an animal than a single cell.

## 11.3 LIFE CYCLE

During its life cycle, a protozoan generally passes through several stages that differ in structure and activity.

- Trophozoite (Greek for "animal that feeds") is a general term for the active, feeding, multiplying stage of most protozoa. In parasitic species, this is the stage usually associated with pathogenesis. In the haemoflagellates, the terms amastigote, promastigote, epimastigote, and trypomastigote designate trophozoite stages that differ in the absence or presence of a flagellum and in the position of the kinetoplast associated with the flagellum. A variety of terms are employed for stages in the *Apicomplexa*, such as tachyzoite and bradyzoite for *Toxoplasma gondii.* Other stages in the complex asexual and sexual life cycles seen in this phylum are the merozoite (the form resulting from fission of a multinucleate schizont) and sexual stages such as gametocytes and gametes. Some protozoa form cysts that contain one or more infective forms. Multiplication occurs in the cysts of some species so that excystation releases more than one organism. For example, when the trophozoite of *Entamoeba histolytica* first forms a cyst, it has a single nucleus. As the cyst matures, nuclear division produces four nuclei and during encystation four uninucleate metacystic amoebas appear. Similarly, a freshly encysted *Giardia lamblia* has the same number of internal structures (organelles) as the trophozoite. However, as the cyst matures, the organelles double and two trophozoites are formed.

## 11.4 REPRODUCTION

Reproduction in the protozoa may be asexual, as in the amoebas and flagellates that infect humans, or both asexual and sexual, as in the *Apicomplexa* of medical importance. The most common type of asexual multiplication is binary fission, in which the organelles are duplicated and the protozoan then divides into two complete organisms. Division is longitudinal in the flagellates and transverse in the ciliates; amoebas have no apparent anterior-posterior axis.

- Endogony is a form of asexual division seen in *Toxoplasma* and some related organisms. Two daughter cells form within the parent cell, which then ruptures, releasing the smaller progeny that grow to full size before repeating the process.
- In Schizogony, a common form of asexual division in the *Apicomplexa,* the nucleus divides a number of times, and then the cytoplasm divides into smaller uninucleate merozoites. In *Plasmodium, Toxoplasma*, and other apicomplexans, the sexual cycle involves the production of gametes (gamogony), fertilization to form the zygote, encystation of the zygote to form an oocyst and the formation of infective sporozoites (sporogony) within the oocyst.

Some protozoa have complex life cycles requiring two different host species; others require only a single host to complete the life cycle. A single infective protozoan entering a susceptible host has the potential to produce an immense population. However, reproduction is limited by events such as death of the host or by the host's defence mechanisms, which may either eliminate the parasite or balance parasite reproduction to yield a chronic infection. For example, malaria can result when only a few sporozoites of *Plasmodium falciparum,* perhaps ten or fewer in rare instances are introduced by a feeding *Anopheles* mosquito into a person with no immunity. Repeated cycles of schizogony in the bloodstream can result in the infection of 10 per cent or more of the erythrocytes, about 400 million parasites per millilitre of blood.

## 11.5 NUTRITION

The nutrition of all protozoa is holozoic, i.e., they require organic materials, which may be particulate or in solution. Amoebas engulf particulate food or droplets through a sort of temporary mouth, perform digestion and absorption in a food vacuole, and eject the waste substances. Many protozoans have a permanent mouth, the cytosome or micropore, through which ingested food passes to become enclosed in food vacuoles. Pinocytosis is a method of ingesting nutrient materials whereby fluid is drawn through small, temporary openings in the body wall. The ingested material becomes enclosed within a membrane to form a food vacuole.

Protozoa have metabolic pathways similar to those of higher animals and require the same types of organic and inorganic compounds. In recent years, significant advances have been made in devising chemically defined media for the *in vitro* cultivation of parasitic protozoa. The resulting organisms are free of various substances that are present in organisms grown in complex media or isolated from a host and which can interfere with immunologic or biochemical studies. Research on the metabolism of parasites is of immediate interest because pathways that are essential for the parasite but not the host are potential targets for antiprotozoal compounds that would block that pathway but be safe for humans. Many antiprotozoal drugs were used empirically long before their mechanism of action was known. The sulpha drugs, which block folate synthesis in malarial parasites, are one example.

The rapid multiplication rate of many parasites increases the chances for mutation, hence, changes in virulence, drug susceptibility, and other characteristics may take place. Chloroquine resistance in *Plasmodium falciparum* and arsenic resistance in *Trypanosoma rhodesiense* are two examples.

Competition for nutrients is not usually an important factor in pathogenesis because the amounts utilized by parasitic protozoa are relatively small. Some parasites that inhabit the small intestine can significantly interfere with digestion and absorption and affect the nutritional status of the host; *Giardia* and *Cryptosporidium* are examples. The destruction of the host cells and tissues as a result of the parasites' metabolic activities increases the host's nutritional needs. This may be a major factor in the outcome of an infection in a malnourished individual. Finally, extracellular or intracellular parasites that destroy cells while feeding can lead to organ dysfunction and serious or life-threatening consequences.

# Cultivation of Microorganisms

## 12.1 INTRODUCTION

The development of microorganisms in culture media is dependent upon a number of factors such as nutrients, oxygen supply, moisture, pH and temperature.

A satisfactory microbiological culture medium must contain available sources of carbon, nitrogen, inorganic salts and, in certain cases, vitamins or other growth-promoting substances. These were originally supplied in the form of meat infusions, which were, and still are in certain cases, widely used in culture media. Beef extract frequently replaces meat infusions, but the preparation of this substance subjects it to the loss of its heat labile nutrient factors in much the same way as infusions are affected. The addition of peptone provides a readily available source of nitrogen and carbon.

Peptone is used in culture media to supply an available form of nitrogen, since native proteins are not generally attacked by bacteria. Most organisms are capable of utilizing the amino acids and other simpler nitrogenous compounds present in peptone.

Certain bacteria require addition of other food substances such as serum, blood etc. to the culture medium. Carbohydrates may also be desirable at times, and certain salts such as those of calcium, magnesium, sodium, and potassium seem to be required. Dyes may be added to media as indicators of metabolic activity or because of their selective inhibitory powers. Growth promoting substances and vitamins are essential to assist in the development of certain types of bacteria.

The consistency of liquid medium may be modified by the addition of agar, gelatin or albumin in order to change it into a solid or semisolid state. Solid media, which were originally devised for the isolation of organisms in pure culture, are now universally used for almost all general cultural work. The semisolid media are used chiefly for carrying stock cultures or for propagating the anaerobes or semi-anaerobes.

One of the principal landmarks in bacteriology was the preparation of a satisfactory medium by Hesse's introduction of agar as a solidifying agent for bacteriological culture media.

Most bacteria are capable of growth under ordinary conditions of oxygen tension. Certain types, however, are capable of deriving their oxygen from their food substances. The aerobic organisms require the free supply of air, while the anaerobes grow only in the absence of atmospheric oxygen. Between these two groups are the microaerophiles, which develop best under partial anaerobic conditions, and the facultative aerobes and anaerobes, which develop over a wide pH range.

For anaerobic conditions a cotton plug is placed just above the culture medium in tubes and a gelatine capsule containing pyrogallic acid in a Durham fermentation tube is placed on top of this plug. A solution of sodium hydroxide is placed in the Durham fermentation tube and the culture tube is then sealed with a rubber stopper or screw cap. The tube is then ready for incubation, the oxygen being removed by the action of the sodium hydroxide on the pyrogallic acid.

Proper moisture conditions must prevail in the culture media employed for the propagation of microorganisms. A moist medium and a moist atmosphere are necessary for the proper growth of the vegetative cells.

The pH or reaction of the culture medium, expressing its hydrogen ion concentration, is extremely important for the growth of microorganisms. The majority of the microorganisms prefer culture media, which are approximately neutral, while others may require a medium which is distinctly acidic. The pH or reaction of the culture medium is determined by colorimetric or electrometric measurement of its hydrogen ion concentration. It should be noted that additions of acid or alkali, which are insufficient to prevent the growth of bacteria in a medium, may inhibit or prevent them from proceeding with the normal functions of their metabolic processes.

The usual range of temperature suitable for the growth of microorganisms lies between 15° and 40°C. Microorganisms have, however, been known to develop at 0°C, and others, such as the sporulating organisms, may grow up to 80°C. The pathogenic organisms, in general, are limited by a comparatively narrow range of temperature, around 37°C, while the saprophytes usually have a much broader requirement.

The media upon which microorganisms are grown must be sterile or free from all other forms whose development might influence or prevent the normal growth of the inoculated type. The usual method for immediate sterilization of culture media is by means of autoclave in which steam under pressure is the sterilizing agent. The proper operation of the autoclave to insure sterilization of media requires careful manipulation. Autoclave sterilization for 15 minutes at 15 pounds pressure (121°C) is recommended for quantities of liquid media up to one litre. Agar media on prolonged sterilization or heating are apt to show a precipitate. Repeated melting of solidified agar, or long holding of melted agar at high temperature, may likewise cause a precipitate to form in the media. Media containing agar may also form a flocculent precipitate if the liquid medium is held in the water bath at 43 to 45°C for longer than 30 minutes. This flocculent agar precipitate, however, may be dispersed by reheating the medium.

External contamination of culture media is prevented by plugging the tubes or flasks with non-absorbent cotton before sterilization. Plugs should fit neither too loosely nor too tightly and should protect the lip of the container against the accumulation of dust.

Media should always be stored in a cool moist atmosphere to prevent evaporation. Prolonged storage of sterile media cannot, however, be recommended. If tubes of media have been kept for any length of time, they should be reheated just before use. Liquid media should be heated in a boiling water bath or in flowing steam for a few minutes to drive off dissolved gases, and then cooled quickly in cold water without agitation just prior to inoculation. Agar tubes should be melted and allowed to solidify in order to secure a moist surface, which is desired by most microorganisms. These precautions for both liquid and solid media are extremely important for the initiation of growth of highly parasitic organisms such as those encountered in blood culture work.

Blood or other body fluids to be cultured should always be taken prior to the administration of the therapeutic agent.

## 12.2 CULTURE MEDIA

Liquid or solid preparation made especially for the growth, storage and transport of microorganisms or other types of cells is known as medium. In order to study a microorganism, it must be isolated from its natural environment and grown in a culture medium designed for optimum growth of the microorganism.

A culture medium is an artificial environment of well-defined conditions for the growth of microorganism. It should contain sources of nitrogen, phosphorus, carbon and other sources and also micronutrients. Sometimes growth factors are used for securing a good growth.

Definitions of some typical cultures are given below:

1. **Pure culture:** A microbial culture consisting of population of cells derived from a single cell is called pure culture. It differs from axenic culture in the sense that the pure culture implies genetic purity, which an axenic culture does not.
2. **Slant culture:** It consists of a microorganism growing on the inclined surface of a solid medium.
3. **Broth culture:** Broth or liquid culture is one that consists of active growth of a microorganism in liquid medium.
4. **Mother culture:** It is the primary culture obtained at the first point of pure culture.
5. **Two-membered culture:** A culture consisting of two types of microorganisms growing together is denoted as two-member culture. It is useful in some cases to derive maximum benefits.

Synthetic, semi-synthetic and non-synthetic or natural media are the three types of media generally used for culturing microorganisms.

### 12.2.1 Routine Laboratory Media for Cultivation of Bacteria and Fungi

Growth media is used for the initiation of a culture/sub-culture for enrichment or diagnostic purposes. Specialised media are used for the growth and maintenance of tissue cultures. The medium should be sterile before inoculations.

The chemolithotrophic bacteria can be grown in a simple media containing inorganic salts. Heterotrophs are grown in a range of culture media containing organic substrates. Anaerobes are generally cultured in the presence of organic substrates under oxygen-free conditions.

Solid medium usually consists of liquid medium, which has been solidified with agar. This is used to obtain colonies of a given organism. Colonies are required for identification purpose and also for the determination of viable (living) cell counts.

Semi-solid medium is used for the isolation of microaerophilic (cultures grown under low $O_2$ conditions) organisms. To create microaerophilic conditions, the agar concentration in the liquid medium should be 1.5 to 2.0 g per litre of medium. Different types of media used for the cultivation of microorganisms are given below (composition of various media are given in appendix).

## 1. Minimal Medium

This is a type of culture medium that lacks certain growth factors. This does not support the growth of auxotrophic strains i.e., a strain of microorganism that lacks the ability to synthesize one or more essential growth factors. It arises by mutations of a wild type strain (prototype).

## 2. Basal Medium

This is a type of medium that supports the growth of various nutritionally selected organisms. It includes peptone and nutrient agar, potato and nutrient agar, potato dextrose agar (PDA) and Saboraud's agar media.

Peptone is a soluble product of protein hydrolysis and is not coagulated by heat or saturated ammonium sulphate. It is precipitated by phosphotungstic acid. Peptone water agar medium consists of 1.2% peptone, 0.5% sodium chloride (NaCl) and 1.5% of agar in 1 litre of water.

This is a simple medium and has been most generally used in bacteriological examination of water, sewage and food products and for isolating organisms in pure culture. This medium is recommended for the cultivation and growth of fungi.

## 3. Enriched Medium

This is a basal medium, which is supplemented with serum, blood etc, or with any range of nutrients or growth factors in order to enable it to support the growth of specific type(s) or organisms.

**Examples:** Blood agar, serum agar, chocolate agar, Jensen's medium etc. This medium is proved to be satisfactory for the isolation of the organisms from all types of gonococcal infections.

### Serum Agar

Bacto-beef blood serum is a dehydrated powdered form of fresh beef blood serum. In a 10% solution, it is equivalent to fresh serum. It is a useful aid to bacteriologists, who have occasional use for serum media. The use of this media eliminates all disagreeable and time consuming procedures when compared to the preparation of fresh serum. When it is used in plain 10% solution or as an ingredient of Loeffler medium, it forms firm, smooth, white slants upon heat coagulation.

## 4. Selective Media

Selective medium is one which allows or supports the growth of certain types of organisms in preference to others. Most of the media are selective in nature. Mannitol salt agar medium is a selective medium for the isolation of pathogenic staphylococci. Growth of most bacteria other than staphylococci is inhibited on this medium. The staphylococci grow luxuriantly producing colonies with yellow zones.

MacConkey agar is a differential plating medium recommended for use in the detection and isolation of all types of dysentery, typhoid and paratyphoid bacteria from stool specimens, urine and other materials harbouring these organisms. It is generally suggested for direct plating of water samples for coliform counts, particularly for differentiating strains of *Salmonella typhosa* from members of the coliform group. Mac Conkey agar, as modified by the addition of 0.5% NaCl, decreasing the agar content to 1.35%, and by altering the concentrations of bile salts and neutral red, has the added advantage of supporting excellent growth of *Shigella* and *Salmonella* strains. It also provides a clear-cut differentiation between these enteric pathogens and the coliform group, making it easier to enumerate the organisms.

Jenson medium is used for isolating free-living nitrogen-fixing microorganisms, particularly *Azotobacter*. The organism uses atmospheric nitrogen as source of N for its growth.

### Quaternary Ammonium Compounds (Qac)

These are a group of cationic surfactants used as antiseptics, disinfectants and preservatives. They appear to act by disrupting the cytoplasmic membrane and by denaturing proteins at high concentrations. A QAC may be regarded as an ammonium halide with four constituents. They are bacteriostatic at low concentrations and bactericidal at high concentrations. They are more active against Gram-positive than Gram-negative bacteria and are reported to be fungistatic, typanocidal and active against certain viruses. They have little or no activity against bacterial endospores. These compounds are used for preoperative skin cleansing and surgical dressings and for the disinfection of equipment used in the food and dairy industries. They include Zephiran, Cetrimide or Cetavlon, commonly used as a preservative in ophthalmic solutions.

## 5. Enrichment Medium

Enrichment medium is a medium containing substances that encourage the growth of the required organism and/or inhibit the growth of other types of organisms. Some of the enrichment media are described below:

### Selenite F-Medium

This medium is used for the enrichment of the strains of *Salmonella* in samples of faeces. Selenite inhibits many of the common enteric bacteria. The medium consists of an aqueous solution of peptone (0.5%), lactose or mannitol (0.4%), sodium selenite (0.4%) and either di-sodium hydrogen phosphate (1.0%) or a combination of di-sodium hydrogen phosphate and sodium di-hydrogen phosphate to give a final pH of 7.0. The medium should not be autoclaved, but may be steamed for 30 minutes. Incubation of the inoculated medium should be around 12 to 18 hours at 30°C.

### Alkaline Peptone Water

This medium is used for isolating *Vibrio cholerae.* It consists of peptone (1%) and NaCl (1%) in distilled water and pH adjusted to 8.6 to 9.0.

## 6. Differential Medium

It is a solid medium on which different types of organisms are distinguished by their different forms of growth. The examples of differential media are blood agar, MacConkey's agar, etc.

### A) Blood Agar

This medium contains oxalated blood. It is prepared by mixing 510 g blood drawn aseptically with a pre-autoclaved molten agar base such as trypticase-soy agar at 50°C. The medium is dispensed in Petri dishes and allowed to set. Blood agar is used for culturing fastidious bacterial parasites of man and animals. It is also used for detection of haemolysins, since a given bacterial haemolysin may be active against the red blood corpuscles of only some animals like horse, sheep etc. The choice of blood plays a very important role in formulating the medium.

### B) Chocolate Agar Medium

This medium is used for the isolation and culture of fastidious bacteria like *Haemophilus* sp., *Neisseria gonorrhoeae* etc. A medium like blood agar base or brain-heart infusion agar is autoclaved and cooled to 80°C. Bovine blood (1%) is added and the medium is shaken gently until chocolate-brown colour is obtained. It may be dispensed into plates or tubes. Alternatively, a chocolate agar plate may be prepared by heating blood agar medium to around 80°C.

## 7. Biochemical Media

### A) Glucose Phosphate Peptone Water

Carbohydrates like glucose, galactose, arabinose, lactose, maltose, mannitol, mannose, rhamnose, sucrose, xylose etc, are added to peptone water composition given in page No. 134–135 to determine the metabolism of the respective carbohydrate by a given strain of microorganism. Bacterial growth in a given peptone-water-sugar may give rise to acid, or acid and gas, as breakdown products of the sugar. Acid production is detected by pH indicator and gas production by a Durham tube. For nutritionally fastidious organisms, peptone-water-sugars are supplemented with serum particular growth requirements in order to permit the growth of the organism.

### B) Tryptone Water

Tryptone is a tryptophan-rich peptone (a pancreatic digest of casein). Tryptone water medium consists of 1–2% tryptone and 0.5% NaCl in water.

### C) Oxidation-Fermentation Medium

Oxidation-fermentation test (Hugh and Leifson's test; O–F test) is used to indicate whether a given carbohydrate (usually glucose) is attacked oxidatively or fermentatively by a given bacterial strain. The peptone-agar medium includes NaCl, $K_2HPO_4$, the carbohydrate and pH indicator (bromothymol blue). The medium imparts green colour and has a pH of 7.1. In the test, each of the two tubes of the medium is stab inoculated with the test organism, and the medium in one of the tubes is immediately covered with a layer of sterile liquid paraffin (about 1 cm. depth). Both the tubes are then incubated and subsequently examined for evidence of carbohydrate utilization. Acidification (yellowing of the medium) in the uncovered medium indicates oxidative attack on the carbohydrate by the bacterium, while acidification in the covered tube indicates that the carbohydrate can be attacked fermentatively. No reaction in either tube usually indicates that the particular carbohydrate is not utilized by the test organism.

### D) Simmon's or Koser's Citrate Medium

Simmon's citrate agar is a medium capable of differentiating between faecal coli forms and members of aerogenes group on the basis of citrate utilisation. It may also be used to differentiate certain members of *Salmonella* group. On such a solid medium, growth is indicated very clearly by colony formation and is usually accompanied by a change of colour of the indicator (bromothymol blue) due to acid or alkali production. Strains producing alkali change the medium from its initial green colour to deep blue in 24-48 hours of incubation.

This medium is also recommended for isolating certain fungi and fungi imperfecti.

## 8. Transport Media

Any medium used specially for the transportation and/or temporary storage of material (swabs) from which the isolation of particular organism(s) is attempted is known as transport medium. The purpose of such medium is to maintain the viability and/or infectivity of the organisms during the delay between collection and culture of the specimen.

The commonly used transport media are Cary-Blair transport medium and Ames transport medium. In the later, charcoal is incorporated into the medium. Charcoal containing media obviate the need for collecting the specimens on charcoal impregnated swabs since the charcoal serves to adsorb inhibitory substances in the medium and/or specimen. Some transport media have been devised for particular organisms, e.g., Monsur transport medium (see P. 340).

## 12.3 ISOLATION OF MICROORGANISMS

Our knowledge of the diversity of the microorganisms is based on two kinds of procedures. Some microorganisms are recognizable by their colony morphology, accumulation of stored substances or changes in their environment. By visual evidence of their occurrence, many of these microorganisms are suitable for direct isolation. Suitable growth media and conditions for their growth can be assessed. Microorganisms can also be isolated by enrichment culture techniques.

Natural populations usually consist of a mixture of different types of microorganisms. For many of the microbiological investigations, a pure culture is very essential. A pure culture is defined as the progeny (clone) of a single cell. A pure culture should be free from contaminants. The procedure for obtaining the pure culture is as follows.

A cell must be separated from a cell population and the colony that results from its multiplication must remain separate from other cells and other colonies. When cells of a single colony are spread out on a fresh surface, many colonies are developed, each with the same shape and colour as the original (Fig. 12.1). The most commonly used methods for obtaining the pure cultures of aerobic bacteria as proposed by Koch are streak plate method, pour plate method and spread plate method.

In the streak plate method, a sterilized inoculation needle is dipped into a suitable diluted suspension of the culture(s) and then streaked on the surface of solidified agar medium in different directions and incubated at ambient temperature. Individual colonies develop on plates after 2–3 days (Fig. 12.2).

In the spread plate method, a small sample (usually 0.1 ml) is spread over the surface of an agar plate containing an appropriate medium and the plate incubated until the colonies are visible to the eye. It is inferred that each colony arose from a single cell, and each colony represents a living cell.

In the pour plate method, the sample is mixed with melted agar and the mixture poured into a sterile plate. The organisms are thus fixed in the agar medium and form colonies.

The above methods allow for the positive identification of the organism(s).

## 12.4 ISOLATION OF MICROORGANISMS BY MICROMANIPULATOR TECHNIQUES

A device named micromanipulator can be used in conjunction with a microscope to pick a single bacterial cell from a mixed culture. The micromanipulator permits the operator to control the movements of a micropipette or microprobe so that a single cell can be isolated. This technique requires a skilled operator and is meant for studies in which a clone must be obtained unequivocally.

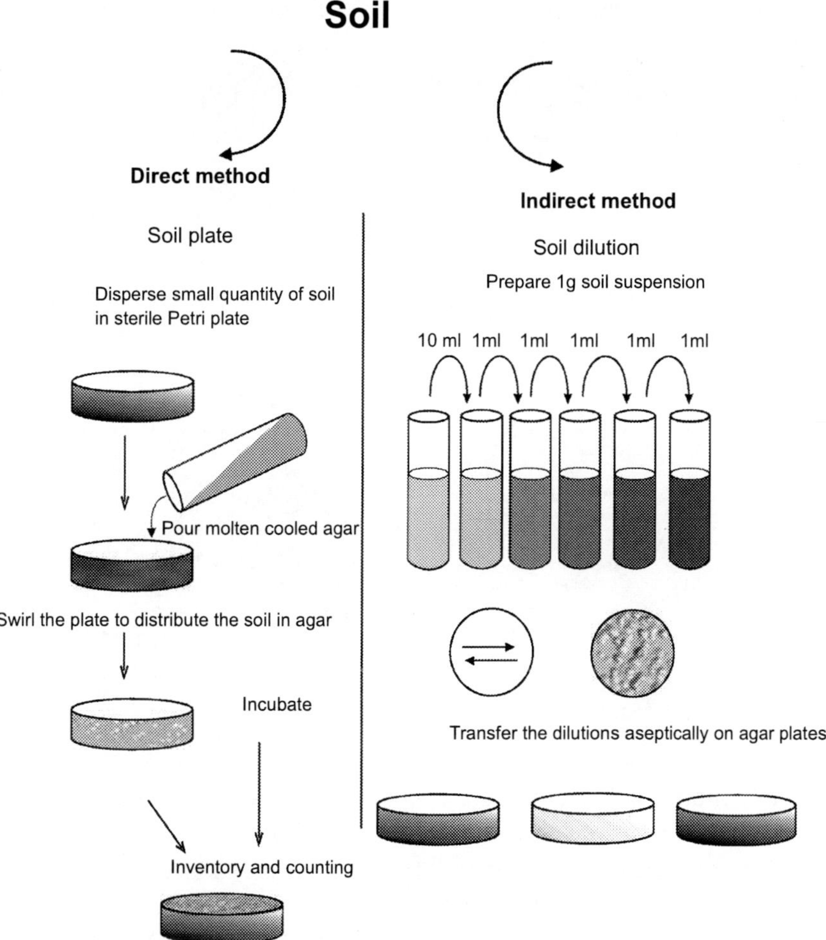


**Fig. 12.1.** Isolation of microorganisms by dilution method.

### 12.4.1 Isolation of Anaerobic Bacteria

Few bacteria are strict anaerobes. The isolation of pure cultures of these forms requires certain modifications of the used plating technique. Provided that their sensitivity to oxygen is not too great, they can be isolated by streaking or pouring plates in the usual fashion and then placing the inoculated plates in an oxygen-free atmosphere. Some strict anaerobes are, however, very rapidly killed by exposure to air. For such forms, the method of choice is the use of dilution shake cultures. A series of sterile tubes, about two-thirds filled with melted agar of the appropriate composition, are cooled to a temperature just above the solidifying point, and the inoculum is added to the first tube. This is then well mixed, and a little of the mixture (about one-tenth) is poured into the second tube, which is mixed in turn and used to inoculate the third tube. The

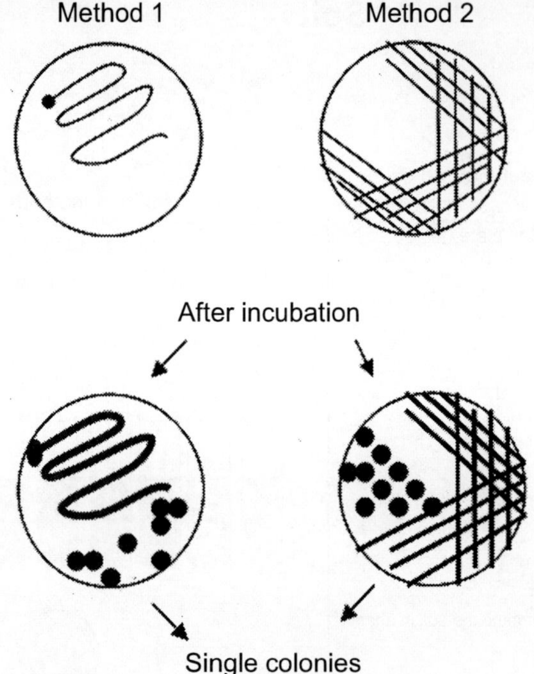


**Fig. 12.2.** Streak plate method for isolating bacterial colonies.

agar is allowed to solidify, and the surface is covered with a seal of vaseline and paraffin in order to minimize the access of air to column of agar.

When a series of ten tubes is prepared in this fashion, the dilution is sufficient to yield one or more tubes containing well-isolated colonies (Fig. 12.3).

When a shake culture has developed, the tube is warmed to loosen the plug of vaseline and paraffin, which is then removed with a sterile wire. A very fine sterile capillary pipette, prepared by drawing out a piece of glass tubing in the flame of a burner, is then introduced between the column of agar and the glass wall of the tube. By attaching the wide end of the pipette with a piece of rubber tubing to a source of compressed air, it is possible to blow out the entire column of agar into a sterile Petri dish, and then it can be sectioned into discs with a sterile scalpel for the examination of individual colonies. By careful separation of single colonies, their suspension in suitable liquid and repeated streaking out or plating on nutrient agar of the same composition, it is possible to obtain pure cultures of the majority of microorganisms under oxygen-free conditions.

### 12.4.2 Enrichment Culture Methods

For the study of the various physiological types of microorganisms that exist in nature, Winogradsky and Beijerinck proposed a new and profoundly important technique: the enrichment culture technique. It is essentially an application on a microscale of the Darwinian principle

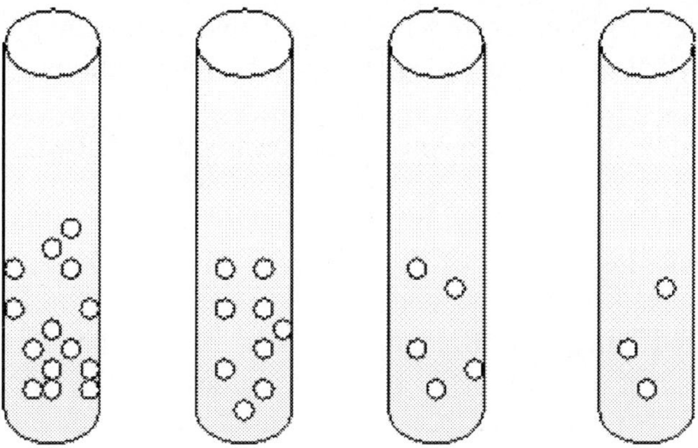

**Fig. 12.3.** Serial dilution method for isolating anaerobic bacteria in soft agar (0.8%).

of natural selection. The investigator devises a culture medium of a particular defined chemical composition, inoculates it with a mixed microbial population, such as can be found in a small amount of soil or mud, and then ascertains by examination what kind of microorganisms comes to predominance. Since their predominance is caused by their ability to flourish in the enrichment medium, they can be readily isolated by plating on a medium of same composition. To take a specific example, if we wish to isolate microorganisms, which can use atmospheric nitrogen, we prepare a medium that is free of combined nitrogen but contains all other nutrients, an energy source, a carbon source and minerals necessary for growth. This is then inoculated with soil, placed in contact with $N_2$, and incubated under any desired set of physical conditions. Since nitrogen is an essential constituent of every living cell, the only organisms that will multiply in such a medium are those that can fix atmospheric nitrogen. If such types are present in the soil sample, they will grow. Such experiments can be varied in innumerable ways: by modifying such factors as the carbon source, the energy supply, the temperature, and the hydrogen ion concentration. For each particular set of conditions, a particular kind of microorganism will come to predominance, provided the organism exists in the inoculum, which can grow under such particular conditions. These experiments also define the medium that should be used for the isolation of the organisms concerned, since it is clear that the enrichment medium, whose exact composition has been set beforehand, provides the necessary nutrient environment for their development. The enrichment culture method is thus one of the most powerful experimental tools available to the microbiologist. By its use, he can isolate microorganisms with any desired set of nutrient requirements, provided that such organisms exist in nature. To isolate microorganisms capable of developing at the expense of cellulose as a source of carbon and energy, one has to prepare an enrichment medium containing cellulose as the only available source of carbon and energy, together with ammonia or nitrate as a nitrogen source, and the other necessary minerals. This is inoculated with soil and incubated until growth occurs. One can further limit the type of cellulose-decomposing microorganisms that develop by altering the physical conditions under which the enrichment culture is maintained. For example, if we wish to obtain anaerobic cellulose-decomposers, the enrichment culture has to be incubated in the absence of air. In case of cellulose-decomposers that are capable of growing at high temperatures, the cultures have to be incubated at 50–60°C.

Once the outcome of a given enrichment culture technique is known, the method also permits the isolation of a given microorganisms at will from nature.

The enrichment of various organisms can be achieved by the manipulation of such environmental factors as the composition of the medium, aeration, pH, and illumination. Other factors, including temperature, osmotic pressure, and surface tension can also be varied to give special selective advantage to different organisms. By incubating enrichment media at very high temperatures (for example, 55°, or 60°C), cultures of thermophilic bacteria can be obtained. Even at moderate temperatures, selective enrichment by the use of varied temperature is possible.

The choice of a suitable inoculum in which a given type of organism is already enriched in nature is often helpful, and in some instances essential to the success of an enrichment. Furthermore, the inoculum can be pasteurized (two to five minutes at 80°C) to destroy bacteria that form vegetative cells and to enrich endospore-forming members of the genera, *Bacillus* and *Clostridium.*

An almost infinite number of permutations and combinations of environmental variables can be devised for the enrichment of different bacteria from nature. Thus, the enrichment culture technique offers a means, not only of isolating known bacterial species, but also for studying undescribed organisms that are capable of flourishing in any given environment.

Enrichment cultures of fermentative bacteria must be incubated under anaerobic conditions, because some of these organisms are obligate anaerobes, and the presence of oxygen inhibits the growth of anaerobic bacteria. The lactic acid bacteria illustrate this point. These organisms are characterized by their remarkable resistance to lactic acid, which they themselves produce in the fermentation of sugar. To enrich lactic acid bacteria, a poorly buffered medium containing glucose and a rich source of growth factors is used (for example, twenty grams of glucose and ten grams of yeast extract per liter). After inoculation, preferably with natural materials that are rich in lactic acid bacteria (vegetable matter, raw milk, sewage etc.), the medium is incubated under anaerobic conditions. The first enrichment culture to develop usually consists of bacteria that carry out the formic acid type of fermentation (*Aerobacter, Escherichia,* and related genera), since these organisms grow very rapidly. However, as lactic acid gradually accumulates, conditions become less and less favourable for these bacteria, whereas the lactic acid bacteria continue to grow. Eventually, the acidity of the medium becomes so high that the lactic acid bacteria become predominant and most other organisms are killed.

Another good example is the enrichment of the propionic acid bacteria. These organisms produce propionic acid, acetic acid, and $O_2$ in fermentation. If a neutral medium containing twenty grams of sodium lactate and ten grams of yeast extract per litre is inoculated with natural materials containing propionic acid bacteria like Swiss cheese and incubated at 30°C under anaerobic conditions, enrichment of these organisms is obtained.

The acetic acid bacteria (*Acetobacter*) are adapted to environments that contain high concentrations of alcohol. They are also less susceptible than other bacteria to inhibition by acetic acid, which they produce from alcohol in respiration. To enrich *Acetobacter*, a complex medium containing alcohol is inoculated with materials containing these bacteria and then incubated under aerobic conditions. Fruits, flowers, and unpasteurized beer are among the favourite sources of inoculum. A medium containing forty millilitres of alcohol and ten grams of yeast extract per litre, adjusted to pH 6.0, can be used for enrichment.

## 12.5 IDENTIFICATION OF MICROORGANISMS BASED ON CULTURAL CHARACTERISTICS

One of the major features of a bacterial strain is its appearance following growth on different media. To determine growth characteristics of a strain, it is vital to observe the features of colonies and broth cultures. After inoculation of medium and subsequent incubation, the following cultural characteristics can be determined:

Growth on agar plate
Growth on agar slant
Growth in broth
Growth in gelatine stab

### 12.5.1 Growth on Agar Plate (agar plate colonies)

a) *Size:* Colonies range in size from extremely small (pin point), measuring only a fraction of a mm in diameter to large colonies measuring 5-10 mm in diameter.
b) *Margin or edge:* The periphery of bacterial colonies may take one of several different patterns depending on the species. It may be evenly circular-like edge of a droplet, or it may show irregularities such as rounded projections, notches, and thread-like or root-like projections.
c) *Surface texture:* Depending on the species, the colony surface may be smooth (shiny, glistening), rough (dull, granular or matte), or mucoid (slimy or gummy). Certain species have colonies possessing a highly wrinkled surface.
d) *Elevation:* Depending on the species, colonies may be thin or thick and surface may be flat or it may exhibit varying degrees of convexity.
e) *Chromogenesis or pigmentation:* Red, yellow, brown and sometimes violet etc. characterize certain species. Colonies may be non-coloured (non-pigmented) also.

**Pigmentation remaining inside the cell:**

Examples:
Red: *Serratia marcescens*
Violet: *Chromobacterium violaceum*
Golden yellow: *Staphylococcus aureus*
Lemon yellow: *Micrococcus flavus*
Brownish black: *Micrococcus niger*

**Pigmentation excreting into the medium:**

Examples:
Greenish to yellow: *Pseudomonas chlororaphis*
Greenish to olive brown: *P. fluorescens*
Green, blue to dark brown: *P. aeruginosa*
*Optical features:* May be opaque, transluscent or opalescent.

### 12.5.2 Growth on Agar Slant (agar streak)

*Amount:* Scanty, moderate or abundant
*Margin or edge of growth:* Uniform or even or exhibiting various irregularities somewhat similar to irregularities described for colonies.
*Consistency of mass of growth:* Butyrans or butter-like consistency, easily removed with transfer needle, viscous or stingy, dry and brittle.
*Chromogenesis or pigmentation:* Similar to that described for colonies.

### 12.5.3 Growth in Nutrient Broth

*Amount of growth:* Scanty, moderate or abundant
*Distribution of growth throughout the broth:* Uniformly distributed (evenly turbid) growth confined to surface of broth as a scum or film (pellicle) or growth accumulated as sediment, which may be granular or viscous.
*Odour:* May be turbid, fruity or aromatic or negligible

### 12.5.4 Growth in Gelatine Stabs

*Growth (no liquefication) along the line of inoculation:* Growth may be confined to the zone of inoculation streak or may exhibit varying degrees of spreading away from the streak.
*Liquefication of gelatin:* May proceed evenly from top, or various funnel-like designs of liquefied medium occur.

## 12.6 IDENTIFICATION OF MICROORGANISMS BASED ON MOLECULAR CHARACTERISTICS

### 12.6.1 On the Basis of G+C Content

For identification of microorganisms is the determination of the relative proportion of guanine (G) and cytosine (C) compared to total number of nucleotide base position in DNA because there is pairing between complementary bases in DNA specifying G+C and also specifies adenosine + thymine (A+T). The composition of DNA is normally described by specifying moles % G+C, i.e., the proportion of the DNA macromolecules composed of G+C base-pairs. Closely related organisms should have similar proportions of G+C in their DNA, and organisms with vastly differing proportions of G+C in their DNA can safely be said to be unrelated. Microorganisms that are otherwise similar, but differ in G+C content in their genomes, represent different species.

To determine the moles % G+C, DNA can be slowly heated and its ability to absorb UV light at 260 nm measured spectrophotometrically. When DNA is denatured, there is a change in its absorption characteristics. This change in UV absorption is known as 'hyperchromatic shift'. The mid-point temperature of the denaturation curve is called 'melting temperature' (Tm) of the DNA because certain ions stabilize the H-bands in DNA. Tm is influenced by chemical composition of DNA solution.

The empirical relationship is:

$$\text{Tm} = 0.41\% \text{ (mole \% G+C)} + 16.6 \log \text{M} + 81.5$$

where, M = Concentration of monovalent cations within range 0.0001–0.2 molar.

The mole % G+C of DNA may also be determined by measuring buoyant density of DNA.

### 12.6.2 On the Basis of DNA Homology

As compared to mole % G+C, a more precise measure of relatedness is "DNA homology" between two organisms. A measure of DNA homology accounts for the order of nucleotides as well as overall composition of the genome. It is a powerful tool for the identification of microorganisms, especially for establishing affinities of strains within the species and species within the genome.

### 12.6.3 On the Basis of Serological Reactions

Certain serological tests are used for the identification of microorganisms where they are behaving as antigens against certain antibodies. Some such tests are precipitation reactions (single diffusion method, double diffusion method, immunoelectrophoresis), radioimmunoassay, enzyme-linked immunoabsorption assay (ELISA), immunofluorescence (direct fluorescent antibody staining i.e., FAB), agglutination reactions (Weil-Felix test), complement fixation etc.

### 12.6.4 Computer-Assisted Identification

Keys and diagnostic tables can be readily programmed performing large number of calculations and comparisons for unknown isolates, also with statistical probability to a particular taxonomic group involving development and use of probabilistic identification matrices.

## 12.7 ISOLATION OF CONIDIAL FUNGI FROM WATER

The conidial fungi are known to colonize the submerged decaying leaves of angiosperms and others. Under natural conditions, submerged dark brown, softened and skeletonized leaves of well aerated streams have to be collected in plastic bags. On return to the laboratory, the leaves are washed and placed in a separate sterilized Petri dish having sterilized distilled water. The dishes are left for a day or two at 20°C. The incubated leaves can be scanned for the fungal growth.

Foam samples can be collected below waterfalls and can be scanned for conidial fungi quantitative techniques involving sucking the water sample through a millipore filter and staining the conidia by cotton blue in lactic acid followed by their counting.

Conidial fungi can be cultured on malt extract agar medium. Conidial fungi in water help in recycling of organic matter, serve as bioindicators and contribute to food chain cycle.

## 12.8 ISOLATION OF *TRICHODERMA*

*Trichoderma* is a very important soil fungus and is involved in the biological control of soil-borne and root-borne pathogens.

This fungus can be isolated from soil, seed, root, leaf surface etc., using serial dilution plate technique and direct plates of the substrate on agar media. The media such as commercial dextrose agar, special nutrient agar, cellulose agar, malt extract agar, oat meal agar, potato dextrose agar, special nutrient agar, *Trichoderma* selective medium etc. are used in the isolation of *Trichoderma.*

## 12.9 ISOLATION OF ZOOSPORIC FUNGI

Zoosporic fungi mostly found in fresh water and marine water can be isolated by baiting half boiled hemp seed, maize seed, chitin and keratin substrates, cockroaches, flies, snake skin, peelings of fruits, tender grass blades, seeds of brassica, bhendi, etc. The baits can be placed in tumblers having water sample. After 24-48 hrs., the baits are scanned under binocular research microscope.

The colonized baits are washed in sterile water and baited with fresh bait material. These fungi grow very well on potato dextrose agar media, maize meal agar medium, hemp seed agar and oat meal agar medium.

## 12.10 COLLECTION OF ARBUSCULAR MYCORRHIZAL FUNGI

The arbuscular mycorrhizal fungi form obligate symbiotic association with roots and other underground parts of the plant. AM fungi play a significant role in the transport of phosphorus and other nutrients.

The AM fungi consisting of spores, sporocarps and hyphal bits can be extracted from soil using modified wet sieving and decanting techniques. Quantification of root colonization is done by detecting the presence or absence of colonization by hyphae, arbuscules, vesicles, resites, spores or sporocarp in the trypan blue stained root cortex.

## 12.11 ISOLATION OF PLANT PATHOGENS

Diseased plant materials have to be collected in fresh polythene bags. After reaching the laboratory, thin transverse sections of the diseased material mounted in lactophenol are observed for preliminary identification of the pathogen. Symptoms such as vascular wilts, rusts, smuts, leaf spot, blights, cankers etc. can be noted. If necessary, microtome sections can be prepared for permanent slide preparation.

The diseased plant tissue is surface sterilized in 0.01%, $HgCl_2$ or sodium hypochlorite followed by repeated washings in distilled water and sterile water. After final washing, the diseased plant tissue can be kept in blotter so that the moisture is removed. Later the diseased tissue is plated on 1% agar medium followed by its transfer onto 2% potato dextrose agar medium.

The inoculated Petri dishes are incubated at room temperature and also under UV light. Sporulation is observed frequently. The vegetative and reproductive structures produced in the culture are compared with the structures observed in thin sections. Thus, the identity of the pathogen is established.

In order to establish whether the same pathogen, which is sporulating in culture, is the causal agent or not, Koch postulates have to be followed. The diseased material is inoculated on the agar medium. The sporulating pure culture is then inoculated onto the specific host. The symptoms produced after inoculation of the culture on susceptible and specific host are compared with the symptoms of field specimen. Thus, the pathogenicity can be established.

## 12.12 ISOLATION OF FUNGI FROM ROOTS

The roots are first washed in running water followed by sterile water. The root pieces are treated with 0.01% $HgCl_2$ or sodium hypochlorite and washed with sterile water. Such treated root species are plated on potato dextrose agar and observed for sporulation.

## 12.13 ISOLATION OF FUNGI FROM AERIAL PLANT PARTS

The aerial plant parts such as stem, bark, petiole, wood, leaves, floral parts, fruits, seeds etc., get affected by fungal pathogens. Such diseased specimens can be collected. The diseased tissues are plated on agar plate having potato dextrose upon medium after treating with 0.01% $HgCl_2$ and sodium hypochlorite and sterile water. The inoculated agar plates are incubated at room temperature and observed for fungal growth.

Purification of fungal cultures of filamentous fungi may be achieved by growing the primary material on plates of plain distilled water agar with a dry surface. To avoid bacterial contamination, an antibiotic such as PCNB, streptopenicillin etc. are used. The agar plates are incubated at room temperature. A part of the growing fungi is cut from agar plate and inoculated into full pledged agar medium. After a period of 8-10 days, the cultures are observed and identified.

## 12.14 SINGLE SPORE ISOLATION

Various techniques such as semi-mechanical and mechanical methods are available. Filtered agar is recommended when microscopical searching for spores is involved. In semi-mechanical method, fungal spore suspension is made on a sterile grid containing sterile water. This spore suspension is streaked across a marked line on a very thin plate of tap-water agar and incubated at 24°C. Germination occurs after 12 hrs. Germinating spores are selected and agar block and such spores are transferred into an agar plate or tube. In the mechanical method, the dummy microscope objective lens is replaced by a sharp edged metal tube. The agar is cut by lowering the cutter until it reaches the bottom of the dish. The cut agar block is reinoculated into fresh agar plate.

## 12.15 ISOLATION OF ACTINOMYCETES

Actinomycetes are isolated from the soil by using differential media. Actinomycetes, in general, are slow growing and hence they are likely to be masked in culture plates when grown in ordinary nutrient media. For isolation of actinomycetes, media containing low nitrogen, such as Ken-Knight's, egg albumin, or Conn's medium (composition given in appendix) are used. The plant pathogenic actinomycetes, like *Streptomyces scabies*, are isolated by using tyrosine-casein-nitrate agar medium. For routine culturing of actinomycetes, yeast extract or Czapek's medium (see appendix) is used.

## 12.16 CULTIVATION AND ASSAY OF VIRUSES

Viruses are obligate intracellular parasites. They are acellular in nature and always require a living host cell for their growth. Cellular organisms are the hosts for viral growth. Viruses do not have the characters of living while they are outside the host cell. While they are in the host cell, they behave as living by making use of cellular biocatalytic machinery of the host cell. While outside (away from the host cell) the host cell, viruses behave as non-living. Therefore, viruses are called acellular entities and not organisms. For a virus to grow, its host cell is required. The host cell can be prokaryotic or eukaryotic. Among eukaryotic host cells, the host can be unicellular or multicellular in nature. A virus infects the specific host cell and replicates to produce its progeny. Result of virus infection and replication in the host cell can be cell death or deviation in normal structure and functioning of host cells. Sometimes host cell infection by certain viruses might go unnoticed because it does not cause any change in the host cell. It is necessary that host cells must be available in actively living state for a virus to grow. Depending on the type of virus and its specific host, the symptoms vary. A virus after infecting a host cell or tissue results in specific symptoms. Such symptoms are normally taken for characterization of virus growth and quantitative assay of virus numbers. Depending on the type of host cell they infect, there are bacterial viruses, animal viruses and plant viruses. Bacterial viruses are also called bacteriophages. Depending on the type of virus to be grown, the hosts can be bacteria, animal cells or whole animal, plant cells or whole plant. Each group of viruses is grown and quantitatively assayed depending on the host cells and type of symptoms they produce. Bacteriophages are viruses that infect the bacterial cells, reproduce, grow in the host cell and come out of the cell as viral progeny. As phages come out of bacterial cell, the cell is destroyed or lysed. Such phages are called lytic phages. Certain bacteriophages after infection do not lyse the host cell but become part of the cell by incorporating viral genome into the bacterial genome. Such phages are called lysogenic phages or temperate phages. Lysogenic phages enter into lytic cycle to produce their progeny and come out of the host cell by cell lysis. There are a number of bacteriophages in nature depending on the type of bacterial host they infect, and genetic material they contain. Bacteriophages are isolated and cultivated in the laboratory by providing living bacterial host cells.

### 12.16.1 Isolation of Bacteriophages

Bacterial viruses are easily isolated and cultivated on young, actively growing cultures of bacteria in broth or agar medium. Optimum growth conditions have to be provided for good growth of host cells and viruses in host cells. While the virus is infecting and growing in bacterial host cell suspended broth cultures, result of virus growth will be observed as clear transparent broth after virus growth. The lysis of bacteria in broth culture causes the turbid culture to become clear and transparent. When viruses grow in bacterial cells, they grow as 'lawn' culture on agar plates. Bacterial lysis results in clear zones, which are called 'plaques'. Formation of plaques is the result of virus and bacterial host cell interaction.

**Plaque assay:** Bacteriophage, when present in a layer of susceptible bacteria on a nutrient medium, will be adsorbed on one of the bacterial cells on the agar medium. Soon after, the infected bacterium will lyse and release large number of phage particles, each of which will be adsorbed onto adjacent bacterial cells. These bacteria in turn will lyse and release phages that infect others bacteria in the vicinity. These multiple cycles of infection continue until, after several hours, the phages will have lysed all the bacteria in a localized area. The result is a clear zone called 'plaque' on the surface of agar medium. One plaque is formed from the action of one initial phage particle. The number of plaques on a lawn of bacteria gives an estimate of the number of phage particles (also called plaque-forming units, or pfu) introduced on the plate. Such a quantitative procedure is called 'plaque assay'. The phages infecting and growing in suspended bacterial host cells in broth will not form plaques but give clear appearance of liquid culture, which otherwise should be turbid. Phage and bacterial host cell interaction is biologically specific. The type of phages or bacterial host strains are also identified by this method of host-phage interaction. Identification of bacterial types with the help of bacterial phages is known as phage typing.

The efficient and most useful way to isolate bacteriophages is from the habitat of the particular host bacterium. For example, coliphages or other phages pathogenic for bacteria found in the intestinal tract can best be isolated from sewage or manure. In practice, the source sample is suspended in saline and centrifuged or filtered to remove suspended particles. A few drops of chloroform are added to the remaining liquid to kill any bacterial cells. A small amount (normally about 0.1 ml) of this preparation is mixed with the host organism (specific bacterial host) and spread onto the agar medium. Virus source sample preparation mixed with host cells together can also be added to melted agar (at about 45–50°C) and poured into sterile plates after mixing and allowed to solidify. Such agar medium plates are incubated at optimum temperature for about two days. Growth of phage is indicated by the appearance of plaques. Number of phage particles in 1 ml of liquid suspension preparation is calculated by the following formula:

where N = Number of phage particles per ml
P = Number of plaques formed
V = Volume of sample (suspension prepared) inoculated
D = Dilution of sample inoculated

For example: A 2 ml sample suspension prepared and diluted up to $10^{-8}$ is inoculated in 0.1 ml volume and the plaques are 20 in a given plate incubated. Then, P = 20, V = 0.1 ml, D = $10^{-8}$. The number of phage particles per ml in 2 ml sample preparation are $2 \times 10^{-10}$ viral particles per ml. In total 2 ml it is $4 \times 10^{-10}$ phage particles. Direct microscopic count of phage particles is also done by observing under electron microscope by taking known volume of sample suspension.

### 12.16.2 Cultivation of Animal Viruses

Animal viruses are isolated and cultivated in the laboratory for various purposes. Large scale cultivation of animal viruses is practised in industries for the preparation of viral vaccines. As viruses are obligate parasites, animal viruses grow in their specific living host cells. Generally, animal viruses are cultivated by three main methods: in living host animals, in embryonated chicken (or duck) eggs, and in cell (tissue) cultures. Depending on the type of virus and its specific host, the method of cultivation is practised. In a given host organism or embryonated egg, specific parts/organs are selected for the growth of specific viruses. Certain viruses exhibit not only host specificity but also organ specificity for their growth.

### 12.16.3 Virus Cultivation in Living Animals

Animal viruses are grown in specific animals maintained in specified animal houses. Using animals for the cultivation of viruses is not the method of choice due to the following reasons: it is expensive, labour-intensive and may add other factors such as contaminating bacteria, into a study. Some animal viruses such as hepatitis B virus can be cultivated only in living animals. In recent times, this hepatitis B virus antigenic genes are cloned in bacteria or yeast cells and grown in large scale for preparation of commercial antigens (vaccines) specific for this virus, which otherwise cannot be grown outside the specific animal host. Studies of host immune responses to a viral infection also demand the use of living animals. In the laboratory, mice, rabbits, guinea pigs, and nonhuman primates are normally used. Animal inoculation is also a good diagnostic tool to determine whether a virus will cause infection, the animal will exhibit typical disease symptoms, and sections of infected tissues can be examined microscopically for the evidence of infection. Certain viruses have organ specificity within the animal host. For example, rabies virus specifically grows in brain tissue in an infected animal. Wherever may be the site of infection (entry) by the virus, it ultimately migrates to brain and establishes its growth to show symptoms. Virus samples are inoculated (injected) into specific organs of host animals for cultivation of animal viruses. Specific pathological changes are observed in the organs/tissues infected by the viruses. This symptom is taken as an indication for identification and quantitation of viruses. It takes certain time of incubation for a virus to show its symptoms of infection and growth depending on the virus and its specific host.

### 12.16.4 Virus Cultivation in Embryonated Eggs

Cultivation in embryonated eggs is one of the most economical and convenient methods widely used for animal viruses. The discovery that viruses could be cultivated by this simple technique was made in 1931. Embryonated chicken eggs can be inoculated aseptically with virus, using a needle and syringe, through a hole drilled in the shell. The opening is sealed with paraffin and the eggs are inoculated at 36–37°C for 2–3 days to allow the viruses to multiply. Generally, 7–13-days-old embryonated eggs are used for virus cultivation depending on the virus. Different organs of embryonated egg are also selected for the cultivation of specific viruses. Chick embryos contain several kinds of cells and tissues in which various viruses will replicate. By using embryos of different ages and using different methods of inoculation, it is possible to cultivate the type of virus desired. Different tissues within the egg are inoculated, depending on the type of virus used. For example, the chorioallantoic membrane (CAM) supports the growth of the herpes virus, smallpox virus (variola), Rous sarcoma virus (RSV) and eastern equine encephalitis virus (EEEV). The amniotic cavity will permit the growth of influenza and mumps virus. The influenza virus also has an affinity for the allantoic cavity. Yellow fever virus is generally grown in yolk sac. The embryonated eggs are used routinely for the production of influenza vaccines, as well as for vaccines against smallpox, yellow fever and a number of other poultry/animal viral vaccines. Certain viruses such as vaccinia virus multiply on one of the embryonic membranes and produce distinct type of lesion called a 'pock'. This symptom can be used for counting the viral particle number in given sample similar to counting of plaques for phages. The embryo itself is used to grow rabies virus and eastern equine encephalitis virus. After virus cultivation in embryonated eggs, the virus culture is harvested and used for quantitative assay by different direct or indirect assay methods.

### 12.16.5 Virus Cultivation in Tissue Culture

Cell culture or tissue culture is the most widely used method for cultivation of animal viruses. Growing of tissue cells is normally practised as a single layer (monolayer) in liquid medium in a flat bottomed container. Once a cell culture has been grown, it is possible to use it as an *in vitro* (in glass) host for a virus. Such tissue cells are living host cells for virus growth. Viruses invade the cells, usually causing some type of visible changes in the growth of the monolayer tissue cells near the site of initial infection (inoculation). This localized change, called cytopathic effect (CPE), is a deterioration of the tissue culture cells caused by the virus after infecting and growing in the cells. The CPE may have different appearances depending on the particular virus and the type of cells in the culture. One type of CPE is an inclusion body, which is an abnormal intracellular structure. Inclusion bodies are important because their presence can help identify the virus causing an infection. There are variety of inclusion bodies depending on the virus and host cell. For example, small particles called Guarnieri bodies can be seen in the cytoplasm of cells infected with the smallpox virus. These are aggregates of virions that have developed in the cytoplasm. Negri bodies occur in the cytoplasm of nerve cells from the brain of animals with rabies, and their presence is important for the diagnosis of this disease. Intra-cytoplasmic inclusion bodies also occur in several other diseases such as sheep-pox, fowl-pox and molluscum contagiosum. Intra-nuclear inclusions are found in cells infected with viruses of chicken pox and herpes.

Groups of lysed cells (plaques) in tissue culture have been used in the quantitative assay of virus concentration. The number of plaques is proportional to the number of infectious virus particles present in a sample similar to phage plaque assay. Each animal virion causes formation of a single plaque on a monolayer of susceptible animal cells, just as a bacteriophage forms a plaque on a lawn of susceptible bacterial host cells. Though tissue culture is the most important method for cultivation of animal viruses, maintaining a tissue culture itself is a difficult job, because it requires utmost aseptic conditions in the laboratory. By and large, embryonated egg cultivation of viruses is the most preferred method of animal virus cultivation in routine teaching laboratories.

### 12.16.6 Cultivation of Plant Viruses

Filtrable agents are the cause of plant leaf mosaic was demonstrated initially by Beijerinck to explain acellular agents as pathogens. Subsequently, in 1930, Wendell M. Stanley obtained purified tobacco mosaic virus (TMV) crystals in purified extract from mosaic leaves. He was awarded Nobel Prize for this work in the year 1946. Viruses need living host and plant viruses grow in living plant cells. Plant viruses cause many diseases of important crop plants. The effects of viral infection on some plants may appear as spots (in tomato spotted wilt virus), wounds (in corn and sugarcane), dwarfness and wrinkles in tomato plants.

Plant viruses can be cultivated by inoculating a plant with a virus suspension through a hypodermic needle or by scratching plant leaves with virus material. Scratching is aided with an abrasive such as carborundum to facilitate the entry of virus into plant cells. This can lead to formation of local lesions as well as general infection. Some plant viruses can replicate to large number in infected plants. For example, a single hair cell of an infected tobacco plant may contain over $10^7$ tobacco mosaic virions. Indeed, as much as 10 per cent of the dry weight of infected leaves may be tobacco mosaic virus. Transfer of virus infection from cell to cell occurs by passage of virions through bridges that connect the cells of plant tissues. In most plant diseases, infected cells can continue to manufacture virus without either disintegrating or drying. Diseased plant material can be crushed to release plant juices containing the virions, which can then be purified. Cell wall of host plant cell is hindrance to the entry of virus into hosts cells. Cell-wall less plant cell protoplasts are directly infected by viruses. Plant tissue cultures of different host plants are made use of for cultivation of specific plant viruses. Quantitative assay of plant viruses can be done by inoculating the virus preparation into specific host plant tissues or host plant itself and observing the symptoms.

## 12.17 ISOLATION OF CYANOBACTERIA

Often initial isolations of cyanobacteria from natural environment give mixed cultures. Therefore, it is essential to purify the individual types of cyanobacteria from the mixture. This could be achieved easily by streak or pour plate techniques. These are essentially dilution techniques resulting in the physical separation of individual cyanobacterium from the mixture allowing them to form distinct colonies, which can then be picked up to make pure cultures. Also, for obtaining clonal cultures of filamentous forms, these techniques prove extremely useful.

There is no accurate way of predicting the number of viable cells in a given sample by the above-mentioned methods. Therefore, one must always make several dilutions of the sample and pour several plates. However, experience with certain type of samples will minimize wasteful dilutions. The success of this method depends upon thorough mixing and even distribution of cells.

The cultures are normally incubated at 29(+l°C) under continuous illumination (2,000-3,000 lux) unless otherwise required. Because of the high concentration of water in agar, condensation of water may occur in Petri dishes during incubation and moisture is likely to drip from the cover onto the surface of the agar and spread out, resulting in a confluent mass of growth and ruining individual colony formation. To avoid this, the Petri dishes are incubated bottom side up. The slant cultures should be placed in a slanting position, with the streak towards light source. All Petri dishes, tubes, flasks and culture bottles should be labelled with name, date and identification of contents.

### 12.17.1 Purification Methods

It is often possible to obtain axenic cultures from field material by performing isolation and subsequent purification steps on plates as described. Sometimes, however, this is not the case, particularly if the natural sample is highly contaminated by bacteria and the cyanobacterial species are immotile and incapable of self-purification by gliding away from their contaminants, or if the single filaments or microcolonies do not grow on solid media. Purification of such cyanobacteria should be attempted after the establishment of unicyanobacterial isolates in liquid cultures. This technique is successfully employed to isolate a cyanobacterium which is found in abundance in a particular habitat or substrate.

It may be difficult to remove certain contaminants by repeated subculturing and / or ultrasonic treatment. In such instances, use of a chemical method rather than a physical method is preferred, but physical methods such as washing and ultrasonic treatment may precede. One such chemical method is the use of antibiotics singly or in combination to kill or inhibit the growth of tenaciously attached contaminants. Purification through antibiotic is a natural practice.

Certain bacteriostatic compounds like potassium tellurite ($K_2LeO_3$ – 10 μg $ml^{-1}$) have also been used to obtain bacteria-free cyanobacterial algal cultures. Purification can also be done by treatment with cycloserine (5μg $ml^{-1}$).

### 12.17.2 Ultraviolet Irradiation

This method has been widely used to obtain pure cultures of cyanobacteria. The cyanobacterial suspension is placed in quartz-windowed chamber and irradiated for 20–30 min. at 275 nm. The suspension should be agitated by continuous stirring during the irradiation period. The cyanobacterial samples are removed from the irradiation chamber at 5 min intervals and a large number of dilution cultures are prepared from each. The diluted cyanobacterial suspension is inoculated into fresh medium and incubated in a culture room. Usually cultures from long exposure times fail to grow, while those from shorter exposure are contaminated. However, certain subcultures, particularly from samples with intermediate exposures, will be free of bacteria.

Cyanobacterial cultures grown in liquid medium under stationary phase are examined critically under the microscope using phase contrast objectives and oil immersion. For additional macroscopic tests, small drops of the liquid culture are inoculated onto plates prepared with the appropriate mineral medium supplemented with casamino acids (0.020.05% w/v) and glucose (0.5% w/v).

The plates are then incubated in the dark for several days at a temperature typical for the growth of the cyanobacteria to be tested, since any contaminants still present should grow at that temperature. The contaminants will then be easily detected by their superior growth, colour and other typical appearances (e.g., fungal hyphae). If the culture is heavily contaminated, the entire area of the test drop will be covered with bacterial or fungal growth while a low degree of contamination will reveal itself in the form of microcolonies on the cyanobacterial lawn. Some contaminants (e.g., gliding bacteria) do not form discrete colonies on the surface of the agar. Their presence can be confirmed by microscopy after screening small sections from the agar occupied by the cyanobacterial deposit.

### 12.17.3 Culture Media

To isolate and maintain the cyanobacteria, appropriate media are used and congenial environmental conditions are maintained. Phosphate, nitrate, magnesium and calcium are the macronutrients generally required by them. The essential micronutrients are iron, zinc, manganese, copper and molybdenum. Good growth is obtained at 2000–3000 lux light intensity. The optimal temperature required for their growth is 29°C (+1°C). A number of culture media for growing cyanobacteria are available. For the isolation and maintenance of heterocystous cyanobacteria, the source of combined nitrogen ($NaNO_3$ or $KNO_3$) in the synthetic medium should be omitted and possibly be replaced by the corresponding chlorides.

## 12.18 CULTURE OF FREE-LIVING PROTOZOA SPECIALLY CILIATES

Freshwater protozoa are cultured successfully in hay infusion medium. Initially water sample is collected for zooplanform in a sample and 20–25 l of water is filtered at the site of water collection. The protozoans collected in the cylindrical jar are observed under microscope and separated with a painted pipette. The isolated and identified protozoa are inoculated in a Petri dish containing equal volume of hay infusion and distilled water. A pinch of meat extract is added to boost the bacterial multiplication and allowed to stand at room temperature after 5 to 6 days. The culture is ready for further use.

## 12.19 MAINTENANCE AND PRESERVATION OF CULTURES

Pure cultures of microorganisms need to be preserved so that they can be used as and when required for research purpose. The preservation can be short term or long term. Different methods are employed for the preservation of microorganisms, and all of them conserve all the characteristics of the organism. Some of the methods of preservation are:

1. **Transfer on Fresh Media:** Microbial cultures can be maintained by periodic transfer on fresh, sterile media in tubes. The frequency of transfer, however, varies with an organism. The agar-based slants should be stored at 4°C for long term storage.
2. **Use of Mineral Oil:** Many bacteria and fungi can be preserved by covering the fresh growth in agar slants with sterile mineral oil. The oil must be above the tip of the slanted surface. Mineral oil covered cultures are stored at room temperature or preferably at 4°C. Some microorganisms have been preserved satisfactorily for more than 15–20 years by this method.
3. **Freeze Drying (Lyophilisation):** It is a process used for long term preservation of cultures of certain types of microorganisms. In this process, the cell suspensions are placed in small vials, which are then frozen by immersing in a mixture of dry ice and acetone or liquid nitrogen. The vials are then evacuated and dried under vacuum, sealed and stored at low temperature. The cells are dehydrated under vacuum (water passing directly from the solid to the vapour phase). Freeze dried cultures can be stored for a long time at room temperature. These cultures can be reconstituted with distilled water.
4. **Storage at Low Temperature:** In this method, cultures are frozen in the presence of cryoprotectants added to the liquid or medium in which cells are suspended. Some of the cryoprotectants used are glycerol, ethylene glycol and DMSO (Dimethyl Sulphoxide). Glycerol and ethylene glycol appear to be more widely used for bacterial preservation, while DMSO has been extensively used for the cryopreservation of protozoans. The cultures are preserved in 80°C deep freezers or in liquid nitrogen containers (–160°C). In nature, microbial populations coexist with a mixture of many other cell types. Therefore, a soil dilution will give many different types of microorganisms on a suitable medium. Population of one cell type can be segregated from the mixed population to obtain a pure culture. These cultures contain only one type of microorganism and are suitable for the study of their cultural, morphological and biochemical properties.

## 12.20 FUNGAL CONSERVATION

Threats to fungi throughout the globe lead to their extinction and such acts have prompted debates about whether the fungi can be conserved. It involves the conservation of site, substrate or the habitat, host and artificial culturing of fungi and their maintenance. Fungi are not only beautiful but also play a significant role in human welfare. Conservation is a major concern at the moment. In the last 40 years, mycologists became aware of a general decline in natural habitats resulting in the dwindling of a number of genera and species in their occurrence. This may be due to climatic / environmental changes, pollution, habitat degradation, desertification, draining of soil nutrients, contamination of water bodies and over-exploitation of habitats by human beings. Moore et al., (2001) stated that some important steps to be taken for fungal conservation are:

1. Conservation of habitats
2. *In-situ* conservation of non-mycological reserves / ecological niches
3. *Ex-situ* conservation, especially for saprotrophic species growing in culture could be developed. This possibly might be of great importance for species, which are on the verge of extinction or those strictly confined to very limited or threatened habitats. Fungi are very seldom legally protected in various parts of the world, but for Slovakia, where 52 species have a special legal status, enabling managers to prevent damage to their habitat (Lizon, 1999). In the absence of legal protection, some efforts should be made to have codes of practice or to have suggestive documents published stressing the importance of fungal conservation, a practice adopted in U.K. and Switzerland. One of the tools that help in the conservation of fungi being inventorization.

*Ex-situ* conservation of fungi is by growing them on agar media and through other methods. It involves artificial synthesis and lists of fungi are not available. Such projects are now initiated all over the world by the IUCN and are currently being funded by ECCF. In order to achieve this, one has to promote the conservation of taxonomists, strengthen inventorization, and use a reasonable taxonomy, which is discerning enough to maintain the ecological bioindicative value of the taxa recognised. Mycologists must be able to prepare red data lists for various geographical or political units and mapping be taken up. If a tropical rainforest is protected for the rhinoceros or tiger, fungi will also automatically profit from the maintenance of the ecosystem. Future of mycologists depends on their input in conservation. Computerisation of information regarding fungal genera and species, ecological niches, occurrence of fungi, correct scientific name, population studies, utility, host / substrate association, growing season, edibility, medicinal importance, micro or macroscopic nature and other relevant characters may be incorporated into database followed by CD RAM preparation.

The future of fungal conservation is counter-balanced by the global threat to the environment, particularly to the forests worldwide. Building or maintaining ecological niche or habitat is a beneficial approach for fungal conservation including single species management followed by habitat conservation. Conservation of fungi can be effected both *in situ* and *ex-situ*. *In situ* conservation of fungi is best effected by ensuring the preservation of the widest range of disturbed habitat types. Safeguarding plant diversity at respective centres in nature throughout the world would be a major step in securing the associated fungi as per the international union for the conservation of nature and natural resources. There is a fear that many fungi will become extinct of single spore cultures and such cultures will be genetically pure. Saprophytic and biotropic fungi can be cultured. Fungi like *Aspergillus, Penicillium, Rhizopus, Mucor, Alternaria, Curvularia, Trichoderma, Fusarium* have been found cultured on artificial media. Fungi like *Erysiphe, Albugo* and obligate symbiotic, arbuscular mycorrhizal fungi multiply on living hosts and unsuccessful attempts are made to grow them in axenic culture: Partial success has been achieved in culturing fungi like *Cercospora, Puccinia* and others. The potential for securing and preserving fungal genetic resource for future exists if isolates of the species are obtained, but this objective can be fulfilled only through international collaboration and availability of massive additional resources (Hawksworth, 1991).The basic principle that centre has to satisfy is that the fungus culture supplied must be correct, pure and viable besides maintaining good quality service. Registration of collection must ensure stability of the host organizations and its funding. Cultures useful in industry, medicine, agriculture and patent strains should remain available at deposited centres. Duplication is necessary particularly for important strains and collections. The selection of preservation techniques for the fungi not only depends upon the success of the method, but also upon the use of the organism, time, facilities and resources available. Long term stability should be considered together with the required availability of the culture without delay, then a collector may select a continuous growth method and this should always backup with one that reduces the possibility of change during storage. Growth techniques allow strain drift; the synthetic medium and conditions provided places selective pressure on the organism, allowing those variants growing best to dominate. Desiccation, for example, freeze drying and freezing methods can prevent such changes Cryopreservation is more widely applicable and methods can be developed for optimum survival of individual strains. Liquid nitrogen can be backed up by storage in freezers that can maintain temperatures of –140°C or below:

1. Mineral oil storage is a very simple method of storage that will retain viability of fungi for many years, but places strains under selective pressure because of the special conditions of storage. Cultures are usually grown on short slopes of agar (30°C to the horizontal) and are covered to a maximum depth of 10 mm with sterile medicinal quality liquid paraffin (specific gravity 0.83 to 0.89). The paraffin is sterilized by autoclaving twice at 121°C for 15 min leaving the oil to cool and settle for 24-hr between each procedure. Storage in the laboratory at temperature a little below those for optimum growth is normally selected, although cultures are often placed in the refrigerator.
2. Water storage technique may allow the growth depending upon the method adopted. The procedure is to cut agar plugs from the edges of actively growing cultures and placing them in sterile distilled water in screw cap bottles. The nutrients available in the agar will allow growth until oxygen is depleted in the storage container.

### 12.20.1 Methods Employed for Preservation of Fungal Cultures

1. Soil storage involves inoculation of spores or mycelium suspended in 1 ml of sterile distilled water into sterile soil of approximately 20% moisture content by dry weight can provide a method of storage that retains viability for 10 to 20 years.
2. Freeze making entails the freezing of the organism and its desiccation by the sublimation of ice under reduced pressure. Cryopreservation is the method of storage at ultra low temperatures, which is the most successful method for the retention of both the viability and characteristics of fungi.
3. Preservation of agar slants at low temperatures.

Identification is an important aspect of biotechnology, since fungi form an important component of biodiversity and biotechnology. Correct identification, removal of confusion in the taxonomy and maintenance of cultures are integral part for any biotechnology venture. The fungi isolated from various habitats and reported are basically the components of nature. But some of them are likely to take part in various biotechnological processes.

# Control of Microorganisms

## 13.1 INTRODUCTION

Creating healthy environment of living for good health has been practised by man since times immemorial. Health of the environment, food and human beings is influenced by the associated microorganisms and their activities. Preservation of food by drying and salting was practised by man much earlier to the knowledge on science of microbiology. Man has been practising the methods of cooking foods and their preservation much early before having the knowledge of microorganisms or their cultivation. Now we know the presence of microorganisms in given environment and try to control them by different kinds and means as required. Different kinds of microorganisms are known to be present in different environments such as in home, laboratory, hospitals, industries etc.

With the advancement of science of microbiology, we are trying to understand a number of new microorganisms, which were not known earlier. Most of the microorganisms are beneficial to us and some of them are harmful. It is necessary to know about the microorganisms for the purpose of propagating the useful ones and controlling the harmful ones. Effective propagation of microorganisms is done with the knowledge of their nutrition and growth requirements. Effective management of controlling microorganisms at various places like laboratory, home, hospital and industry etc., depends on the knowledge of how to control them in a given environment. The control of microorganisms may be by killing, inhibiting or removing them in the given environment. This aspect of microbiology dealing with ways and means of controlling microorganisms is studied as "control of microorganisms".

Various agents of physical, chemical and biological nature, which become detrimental to the survival and growth of microorganisms, are used as antimicrobial agents to control them. Selection and use of the best agent depends in part on whether we want to kill, inhibit or remove all of the microorganisms present, or kill only some of them. While practising the methods of controlling microorganisms, we come across certain terms frequently used such as sterilization, sanitation, disinfection, antisepsis etc. Therefore, it is necessary to know and understand these terms for their meaning and application.

## Sterilization

The term 'sterile' means free of living or free from contamination. The process of making sterile is called "sterilization", i.e., the killing of all forms of life (including spores). The method of sterilization is practised in microbiology for creating an environment, an agent, an article, a medium etc., which is free of any living organism. Sterilization is done to create a sterile growth medium, sterile water, sterile air, sterile animal fluids etc. It is necessary to have a sterile medium to grow the microorganism of interest so that it can grow and give population of its kind.

## Contamination

Contamination is the entry and/or presence of undesirable organisms/materials into some environment/material/object. For example, a medium, which is supposed to be sterile, but entry of a microorganism (such as *E. coli*) into it makes it un-sterile and undesirable. This is called contamination of medium. Similarly, contamination of microbial pure cultures, water, air, foods etc. can be explained. It need not be always by living microorganisms. It could be anything whose presence is not desirable. However, more frequently this term is used for the presence/ entry of microorganisms whose presence is not desirable in sterile medium or pure cultures. Contamination also indicates the presence of unwanted organism or foreign organism other than the required organism.

## Sanitation

A method in which the quantity of microbial population is reduced to safe levels in environment, air, water, foods, articles such as eating utensils etc., as judged by public health authorities. The agents used for the purpose of sanitation are called sanitizers. Sanitation is practised in habitats of living (human, animal etc.) such as houses, hospitals, laboratories, production and processing industries, food processing plants etc.

## Disinfection

This is a process of using chemical or physical agent to destroy infectious microorganisms present in a given environment. A disinfectant refers to the use of chemical agents for the treatment of inanimate objects and surfaces. These substances do have their harmful effects on the users.

## Antisepsis

Sepsis is a phenomenon in which the infected tissues are poisoned by the products of putrefaction. This is a severe toxic state resulting from an infection with pyogenic (pus forming) microorganisms. Result of sepsis is death and degeneration of infected cells and tissues. Antisepsis is a phenomenon in which sepsis is prevented or inhibited. Antiseptics are the chemical agents, usually applied to the surface of the body, which prevent microorganisms from multiplying. Antiseptic substance may kill the microorganisms or inhibit their growth and metabolic activities. These are used on living tissues, often on cuts or abrasions to prevent infection and sepsis.

## 13.2 METHODS OF STERILIZATION

Agents that either kill microorganisms or prevent their growth are called antimicrobial agents. Based on the type of microorganisms that are affected, these agents could be antibacterial, antiviral, antifungal, antiprotozoan etc. The antimicrobial agents that kill microorganisms are called microbicidal agents. These may be bactericidal, viricidal and fungicidal indicating the type of microorganisms they kill. Certain agents that merely inhibit the growth of microorganisms but do not kill them are called microbiostatic agents. These can be bacteriostatic or fungistatic etc. The microbicidal or microbiostatic activity of a given agent depends on various factors influencing its activity. These include the nature of agent (physical, chemical, biological) and its concentration, type of microorganisms and their number, physiological state of microorganism, environment in which it is acting, time of exposure, state of microorganisms etc. These factors influence the antimicrobial activity of an agent individually or collectively in combinations.

Antimicrobial effect of an agent is exploited and made use of for the purpose of sterilization. Every living organism is influenced by a physical, chemical or biological agent for its life and growth. Life and growth in living organisms are the resultant products of chemical reactions carried out in its cell(s). Every chemical reaction has its reactants and product(s). Such reactions may be influenced for their rate of reaction by agents called catalysts. A catalyst enhances the rate of reaction. The nature of catalyst can be physical, chemical or biological. Biological catalysts are called biocatalysts or enzymes.

Most of the physical and chemical catalysts work at extreme conditions such as high temperature and pressure, whereas most of the known biocatalysts (enzymes) work at normal temperature and pressure. Catalytic activity of biocatalysts is influenced by physical environment such as temperature, pressure, pH etc. The physical agent has its effect on enzyme activity with its minimum and optimum conditions. For every living system, it is necessary to have the environment to work by the biocatalysts at its optimum. Growth rate of cells will be high at their optimum environmental conditions. Taking the temperature of growth, environment and its effect on microbial growth, its effect can be explained. At low temperature, microbial cells do not multiply but remain static. At optimum temperature, the growth rate is highest. As the temperature increases from its minimum to optimum, the growth rate also increases till it reaches its maximum at optimum temperature. The temperature becomes detrimental for cell growth when it is more than its optimum and leads to cell death at a point where cells do not survive. This temperature is called lethal temperature. Such lethal effects of physical agents are used for killing living systems for the purpose of sterilization.

Different methods of sterilization are practised by using physical or chemical agents for the purpose of killing microorganisms to sterilize materials. Physical separation of living cells is also practised for sterilization of certain materials, which tend to lose their nature for intended purpose if sterilized by methods to kill the cells. The general methods of sterilization for practical application are broadly classified as:

1. Physical methods
2. Chemical methods

Whatever may be the agent used for sterilization purpose, the interaction with the agent will result in the death of the microorganism creating sterile condition. The fundamental features that apply to both these antimicrobial agents used for sterilization are:

(i) Pattern of death of microbial population after exposure to a given antimicrobial agent.
(ii) Conditions influencing effectiveness of an antimicrobial agent.
(iii) Ways in which the microbial cells can be damaged by a given microbicidal agent.

The pattern of death of a microbial population on exposure to a given microbial agent follows a progressive decrease in its population (viable count) when studied with time vs. population by keeping the lethal agent constant. The probability of killing the targeted cell with a lethal agent becomes less as time increases due to decreased population of viable cells in a given system. All such experiments for assessing the effectiveness of an antimicrobial agent are done by taking certain population of microbial cells but not individual cell for its exposure to the lethal agent in a given environment.

Various conditions influence an antimicrobial agent for its effect on a given microbial population. Important among them are:

(a) Size of microbial population in a test system where largest populations take longer period to get killed than smaller populations.
(b) Concentration of microbicidal agent, where higher the concentration, shorter the time it takes to kill a microbial population or vice-versa.
(c) Nature and physiological status of microbial cells, where vegetative cells are more sensitive to a given antimicrobial agent than its spore. Actively growing, physiologically active cells are more affected than the non-dividing cells.
(d) Temperature of the exposure to an antimicrobial agent. Higher the temperature, faster the population killed. At higher temperatures, there will be enhancement of antimicrobial activity due to a sort of synergistic action.
(e) Time of exposure to the inhibiting agent is very important. Longer exposure times will kill greater number of cells.
(f) In addition to the above factors, type of microorganism is very important on which a given antimicrobial agent can have inhibitory effect. There may be differential effect of a given antimicrobial agent on fungal and bacterial organism. Similarly, Gram positive bacteria are more resistant to heat than Gram negative bacteria. Type of microorganism, its chemical makeup and physiological status influence its sensitivity to a given antimicrobial agent, which could be physical or chemical in nature.

Microbial cells get killed by a given antimicrobial agent by one or more mechanisms in which a given agent acts. Damage could be to the structure and function of a given microbial cell. Damage by high temperature may be by oxidation or coagulation of cellular components, and moist heat can coagulate cellular proteins to denaturation resulting in cell death. Similarly, chemical antimicrobial agents kill the cells in different ways. It could be due to direct reactions with biomolecules to make them non-functional or by creating a damage in cell structures so that the cell loses its functions. Membrane damage by a number of chemical antimicrobials leads to leakage of cellular constituents resulting in cell death.

Various cellular structures like cell wall, membrane, nuclear materials and other structures could be targeted by antimicrobial agents. Some of the functional enzymes of microbial cells can also be targeted by chemical antimicrobial agents. Structures and their functions, which are common to the host and pathogenic microorganisms, if targeted by antimicrobial agents, become non-specific. However, certain structures and functions that are specific only to microorganisms, if targeted, become selective targeting by a given antimicrobial agent. Such criteria are very important in developing chemical molecules as selective toxic agents for the treatment of infectious diseases.

### 13.2.1 Physical Methods of Sterilization

The method of sterilization depends on the nature of material to be sterilized, purpose for which it is to be used and the likely microflora associated with it. There are different kinds of physical agents used for sterilization. The agents like temperature or radiation are used to kill microorganisms.

**Sterilization at high temperature:** High temperature is commonly used and is the effective means of sterilization. Exposure to high temperature is mainly based on the nature of microorganisms and their sensitivity to highest temperatures. It depends on the thermal death point and thermal death time of the organisms. These characters are interdependent and commonly expressed as time-temperature relationship.

**Thermal death point (TDP):** This is the temperature required to kill microorganism(s) at a given time of exposure under standard (defined) conditions. This may be different for different types of microorganisms or same microorganisms under different conditions and its physiological status.

**Thermal death time (TDT):** It is the shortest time required to kill all the microorganisms in a sample when exposed to a specific temperature under standard (defined) conditions. In experiments to determine TDT for a given microorganism, a specific temperature is selected as a fixed point. Then the microbial population is sampled at various time intervals of heat treatment and cultured for growth. This method helps to determine the minimum exposure time that is required for destroying all the organisms present under test conditions. Normally, the TDT is inversely proportional to the temperature of exposure. Higher temperature fastens the death rate and lessens the time of exposure for sterilization.

**Decimal reduction time (DRT):** It is the time required to decrease the microbial population in a given sample by 90 per cent (i.e., by logarithmic value of 1) at a prescribed temperature. The temperature is kept at fixed point and population is sampled at various intervals. In such experiments, the exposure time that gives 10 per cent survival is determined rather than that results in 100 per cent kill.

While determining the TDT or DRT values, all other factors except time of exposure must be strictly controlled. The factors such as the nature of microorganisms, their concentration, nature of the medium, the pH can influence the susceptibility of microorganisms to heat. The measurements of TDT or DRT are important in the dairy and food processing industries where optimum processing times and temperatures must be established for various dairy products and canned foods. Excessive exposure to heat is likely to affect the quality of food, and it is important to know the lowest temperature and shortest time that will effectively treat the product.

**Dry heat sterilization:** Sterilization at high temperature is practised with dry heat or moist heat. As explained earlier, use of dry heat or moist heat for sterilization depends on the nature of material and its intended purpose of use. Each method has its own advantages and limitations. In this method of sterilization, the microorganisms are subjected to burning at high temperature under dry conditions. Dry heat or hot air at high temperature will kill microorganisms. The principle of killing microorganisms in dry heat sterilization is by oxidation of biomolecules where cellular materials get oxidized to give ash (oxides).

**Incineration:** Destruction of microorganisms or any other material by burning is called incineration. It is the most practised method of sterilization in the laboratory for culture inoculation/ transfer loop/needle. Bacteriological loop/needle is sterilized by exposing it to Bunsen burner flame prior to its use and also after its use. There are also certain designed incinerators which are fixed in the laboratory for routine use. Dry (red hot) heat is created into it for sterilization. While sterilizing by incineration, care has to be taken as the heated material (on loop/ needle) may boil and spatter. Such droplets may carry viable microorganisms and contaminate other materials on the work bench.

Incineration is also used for disposal of contaminated laboratory materials such as swabs, sponges, plugs, dressing material and also carcasses of infected animal/ human body parts. Burning the dead bodies is an age old practice for protecting the environment from putrefying biological materials and prevents diseases. Care should be taken to ensure that exhaust fumes coming from incinerator do not carry matter containing microbes into atmosphere.

**Hot air sterilization (hot air oven):** Materials impermeable to or damaged by moisture (e.g., oils, glassware, sharp instruments and metals etc.) are sterilized by dry heat in hot air oven. Various laboratory glassware like test tubes, flasks, Petri dishes etc., are sterilized in hot air oven with care and effectiveness. Material to be sterilized is exposed to dry heat/hc air in the oven at high temperature for specified times. Loading of oven or arrangement glassware in the oven is important for effective and uniform distribution of heat in the (Fig. 13.1). Normally, glasswares are sterilized in laboratory hot air oven at 160°C to for 2 to 3 hours. However, this varies depending on the type and size of glassware to ized. Smaller size glassware (tubes, flasks etc.) require less time of exposure than

**Fig. 13.1.** Hot air oven.

ones. Dry hot air fumes are let off from the oven through small openings on top or sides. Care should be taken while operating the oven at high temperature. Hot air oven with high temperature should not be opened as it might burst due to differential temperatures inside and outside the oven. This results not only in the breakage of hot glassware but also injuries to the operating persons.

**Moist heat sterilization:** Moist heat sterilization is a more effective method for killing microorganisms than dry heat. Moist heat causes denaturation of vital proteins of cells such as enzymes by coagulation, whereas dry heat causes oxidation of organic constituents (i.e., slow burning). Denaturation of cell proteins occurs with lower temperatures and shorter exposure times than those required for oxidation. For example, bacterial endospores are destroyed faster at lower temperature with moist heat than with dry heat (Table 13.1). Bacterial endospores are more resistant to high temperatures than their vegetative cells. For example, vegetative cells of bacteria are usually killed within 5 to 10 minutes by moist heat at 60 to 70°C, whereas bacterial endospores require 10 to 20 minutes at 100 to 120°C in routine sterilization practice by autoclaving. Vegetative cells of yeasts and other fungi are killed within 5 to 10 minutes by moist heat at 50 to 60°C, but their spores require higher temperature (70 to 80°C) in the same time. Protozoa and most viruses are killed like vegetative cells of bacteria and fungi with moist heat. Moist heat is used for killing microorganisms as steam, boiling water or sometimes at sub-boiling temperatures.

**Steam sterilization:** Use of steam under pressure is the most practical and dependable way of sterilization with moist heat. The temperature of pure boiling water or steam is 100°C at 0 (lb/in$^2$) pressure. However, boiling point of water depends on pressure acting on it. In general,

**Table 13.1.** Reported killing times of bacterial endospores by moist and dry heat

| Species of bacteria | Moist heat | | Dry heat | |
|---|---|---|---|---|
| | Temperature (°C) | Killing time (min) | Temperature (°C) | Killing time (min) |
| ***Bacillus anthracis*** | 100 | 2–15 | 140 | Up to180 |
| | 105 | 5–10 | 160 | 9–90 |
| | 180 | 3 | | |
| ***Clostridium botulinum*** | 100 | 300–530 | 120 | 50 |
| | 110 | 32–90 | 130 | 15–35 |
| | 115 | 10–40 | 140 | 5 |
| ***Clostridium perfringens*** | 100 | 5–45 | 120 | 50 |
| | 105 | 5–27 | 130 | 1–35 |
| | 115 | 4 | 140 | 5 |
| | 120 | 1 | | |
| ***Clostridium tetani*** | 100 | 5–90 | 130 | 20–40 |
| | 105 | 5–25 | 140 | 5–15 |
| | 160 | 12 | | |

boiling point of water is directly proportional to the pressure acting on it. If water is subjected to heating in a closed system of constant volume, steam pressure acts on it and increases the boiling point of water resulting in increased steam temperature (Table 13.2). This principle is made use of in steam sterilizers (autoclaves) for sterilization using moist heat. Steam under pressure has the advantage of rapid heating and greater penetration. The laboratory apparatus designed to sterilize with pressurized steam is the autoclave (Fig. 13.2). It is an important and essential unit of equipment in every microbiology laboratory. Autoclave was developed by Charles Chamberland in the year 1884. Many culture media, solutions, contaminated materials, discarded materials and cultures are sterilized routinely with autoclave. Autoclave(s) resemble

**Table 13.2.** Temperature of steam (boiling point of water) under pressure normally used in steam sterilizer (autoclave)

| Steam pressure (lb/in$^2$) | Temperature (°C) |
|---|---|
| 0 | 100.0 |
| 5 | 109.0 |
| 10 | 115.0 |
| 15 | 121.5 |
| 20 | 126.5 |

**Fig. 13.2.** Autoclave.

home pressure cookers in that both use steam under pressure, but the temperature and pressure inside an autoclave can be more closely controlled. The double-jacketed chamber of the autoclaves is first flushed with free-flowing (non-pressurized) stream to remove any air. It is then filled with pure steam and maintained at a designated temperature and pressure for a specified period of time. It is essential that all the air initially present in the chamber be completely replaced by pure steam. If air is present, it will reduce the temperature. It is the temperature and the pressure within the chamber that kills the organisms.

An autoclave is usually operated at a pressure of 15 lb/in$^2$ at which the temperature of pure steam is 121°C. The length of time needed to sterilize at this temperature depends on the material being treated. It takes longer for heat to penetrate viscous or solid material than fluid. The necessary time also depends upon the volume of material being sterilized (Table 13.3). For example, 1000 test tubes containing 10 ml each of a liquid medium can be sterilized within 10 to 15 minutes at 121°C, whereas the same amount of liquid (10 litres) in a single container would require 1 hour or more at the same temperature. More time is required for the heat to penetrate to the centre of bulk material. While operating an autoclave, either in the laboratory or at large scale industrial level, the handling persons should be trained. Since it works with high pressure at high temperature, accidents could be possible. One should not face the steam coming out of the autoclave immediately after opening it, as it becomes dangerous. Lid of an autoclave should be opened in such a way that direction of steam coming from it goes away from the person handling it (Fig. 13.2).

**Testing of steam sterilizer:** Efficiency of steam sterilizer (autoclave) for its effective functioning is tested by using a chemical dye or biological indicator. A chemical dye that is heat sensitive and loses its colour on exposure to a given temperature is kept in the autoclave along

**Table 13.3.** Exposure time required for liquids or aqueous solutions in various containers for sterilization by autoclaving

| Container at 121–123°C | Minutes of exposure |
|---|---|
| Test tubes | |
| 18 × 150 mm | 12–14 |
| 32 × 200 mm | 13–17 |
| 38 × 200 mm | 15–20 |
| Conical flask | |
| 50 ml | 12–14 |
| 500 ml | 17–22 |
| 1000 ml | 20–25 |
| 2000 ml | 30–35 |
| Serum bottle, 9000 ml | 50–55 |
| Milk dilution bottle, 100 ml | 13–17 |

Source: Perkins, J. J., Principles and methods of sterilization in health sciences, Charles C Thomas, Springfield, III., 1983.

with the material to be sterilized. After sterilization process, it is observed for its change in colour. However, it does not tell the time for how long it is maintained at a given sterilization temperature. Therefore, biological indicators are more dependable for testing of autoclaves. Spores of heat resistant bacterium, *Bacillus stearothermophilus*, are commonly used as biological indicator. Paper strips impregnated with spores of *B. stearothermophilus* are kept in a tube of nutrient medium in the autoclave along with other materials for sterilization. After sterilization, the tube with bacterial spores is incubated for overnight and observed for the presence or absence of growth. Absence of growth indicates complete sterilization and perfect functioning of autoclave. However, the disadvantage with this method is that it requires one day to confirm the result of sterilization.

**Boiling water sterilization:** Boiling water temperature (100°C) kills microorganisms present in liquids. At this temperature, the vegetative cells of bacteria and other microorganisms are killed. However, contaminated materials or objects exposed to boiling water for several minutes cannot be sterilized with certainty, because some bacterial endospores can withstand temperatures more than 100°C for an hour or more.

Exposing materials to boiling water for defined periods of time will certainly kill all the vegetative cells present but will not ensure sterilization. Therefore, boiling water temperature treatment is not considered a method of sterilization. However, boiling of syringes, needles and other instruments is often practised. This is not for sterilization but to eliminate any vegetative cells that may be considered to be present and cause harmful effects.

**Fractional sterilizations:** Sometimes while preparing a sterile medium by autoclaving at 15 lb/in$^2$, certain nutrients present as media ingredients which are heat sensitive become inactivated. However, requirement is that the medium has to be sterilized by protecting its nutritive value and killing all the microorganisms present in it. This type of sterilization is mostly required for preparation of solid media having heat sensitive nutrients and which cannot be filter sterilized. In such cases, the medium is subjected to moist heat treatment at 100°C (steaming) for about one hour and then incubated at 30–37°C for one day, then again steamed for one hour next day. Such steaming process is continued everyday for three days (or as required) in which the medium becomes sterile but not subjected to temperature more than 100°C. The vegetative cells are killed on steaming, and endospores, if any present in the medium, are allowed to germinate and grow into vegetative cells on incubation. Such germinated vegetative cells are killed on steaming again at 100°C. Such repeated exposure of medium to steaming on successive days completely kills any vegetative cells growing in the medium, making it sterile. The sterilization process is completed in fractions on successive days, therefore, it is called fractional sterilization or tyndalization. Steaming may also denature some of the highly heat sensitive nutrients. In such cases, alternative methods of sterilization for nutrients or other fluids are practised to sterilize them separately by filtration and mixing to main medium under aseptic transfer. The medium or any other material must be checked for its sterility before use.

**Sub-boiling temperature:** The practice of treating materials with temperatures less than 100°C under controlled conditions was introduced by Louis Pasteur in 1860 for eliminating unwanted microorganisms spoiling the wine and its making process. Higher temperatures are often avoided in the food industry when they are unnecessary, since they affect the taste, appearance, or nutritive value of foods such as milk products, fruit juices and vegetables. Pasteur's method of controlled heat treatment of liquids is known as pasteurization.

Pasteurization kills the vegetative cells of many microorganisms but does not sterilize. Pasteurization eliminates the vegetative cells of pathogenic microorganisms, as well as those of many other microorganisms, and thereby prolongs the shelf-life of the product. Now pasteurization is used for eliminating unwanted microorganisms in many liquids, food items, milk and beverages.

**Pasteurization of milk:** The terms pasteurization, pasteurized, and similar terms shall mean "the process of heating every drop/particle of milk or milk product to at least 145°F (62.8°C), and holding it continuously at or above this temperature for at least 30 minutes, or to at least 161°F (71.7°C), and holding it continuously at or above this temperature for at least 15 seconds in equipment which is properly operated and approved by the health authority". The original time-temperature relationships for pasteurization were determined with *Mycobacterium tuberculosis*, because this species was regarded as the most heat-resistant pathogen likely to occur in milk. This organism is destroyed when exposed to 140°F (60°C for 10 minutes). To be safe, the pasteurization temperature was set at 143°F (61.6°C) for 10 minutes. Subsequently, it was discovered that *Coxiella burnettii*, the causative agent of Q fever transmitted by milk, can survive in milk heated to 143°F for 30 minutes. This observation changed the conditions of milk pasteurization. At present pasteurization of milk is done by one of the two methods: (i) The low temperature holding (LTH) method, which holds the milk at 145°F (62.8°C) for 30 minutes; or (ii) The high temperature short-time (HTST) method, which exposes the milk to a temperature of 161°F (71.7°C) for 15 seconds. In either method of pasteurization, it is essential that the equipment is designed and operated so that all of the milk is heated to the required temperature and held for a specific time. Precautions must be taken to prevent recontamination after pasteurization. The finished product should be stored at low temperature to retard the growth of microorganisms that survive pasteurization because pasteurized milk is not sterile milk (Table 13.4).

**Sterilization by radiation:** Energy in the form of electronic waves is called electromagnetic radiation. It is classified based on its wavelength. Wavelength of electromagnetic radiation ranges from $10^{-5}$ to $3 \times 10^{15}$ nanometres. Energy content of radiation is inversely proportional to its wavelength. Shorter the wavelength, greater the energy content. Wavelength of visible light is between 380 to 780 nanometres. Lower wavelength below visible light begins from ultraviolet, X-rays, gamma rays, cosmic rays in decreasing order. High energy radiation includes gamma rays, x-rays and ultraviolet light in decreasing order. These high energy rays kill living cells including microorganisms. Lethal radiation kills living organisms by either ionizing the molecules or non-ionizing by forming free radicals in biological systems.

**Sterilization by ionizing radiation:** High energy electron beams, gamma rays are commonly used for sterilization. These become lethal to biological systems because they cause ionization of molecules by driving away electrons and splitting molecules into atoms or groups of atoms. Water molecules in cells are split into hydroxyl radicals ($OH^-$), electrons ($e^-$) and hydrogen ions (or protons, $H^+$). Hydroxyl radicals are highly reactive and destructive to normal cell components such as proteins and DNA. Change in or destruction of biomolecules leads to cell death. Microorganisms present in materials to be sterilized are killed by ionizing radiation. But the lethal effect of this ionizing radiation depends on the type of radiation, period of radiation, type of material in which microorganisms are present, type of microorganisms and their number. Gamma and X-rays have high penetrating power; they can penetrate packaged materials. X-rays are more expensive than gamma rays. Gamma rays are emitted spontaneously from certain radioactive materials (isotopes) such as cobalt-60 ($^{60}Co$). They are emitted in all directions and are difficult to control because they cannot be turned on or off like an x-ray machine. Establishment of gamma radiation unit is costly, but once established, cost of gamma rays works out to be cheap in long term high usage. Gamma rays are used for sterilizing packaged food and medical equipment at final stage because of their high penetrating property. Their application is limited only to a few food items such as spices. This limited use is because of their uncertain effect on the product quality and public apprehension on radiated food.

**Sterilization by nonionizing radiation:** Ultraviolet (UV) light is the most commonly used non-ionizing radiation in the laboratories, small work places, hospital operation theatres, sterile rooms etc. UV light has a wavelength ranging between 136 to 400 nm. UV light excites electrons in a molecule causing it to become a radical, which is more reactive than molecule itself. Free radical reacts differently than ions resulting in the formation of different types of newer molecules in cells. Some of such molecules become lethal for survival of living organisms. UV light is absorbed by many molecules in cells, of which DNA and proteins are more

**Table 13.4.** Use of high temperature for sterilization

| Method | Temperature | Application | Limitations |
| --- | --- | --- | --- |
| | | **Moist heat** | |
| Autoclave | 121.6°C at 15 lb/in$^2$ pressure, 15–30 min | Sterilizing media, instruments, liquids | Heat sensitive items |
| Boiling water | 100ºC, 10 min | Killing vegetative cells on instruments, containers; does not give sterile material | Endospores not killed |
| Pasteurization | 62.8°C for 30 min at 71.7°C for 15 sec | Killing veg. cells of pathogens and others in milk and liquids | Does not give sterile material |
| | | **Dry heat** | |
| Hot air oven | 170–180°C for 1–2 hrs | To sterile material impermeable to or that cannot withstand high temperatures | Destruction to materials damaged by moisture |
| Incineration | Swabs, sponges, plugs, disposal of infected carcasses and contaminated objects | To sterilize loops and incinerators, flames etc., to prevent air pollution | Properly designed |

damaged. Maximum absorption of UV light by DNA is at 260 nm and proteins at 280 nm. Greatest bactericidal action by UV light occurs at 260 nm and pyrimidines are the most targeted molecules. After DNA is exposed to UV light, two adjacent pyrimidine molecules in a polynucleotide strand form dimers. Unless these dimers are removed either by light reaction or intracellular enzyme mediated dark repair, DNA replication is inhibited or altered. This causes either mutations or death of cells. UV light is the convenient and most used cheap method of sterilization by radiation. Disadvantage with this method is that UV light cannot penetrate through glass or other hard materials. Therefore, material needs to be exposed directly to UV light for sterilization. The time of exposure needs to be standardized depending on the type of material and microflora present in it. One should not forget to put off the UV light prior to the use of sterile area. Direct exposure of human body to UV light is harmful.

UV light is a component of sunlight, but most of the shorter and more damaging UV wavelengths are filtered out by substances present in atmosphere, such as ozone, water droplets in clouds, smoke etc. UV light that actually reaches the earth has a higher wavelength between 295 to 400 nm and it is not damaging. In the laboratory, UV light is generated through commercially available electrical tubes and can be fitted at appropriate places for effective radiation sterilization. Certain special UV lamps are commercially available, which are used for radiation experiments to develop mutants of microorganisms for research.

**Sterilization by physical separation (filtration):** Microorganisms present in certain liquids, air and others are removed by physical separation. Sometimes liquids containing heat sensitive or radiation sensitive materials need to be sterilized by filtration. Sterile air is required for the growth of microorganisms in growth vessels (tubes, flasks, jars, fermenters etc.) and a number of other living organisms are also sterilized by filtration. Clean air or sometimes sterile air used in sterile operation rooms is made available by passing air through bacteriological filters to remove all bacterial organisms present in the air. Organisms having a size equal or greater than bacteria are physically removed by bacteriological filters. Filters are designed and prepared with different pore sizes depending on the requirements (Fig. 13.3). Routinely used bacteriological filters have a maximum pore size of 0.45 microns, since, on an average, a bacterial cell has a size ranging from 0.5 to 5 microns. These can be conveniently separated by passing material (liquids, air) through filters having pore size of 0.45 microns. Smaller pore size filters are also used for specialized purposes, to separate particles of sizes smaller than 0.45 microns.

Use of filters to remove bacteria from drinking water was described by Charles Chamberland, the developer of autoclave in 1884. He used porous porcelain in funnel shape to purify water by passing through it. This method was practised in Pasteur's laboratory to separate microorganism from culture media. Filters are used in the laboratory and industries to sterilize materials that cannot be sterilized by autoclaving, such as heat sensitive vitamins, proteins, serum components etc. Cotton plug is the traditionally used filter in microbiological labs.

Earlier filters were made from earthenware, asbestos, and sintered glass. Most of these are now replaced by cellulose membrane filters, commonly known as membrane filters. Filter membranes are made with variety of cellulosic materials to give quality and strength with different pore sizes. Filter membrane materials are cellulose esters made into extremely thin (about 150 microns) disks

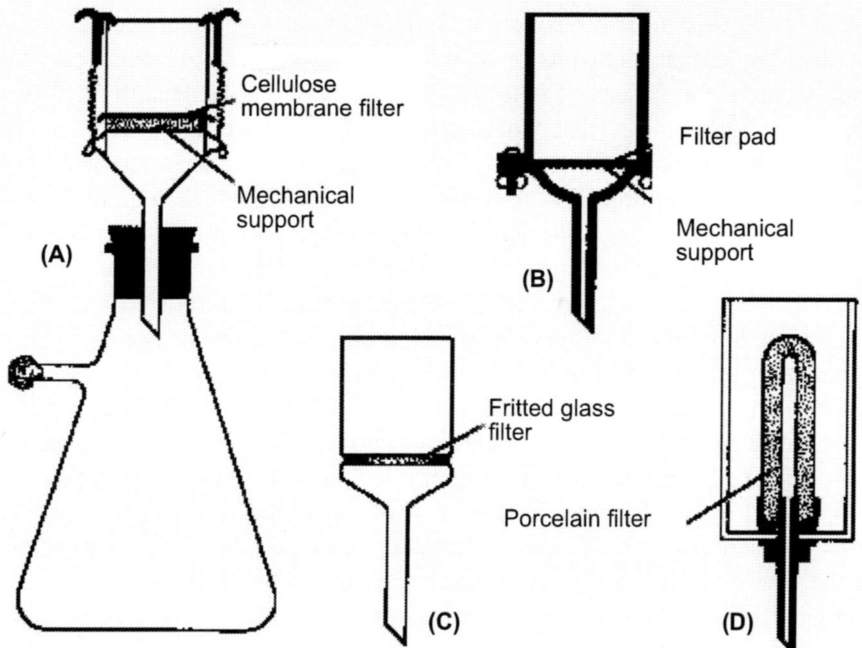


**Fig. 13.3.** Bacterial proof filters. A: Cellulose membrane filter; B: Seitz filter; C: Fritted glass filter; D: Porcelain candle filter.

with pores small enough to prevent the passage of microorganisms. Membrane filters are superior to older types of filters because:

a) The pores of membrane filters are of uniform known diameter.
b) Filter can be manufactured with any desired pore size.
c) They absorb very little fluid being filtered.
d) Filtration through membrane filters is more rapid than with older filters made up of other materials.

Membrane filters are also disposable, which eliminates the need to clean and sterilize the reusable filters. Specially designed equipment are available for holding membrane filters while liquids pass through them. Besides sterilization, membrane filters are used for separating types of microorganisms and for collecting microbial samples. For example, viruses can be separated from other microorganisms by using filter membrane with smaller pore sizes. This is the filtrable character assigned to viruses because they pass through the filter pores, which trap bacteria. Membrane filters are also used frequently for microbiological examination of water in which the function of a filter is to concentrate the bacteria on filter membrane, which are small in number in large volume of water.

Certain microbiological laboratories involve work with hazardous microorganisms like highly infectious organisms, genetically modified microorganisms etc. Release of such microorganisms into environment leads to dangerous situation. It is essential to protect the research persons working with such dangerous organisms or other persons from possible infections. Guidelines issued from the Centers for Disease Control (CDC) of U.S. Public Health Service establish safety procedures for laboratories that work with biological materials. There are four CDC 'biosafety levels' differentiated by the degree of possible danger and characterized by the safety equipment and procedures required. These are, for e.g., level I, which requires only standard open-bench laboratory techniques, while level IV requires maximum containment with special clothing, decontamination procedures and gastight cabinets. A person working in level IV is required to enter the area in completely surface sterile form by wearing sterile clothes to cover his body surface including nose and eyes. After completing the work in sterile room, he has to come out by removing the covering in a designated location for disposal.

There will not be any connection between the environments inside and outside to such containment work place. Most rigorous precautions are built into the biosafety level IV system wherein, the entrance and exit of air in an entire room is controlled and each laboratory worker in the room is completely protected by a full-body suit with its own breathing equipment. This requires that the street clothing be removed and laboratory clothing put on, and that the personnel take shower when existing the facility. Numerous other stringent precautions and practices are also required.

Special biological safety cabinets have been developed that provide protection from contaminated aerosols. Simplest of these is the biological safety cabinet, normally called laminar air flow cabinet (Fig. 13.4). Sterile air comes into such cabinet horizontally or vertically through

**Fig. 13.4.** Laminar air flow cabinet.

a filter called 'high efficiency particulate air' (HEPA) filter. HEPA filter traps particulate matter such as microorganisms coming in and the air becomes sterile. HEPA filter consists of cellulose acetate plated around aluminium foil. It captures 99 percent of the particulate matter from the air passing through it.

### 13.2.2 Chemical Sterilization

Sterilization of certain materials is done by making use of chemical agents. Chemical sterilants are particularly useful for the sterilization of heat-sensitive medical supplies, such as plastic blood transfusion or donor sets, plastic syringes, and catheterization equipment etc. They are also used to sterilize enclosed areas, including the aseptic used for procedures that must be microbe-free. Potable water treatment for elimination of microorganisms is normally practised by treatment with halogen gases or halogen compounds. Chlorination of municipal water is the commonly used method for elimination of microorganisms.

A number of chemicals are used for sterilization of liquids, solid materials etc. Chemicals may act on the microbial cells in different ways to kill them. There may be direct attachment of chemical groups to the microorganisms so as to make them non-functional, or certain chemicals may release nascent oxygen, which kills the microorganisms by oxidation of cellular components. Treatment of water with potassium permanganate is commonly practised for sterilization of small quantities of water in hotels. Municipal water is treated with liquid chlorine or chlorate compounds. In all such cases, it becomes toxic for microorganisms present in water. However, such treatment may not always result in settle water, but it gives sanitary quality water for potable purposes.

Commonly used chemicals for sterilization of medical appliances to make them microbe-free are ethylene oxide, beta-propiolactone, glutaraldehyde and formaldehyde. These are commonly called chemosterilizers. Formaldehyde is the most commonly used chemosterilizer. Operation theatres, sterile operation rooms, aseptic transfer zone area etc., are treated with formaldehyde for sterilization of air and surfaces of materials present. Formaldehyde in gaseous state is allowed to spread all over the space in a given room for standard period of time and neutralized with ammonia vapour before the room is used for aseptic work. Formaldehyde is highly toxic and irritating to eyes even in small concentrations. Care should be taken to prevent any toxic effects with such chemosterilizers. Formaldehyde inactivates enzymes and other active groups in proteins.

Ethylene oxide is another organic chemical compound used for chemosterilization. It is a liquid at temperatures below 10.8°C, but becomes gas at room temperature. Its vapours are highly irritating to the eyes and mucous membranes and also flammable. It is commonly used in combination with carbon dioxide or freon, which are non-flammable gases. These mixtures retain microbicidal activity of ethylene oxide. Ethylene oxide is microbicidal even to bacterial endospores. It is highly penetrating and can sterilize large packages of materials, bundles of cloth, plastics, medical syringes and other packaged plastic materials. The disadvantage of this compound is its slow rate of action against microorganisms, requiring several hours of exposure, sometimes overnight. Ethylene oxide inactivates enzymes and other proteins that have liable hydrogen atoms, such as sulphydryl groups. This reaction gives alkylation by opening the ring structure of ethylene oxide.

# Microbial Genetics

## 14.1 INTRODUCTION

The genetic information of prokaryotic organism is present in a single thread of DNA, the bacterial chromosome. In eukaryotes, the genetic material is DNA, but it is present in chromosomes. Chromosomes form a network inside the nucleus. Nucleus is covered by a nuclear membrane, which has some openings too. Inside the nucleus, nucleolus is also present. DNA was discovered by Oswald Avery, Macleod, Cal Maclyn and Mccarty in 1943. From DNA the genes are made and this forms hereditary material transmitting hereditary characters. Watson and Crick proposed the model for DNA based on X-ray information supplied by Wilkins and Franklin. According to Watson and Crick, the DNA (Fig. 14.1) molecule consists of two long strands wrapped around each other. This structure is called double helix, which is a twisted ladder and nucleotides are the structural units of nucleic acids. Thus each strand of DNA contains many nucleotides. DNA has three components:

(1) a nitrogen containing base
(2) a pentose sugar called deoxyribose
(3) phosphate group (phosphoric acid Fig. 14.1).

The nitrogen bases are cyclic compounds made up of carbon, hydrogen, oxygen and nitrogen atoms. The bases are named as adenine (A), thymine (T), cytosine (C) and guanine (G). A and G have double ring structures called purines, while T and C bear single ring structures called pyrimidine nucleotides and are named according to their nitrogenous base. Thus, nucleotides are known as thymine nucleotide and adenine nucleotide. Nucleotide refers to the combination of a purine-pyrimidine plus a pentose sugar. Phosphate group is absent.

DNA, the double helix, thus contains alternating sugar and phosphate groups, which form the backbone of hereditary material. Purine 'A' always pairs with the pyrimidine 'T' and 'purine G' pairs with the pyrimidine C is held together by two hydrogen bonds while G and C are held by the free 'H' bonds. A segment of a nucleotide chain constitutes

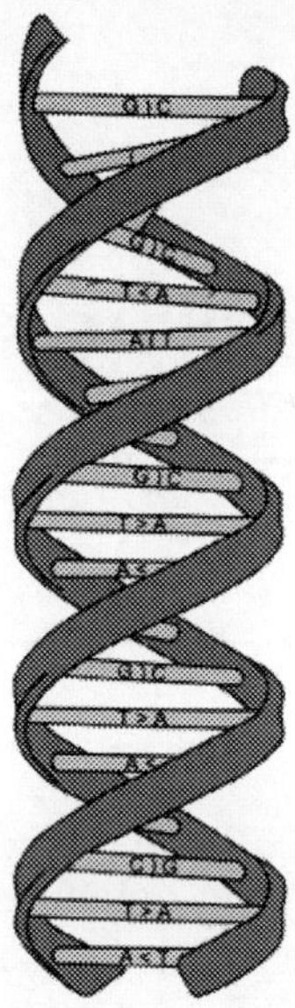

**Fig. 14.1.** Structure of DNA.

a gene and a single DNA molecule may contain thousands of genes. Genes are the determinants of hereditary traits and control all the cellular activities. The strand arrangement is such that ATGC alternates with TA CG, thus becoming complimentary.

Ribonucleic acid (RNA) is the second category. RNA is single-stranded. Ribose is a five carbon sugar of the RNA nucleotide. It has one more oxygen atom than deoxyribose. RNA has adenine, guanine, cytosine and uracil instead of thymine. There are three types of RNA's, namely messenger RNA (mRNA), ribosomal RNA (rRNA) and transfer RNA (tRNA).

A gene is nothing but a segment of DNA and it codes for a functional product. The final product can be a molecule of ribosomal RNA or tRNA. Genes, that code for proteins are called structural genes. Specific proteins are coded by a unique base.

Cell's machinery is concerned with translating the DNA of structural genes into specific proteins. The sequence of DNA is copied to produce a particular molecule of RNA called mRNA. The information encoded in the mRNA is translated into a specific sequence of amino acids that forms a polypeptide chain and protein. The gene expression involves transcription of DNA to produce a specific molecule of RNA, and mRNA encodes information that gets translated into protein.

DNA can change either spontaneously or due to induction of some chemicals, plant extracts, radiation etc. Such changes appear in the base sequence of DNA and are called mutations. The physical or chemical agent present in the environment also brings about mutations. Some mutations are harmful. Many mutations create new characteristics that give the offspring inheriting the mutation a better chance of survival in a new environment. Such DNA mutations are of great advantage that contributes to the successful evolution of species. Phenotype refers to the expressed properties including the ability to perform a particular chemical reaction. Phenotype is the collection of proteins. Genotype is the genetic makeup of an organism that codes for all the particular characteristics of the organism.

DNA replication results in the synthesis of a duplicate DNA strand. From one DNA molecule, two daughter DNA molecules are produced. During replication of DNA, the strands of parental DNA unwind and separate from each other in one small DNA segment after another. Free nucleotides of the cell match up with the exposed base of single stranded parental DNA. This is followed by alignment of purines and pyrimidines, then the newly added nucleotide is joined to the growing DNA strand by an enzyme called DNA polymerase. Soon after this, the parental DNA unwinds a bit further to allow the addition of next nucleotide. This process is called replication. The parental DNA strand and newly formed DNA then rewind. Thus new DNA will have original DNA strand (conserved) and the newly added DNA strand. This is called semi-conservative replication.

The DNA synthesis requires proteins or enzymes. During replication, the parental double helix is unwound by enzymes. DNA polymerase attaches new additional nucleotides to the free 3′ end of an existing DNA or RNA strands to elongate the strand. The leading DNA strand is synthesized by DNA polymerase. RNA polymerase starts the synthesis of each DNA fragment of the lagging strand with a stretch of about 20 RNA nucleotides. The DNA polymerase can then proceed from the RNA primer. Exonuclease digest the RNA. DNA-ligase joins the discontinuous fragment.

## 14.2 BACTERIAL DNA

Bacterial DNA is called nucleoid, which contains a single, long, circular molecule of double-stranded DNA, the bacterial chromosome. It is this genetic material that carries all the information required for cell's structure and functions. Bacterial DNA does not have histone, and is not surrounded by nuclear envelope. The nucleous material can be spherical, elongated or dumbel-shaped.

There are some small circular, double-stranded DNA molecules called plasmids. Plasmids are the extra-chromosomal genetic materials present and are not connected to the main bacterial chromosome. Plasmids replicate independently. Plasmids contain 5 to 100 genes. Plasmids may help in carrying genes responsible for antibiotic resistance, tolerance to toxic metals, toxin production and enzyme synthesis. Plasmids are also known to be transferred from one bacterium to another.

### 14.2.1 Bacterial DNA Replication

Replication starts at a particular site. In *Escherichia coli,* the DNA replication is bidirectional around chromosome, wherein two replication forks move in opposite directions away from the origin of direction. The replication forks meet only after the process is completed because of loop-like structure of bacterial DNA. During the cell division of bacteria, each daughter cell receives one copy of the DNA molecule Replication is a fast process. In *E. coli,* around 1000 nucleotides get synthesized per second at 37°C. DNA replication helps in the flow of genetic information from one generation to the next (Fig. 14.2).

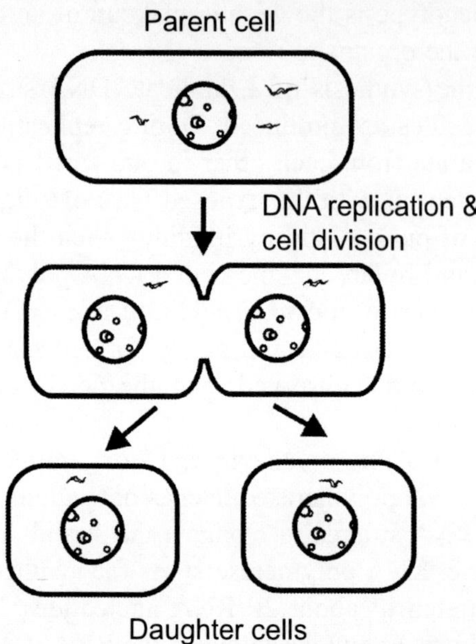


**Fig. 14.2.** Flow of genetic information.

## 14.3 GENETIC MATERIAL IN OTHER MICROBES

Viruses possess either single or double-stranded DNA or RNA molecules, the nucleic acids being in linear or circular form. Plant viruses contain RNA while animal viruses possess DNA.

Viroids have 1000 molecular weight nucleic acid and cannot code for any protein. These differ from viruses in lacking protein coat.

Virusoids are the small circular RNA's, which are similar to viroids, but they are always linked with larger molecules of the viral RNA.

The proteinaceous and infectious agent is called prion. Prions are known to cause diseases such as mad cow disease, scrapie etc. No nucleic acid could be isolated from prions.

Bacteriophages are the viruses that infect bacteria. They contain a single-stranded DNA or ssRNA as genetic material.

Archaebacteria's genome has G + C contents (DNA) with a broad range of 25–68 mol. per cent. Cyanobacteria are also prokaryotic.

## 14.4 GENETIC CODE

The code refers to the relationship of nitrogenous bases of DNA and the corresponding codons of mRNA. Codons determine their base sequence in mRNA. There are 64 possible codons, which code for 20 amino acids. Most of the amino acids are signalled by several alternative codes. Leucine has 6 codons and alanine has 4 codons. Though there are 64 codons, 61 are sense codons and three are non-sense codons. Sense codons code for amino acids. Examples of sense codons are AUC, UUU etc., while non-sense codons are UAA, UAG and UGA. AUG codes for methionine, which is the starting point for protein synthesis. Each codon containing three nucleotides code for a particular amino acid. This is represented by a three-letter abbreviation. Genetic code is the language of cell.

## 14.5 DNA AS A GENETIC MATERIAL

The information of DNA is translated into the proteins that control cell activities. The cell utilizes the information encoded in RNA message to synthesize specific proteins through the process of translation. In transcription, the genetic information for protein synthesis is stored in DNA and passed on to mRNA. The genetic information in DNA first required for the synthesis of a protein is to be transferred to a molecule of mRNA by a process called transcription. In translation, the mRNA directs the assembly of amino acids into a polypeptide. The mRNA gets attached to a ribosome, followed by the delivery of amino acids as mRNA by tRNA. However, there must be complimentarity in pair base pairings in codons and anti-codons. The amino acids are then linked to form polypeptide, and proteins can be synthesized in one or more polypeptide chains. Each chain may have 50 to several hundred amino acids. The DNA of *E. coli* has about 4000 genes and specify about 4000 different kinds of proteins. mRNA, tRNA and rRNA are the nuclear materials involved. RNA polymerase is the enzyme involved. tRNA and rRNA are synthesized from DNA template.

Genes are involved in the synthesis of proteins through transcription and translation. All the enzymes are regulated by the genes. The process of inhibiting the gene expression and decreasing the synthesis of enzymes is called repression. It is the response to the presence of an end product of a metabolic pathway, resulting in the decrease in the production of enzymes. Repressors block the ability of RNA polymerase to initiate transcription from the repressed gene.

A gene acts as an inducer and enzymes are synthesized in the presence of inducers. Induction is nothing but the process that turns in the transcription of a gene or genes. Lactose metabolism of *E. coli* is controlled by inducible system, which is controlled by genes. One of the genes code for the β-galactosidase. This enzyme splits the substrates, glucose and galactose, respectively. *E. coli* failed to show β-galactosidase enzyme when grown in a medium that contains lactose. However, when lactose is added to the medium, the bacterial cells produce a large quantity of the enzyme. Thus lactose is converted in the cell to the related compound, allolactose. This is the inducer for these genes. The presence of lactose indirectly induces the cells to synthesize more enzyme. All this process is under genetic control (enzyme induction). The control of gene expression, induction and repression are described by the operon model. Francis Jacob and Jacques Monad formulated the "Operon model", which accounts for the regulation of protein synthesis. *E. coli* was the model organism used by them. β-galactosidase breaks the disaccharide lactose into glucose and galactose, while β-galactosidase permease is involved in the transport of lactate into the cell. Thiogalactoside transacetylase metabolizes certain disaccharides other than lactose.

The structural genes are responsible for these three enzymes that are arranged next to each other on the bacterial chromosome and regulated together. The genes are transcribed and translated rapidly and simultaneously when lactose is introduced into the culture medium.

There is a regulatory gene called the *I* gene that codes for the synthesis of a repressor protein. Operator, another site located next to structural genes, controls the transcription promoted in an another region of DNA where RNA polymerase binds to initiate transcription. In the presence of lactose, the repressor protein binds tightly to the operator site. In the presence of lactose, a portion of it diffuses into cells and gets converted into inducer allolactose. The inducer binds the repressor protein and alters it so that it cannot bind to the operator site.

The absence of operator bound repressor, the RNA polymerase, enzyme can transcribe the structural genes into mRNA, which is then translated into enzyme. Thus it is proved that lactose induces enzyme synthesis and the lac operon is called an inducible operon. There are small molecules, called co-repressor, which helps to bind to the operator gene. Sixty to 80% genes are constituted and these code for enzymes that the cell needs in fairly constant amounts in its major life processes.

The regulation of the lactose operon depends on the glucose level in the medium and this controls the intracellular level of the cyclic AMF that is derived from ATP, which serves as a cellular alarm signal.

Glucose is more directly metabolized as a carbon and energy source over lactose. Cells grow using glucose as carbon source. Enzymes responsible for that metabolize glucose and constitutive. In the absence of glucose, *E. coli* responds by producing cyclic AMF. Increasing the level of cyclic AMF in the cell triggers a series of events that result in the transcription of lactose. Thus, catabolite repression or glucose effects are controlled by the genes.

## 14.6 TRANSFER OF GENETIC MATERIAL IN PROKARYOTES

The exchange of genes between two DNA strands results in the formation of new combination of genes on a chromosome called genetic recoms. A and B chromosomes break and rejoin. This is called crossing over and chiasma 'X' is formed. Now each portion of the chromosome carries the genes of other chromosome. This is called recombination genes that show recombination during meiosis. In eukaryotes, the recombination is an ordered process. However, in bacteria, genetic recombination can happen in many ways. Genetic recombination contributes to a population's genetic diversity. Diversity is nothing but variation. Recombination results in a valuable new function.

In the genetic recombination process, the transfer involves a donor cell that gives only a portion of its total DNA to a recipient cell. On transfer, part of the donor DNA is usually incorporated into recipient DNA, the remainder is degraded by cellular enzymes. If the recipient cell incorporates donor DNA into its own DNA, it is called recombination. Genetic recombination may not be a frequent process, but it may occur in only 1% or less of an entire population.

### 14.6.1 Transformation

Frederick Griffith of United Kingdom worked on *Streptococcus pneumoniae* in 1928 and established the process of transformation. In the transformation process, genes are transferred from one bacterium to another as naked DNA (Fig. 14.3).

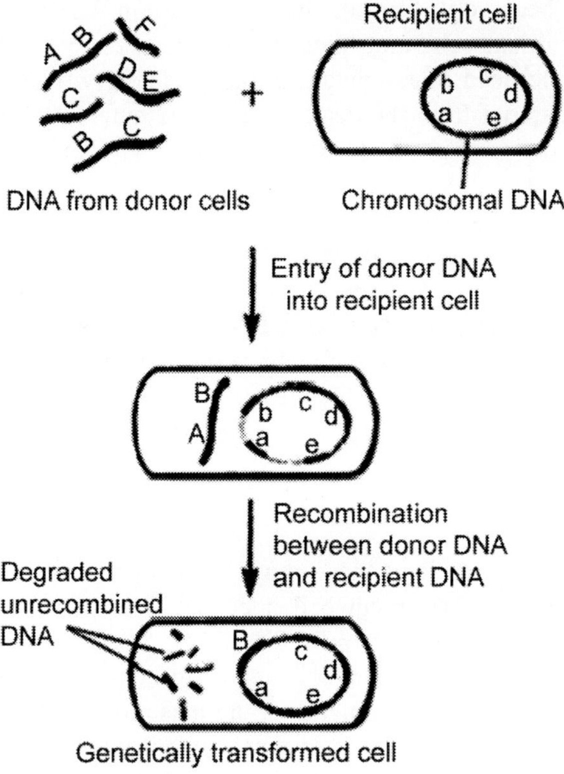


**Fig. 14.3.** Genetic recombination by transformation.

Griffith selected disease causing virulent bacteria. The virulent strain had a polysaccharide capsule and caused pneumonia. The second strain was an avirulent strain that lacked the capsule and did not cause disease. The virulent bacteria were inoculated into mouse, and the mouse died. The blood smear of dead mouse when examined showed virulent bacteria.

The avirulent bacteria were also injected into the mouse and it was healthy. The blood smear showed avirulent bacteria.

When heat killed virulent bacteria were injected into mice, it did not become ill and no bacteria could be observed in blood smear. However, when avirulent and heat killed virulent bacteria were injected, the mouse died. Colonies of avirulent and virulent bacteria were isolated from dead mouse (see chapter 7, Fig. 7.7)

The above experiment demonstrates that some material from heat killed virulent bacteria transformed the living avirulent bacteria into virulent bacteria, which killed the mouse. Avirulent bacteria lack capsules and are readily destroyed by the host, hence few show up as colonies on the medium.

The main objective of Griffith was to determine whether injections of heat killed (at 60°C) virulent pneumococci could be used to vaccinate mice against pneumonia. Griffith found that the injected mice did not become ill from the heat killed strain. Further, when these dead cells were mixed with live avirulent cells and injected into mice, many of the mice died. Griffith found that the blood of the dead mice contained encapsulated, virulent bacteria. Thus it was inferred that hereditary material from the dead bacteria had entered the live cells and changed them genetically so that their offspring were encapsulated, virulent forms. Thus bacteria had acquired a new hereditary trait.

Oswald J. Avery and his associates at Rockefeller Institute in USA tried to extract various chemical components from the killed cells and determine the chemicals responsible for transformation. In 1944 they announced that the chemical component responsible for transforming harmless pneumococci into virulent strains was DNA. *E. coli* is a good laboratory microbe that enables to take up DNA from other *E. coli* cells or even grown unrelated cells. This discovery along with recombinant DNA technology brought about the genetic engineering revolution.

### 14.6.2 Conjugation

Conjugation is cell to cell contact and these must be of opposite mating type. Donor cells have to carry the plasmid while recipient cell need not. The plasmid of donor cell carries genes that code for the formation of sex pili (Fig. 14.4). Sex pili are the projections of the donor cell surface that contact the recipient. Thus sex pili help to bring the two cells to come into direct contact. In this process of conjugation, the plasmid is replicated during transfer of a single-stranded copy of the plasmid DNA to the recipient where the complementary strand is synthesized. Donors carrying F+ factor transfer the plasmid to recipients (F– cells). After conjugation F+ cells are formed. Some bacterial cells carrying the F factor integrate into the

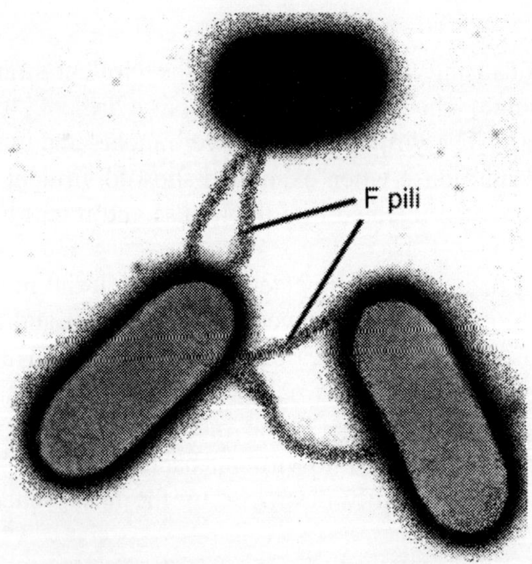


**Fig. 14.4.** Conjugation in bacteria.

chromosome, forming an Hfr (high frequency of recombination cell). In the process of conjugation between an Hfr and F– cell, the Hfr cell's chromosome with its integrated F factor replicates and a new copy of the chromosome is transferred to the receiving cell (recipient). Replicates of Hfr begin within the F factor. Small piece of the F factor leads the chromosomal genes into F– cell. Chromosome breaks before it is completely transferred. Further, the donor DNA can recombine with the recipient cells' DNA. DNA of cell is not integrated but degraded. The plasmids responsible for conjugation are transmissible between cells during conjugation.

### 14.6.3 Transduction

DNA is transferred inside a bacterial virus called bacteriophage from the donor cell to the recipient cell. In this method, the naked DNA is not passed as in transformation or during cell to cell contact in conjugation.

In this process the phage attaches to the bacterial cell wall and injects its DNA into the bacterium. DNA of bacteriophage acts as a template for the synthesis of new phage DNA and also directs the synthesis of phage protein coats (Fig. 14.5). The bacterial chromosome breaks apart inside the phage infected cell. Some fragments of the bacterial chromosome are accidentally packed inside phage protein coats. The resulting phage particles then carry bacterial DNA instead of phage DNA. The released phage particles later infect new population of bacteria. During this process the bacterial genes get transferred to the newly infected cells.

Transduction by a virus can lead to recombination between the DNA of the first host cell and the DNA of the second host cell. New genotypes are acquired by transduction.

## 14.7 PLASMIDS IN GENETIC ENGINEERING

The manipulation of genetic material in the laboratory is called Genetic Engineering. Plasmids are found in bacteria and also in yeast. Plasmid is a small cyclic DNA molecule in bacteria capable of replicating independently of the chromosome. It is an extra-chromosomal material present in the cytoplasm. Conjugative plasmids may have genes for carrying out their transfer to another cell. The cells can survive under most conditions without the use of proteins coded by plasmid gene, thus plasmids are usually dispensable. In some the genes carried by the plasmids can be crucial to the survival and growth of the cell. Dissimilation plasmids code for enzymes, which help in the catabolism of unusual sugar and hydrocarbons. There are different types of plasmids.

### 14.7.1 Resistance Plasmid Factors

These plasmids are of much importance in medicine. R factors carry gene that confer upon their host cell resistance to antibiotics, toxins and others. The resistance transfer factor has genes for plasmid replication and conjugation; r-determinant has the resistance genes that code for the production of enzymes that inactivate certain drugs etc. R factors are known to pose very serious problems for the treatment of infectious disease using antibiotics.

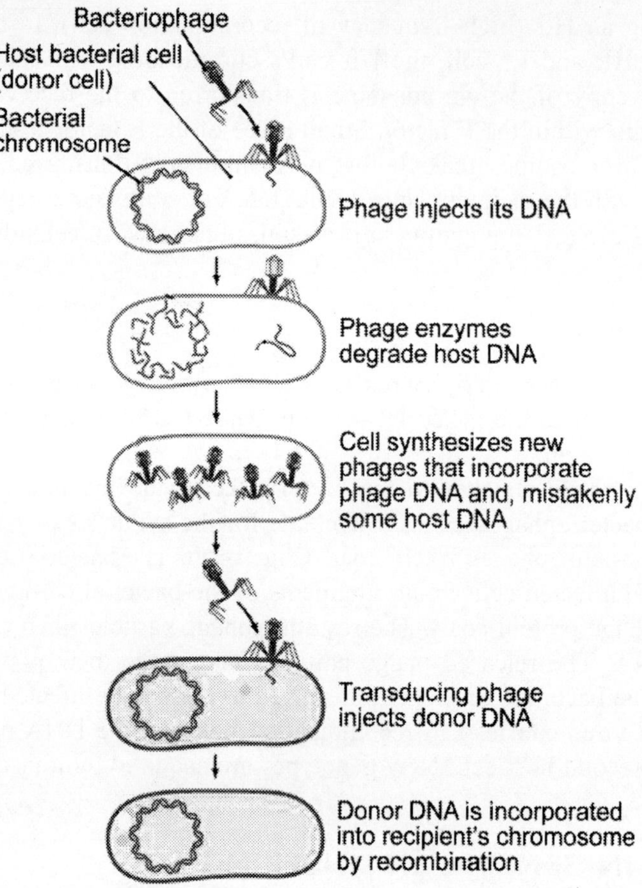


**Fig. 14.5.** Transduction in bacteria.

### 14.7.2 Col Plasmids

Many bacteria produce toxins of proteinaceous nature called bacteriocins. These are lethal to other strains of the same genus. The toxins secreted by *E. coli* are colicins. Synthesis of colicins is managed by plasmids present in *E. coli* cells but not by bacterial chromosome. These plasmids are associated with colicin production, hence called col plasmids.

### 14.7.3 Degradative Plasmids

*Pseudomonas* spp. are found to catalyze a number of complex organic compounds through specific metabolic pathways. 2,4-D, chlorobenzene, phenyls have been reported to be utilized by *Pseudomonas putida.* Anand Mohan Chakraborthy reported that *Pseudomonas putida* degrade urinal complex organic components. The plasmid of *P. putida* encodes enzymes for the degradation of camphor. Similarly, several plasmids are reported that degrade octane, naphthalene, salicylate in *P. putida*.

### 14.7.4 Ti-plasmids

*Agrobacterium tumefaciens,* a bacterium that is commonly used as vector in genetic engineering experiments, bears tumour inducing principle. Ti-plasmid present in the above bacteria encodes enzymes for the synthesis of auxin and cytosine.

### 14.7.5 Ri-plasmid

*Agrobacterium rhizogenes* causes hairy root disease in plants. Ri-plasmids are closely related to Ti-plasmids on the basis of opine production in the infected plant. The Ri-plasmids are put into manopine–Ri-plasmids, agropine plasmids, orcucumopine plasmids.

## 14.8 MUTATION

Any change in the base sequence of DNA is called mutation, generally referred as sudden change in the plant, animal and microbes. The agent in the environment that brings about mutation is called mutagen. The probability of a gene mutating each time a cell divides is known as mutation rate.

Spontaneous mutations are those mutations, which occur without the known intervention of mutation causing agents. Agents include chemicals or radiation, that directly or indirectly bring about mutations as mutagens. Such mutations are called induced mutations.

Commonly found mutation is base substitution or point mutation in which a single base at one point in the DNA is replaced with a different one. Adenine and thymine may be substituted for guanine and cytosine. Such case substitution brings about change in DNA molecule that codes for protein. The mRNA transcribed from the gene will carry an incorrect base at some position. Such incorrect base insertion causes the formation of incorrect amino acid in a protein. This is also called missense mutation.

In non-sense mutation, a base substitution results in a non-sense codon. Changes also occur in DNA called frame shift mutations wherein one or few nucleotide pairs are deleted or inserted in the DNA. The insertion of one nucleotide pair in the middle of a gene causes many amino acids down-stream from the site of the original mutation to change. As a result, non-sense codon will be generated and thereby terminate translation.

Luria and Delbruck (1943) were responsible for demonstrating the spontaneous as non-adaptive mutations. They found that *E. coli* conferred resistance to phage Ti infection. Lederberg and Lederberg (1952) gave an evidence for the origin of spontaneous mutations in T1 cells of *E. coli* without exposing to phage.

### Induced Mutations

**Chemical mutation:** Several chemicals, radiation, pollutants, plant extracts, alkaloids, drugs etc., have been considered as mutagens, thus bringing in change in the DNA of prokaryotes. For example, exposure of DNA to nitrous acid converts the base adenine that cannot pair with thymine but pairs with cytosine. This molecule formed will have a different base. Nitrous acid is considered a base pair mutagen that makes a specific mutational change in DNA. Altered DNA is found in a mutant.

Chemicals like 2-aminopurine and 5-bromouracil are the examples that bring in changes called base analogs. The 2-aminopurine molecule is incorporated into DNA in place of adenine but can sanction pair with cytosine. Similarly, the 5-bromouracil is incorporated into DNA in place of thymine, which often pairs with guanine during DNA replication. Benzopyrene is an effective frameshift mutagen. *Aspergillus flavus,* the fungus that grows on groundnut and stored grains produce aflatoxin. Aflatoxin is also carcinogenic and mutagenic.

**Radiation:** Ionizing radiation such as X-rays, gamma rays etc., are the mutagens. Exposure of microbial materials to such radiation results in change of DNA or causes damage to DNA. Ions and free radicals combine with bases in DNA. Such a combination results in mistakes in base pairing during replication. This leads to base substitution. The breaking of sugar-phosphate of DNA cause physical breaks in chromosomes. Ultraviolet light is a non-ionized component that affects DNA. Thymine of a DNA strand can cross line to form thymine dimers. Several enzymes are used to repair radiation damage of DNA in bacteria and other microbes.

The mutation rate is stated as power of 10. As mutations are rare, the exponent is always a negative number. The probability that any given cell in a growing population will undergo, spontaneously a specific mutation during its division cycle. The concept of mutation rate can be expressed as:

$$a = \frac{m}{d}$$

where $a$ = mutation rate; $m$ = the number of specific mutations which have occurred within the population and $d$ = the number of individual division cycles within the population. DNA replication in spontaneous mutations is very low.

## 14.9 RECOMBINANT DNA TECHNOLOGY

A DNA molecule produced by recombination is known as recombinant DNA. Genetic engineering deals with manufacturing and manipulating genetic material *in vitro*. Recombination is a process of joining pieces of DNA from different sources. Joining of DNA from different sources together under *in vitro* conditions results in the formation of recombinant DNA. Such recombinant DNA is synthesized artificially when introduced into an appropriate microbe. The microbe can often synthesize the products of the new gene it has acquired. Restriction enzymes are employed to make recombinant DNA. The enzymes identify unique site in DNA—the short symmetrical sequence of bases called palindromes. These enzymes cut double-stranded DNA within the palindromic recognition site and cut sites on two DNA strands. This leaves single-stranded self-complementary ends called sticky ends. For example, ECORI (*Escherichia coli* restriction enzyme) acts on the recognition site followed by cut site formation leading to sticky ends formation.

cut site

5′ – GAATC – 3′
3′ – CTTAAS – S′ ⟶ 5′-G 3′-C TTAA + AATTC – 3′
3′-C TTAA G 5

cut site
Recognition site Sticky ends

Recognition site is a nucleotide pair having same sequence on both strands. Thus the pieces of DNA are cut by Eco-RI, and they will have sticky ends of same DNA sequence. These will join in linear or circular form, which has the following arrangement:

AATTC G GAATTC G
G CTTAA GCTTAA
G AATTC
CTTAA

Recombinant plasmid construction is shown in Fig. 14.6. The recombinant is put into bacterial cells by transformation. The structural barriers present on the cell wall of recipient bacteria

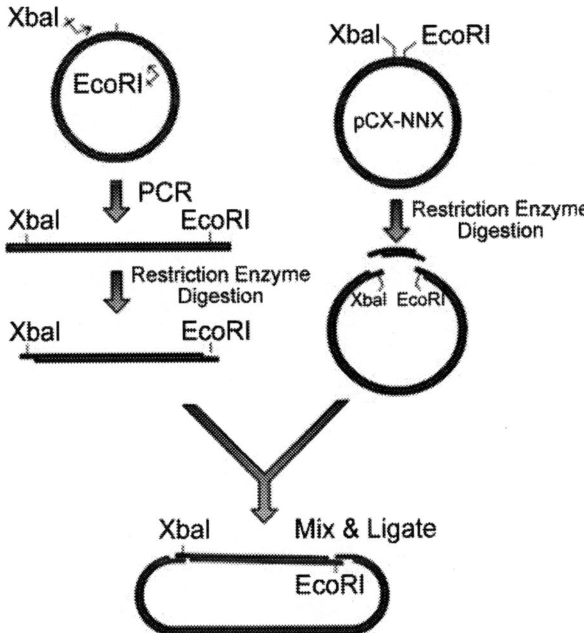


**Fig. 14.6.** Recombinant plasmid construction.

are removed. Thus hybrids having new genotypes are formed. As and when hybrid bacteria divide, the plasmids will also carry information in their new genes. Culturing of recombinant cells that secrete the substances has been done.

Eukaryotic genes have exons and introns. Exons are regions of a gene that code for the protein product of the gene and introns are the regions within the gene that do not code for the protein. The true nucleated cell transcribes the DNA of the entire gene, including both exons and introns. Intron RNA is removed and finally splices together the exons to form the mRNA. This gets translated into the functional protein.

Bacteria do not have enzymes that remove intron RNA, hence no protein can be formed. The DNA of eukaryotic cell cannot make recombinant followed by making a DNA copy of mRNA *in vitro* using viral reverse transcriptase as catalyst. The complementary DNA represents the entire coding sequence of DNA for the protein formation. With this phenomenon, the exons will be in correct sequence but introns are absent on the introduction of cDNA into bacterial cell in a recombinant plasmid. The bacterium then transcribes and translates the eukaryotic cDNA.

Recombinant DNA technology helps in understanding gene regulation. Animal genes can be inserted into bacterial DNA so that bacteria produced substance of therapeutic importance (examples – synthesis of insulin, somatostatin, interferon). The proteins that are produced as a part of recombinant DNA technique can act as vaccine. This results in the production of antibodies against the pathogenic microbes. These vaccines protect animals from infection. Recombinant DNA technology is important in pharmaceutical processes (antibiotics, conversion of ethylene to ethylene glycol, ethanol etc.).

# Molecular Techniques for Monitoring Microbial Systems

## 15.1 INTRODUCTION

A major problem hindering the successful application of microorganisms to improve crop production in the environment is knowing the fate of the bacteria in the field, as the released strains are mostly not distinguishable from the indigenous bacterial population in soil environment. Thus, to study the inoculum, establishment and survival, it is necessary to have a sensitive and reliable means of specifically detecting and quantifying the inoculated strain in the system.

Traditional approaches in studying bacterial establishment employ endogenous molecules as markers. For example, rhizobia may be distinguished by patterns of intrinsic or induced antibiotic resistance, and these always result from molecular change. The other traditional class of molecular markers used in bacterial ecology is antigens expressed on the cell surface that can be recognised by specific antisera. This is commonly achieved through the use of enzyme-linked immunosorbant assay (ELISA), fluorescently labelled antibodies or immunodiffusion.

Molecular methods have been developed that can detect and quantify phylogenetic groups on the basis of rDNA sequences and relevant structural genes. Many of these techniques rely on PCR for the amplification of DNA sequences that might be in low abundance in a mixed microbial community. Reverse transcriptase can be coupled to PCR for measuring gene expression on the basis of mRNA abundance. Microbial diversity and community structure can be addressed by further examination of PCR products by various separation techniques and restriction analyses.

## 15.2 MOLECULAR MARKERS

Over the past years, the use of marker genes, such as *gusA, lacZ, celB, xylA, luxAB*, has become an important tool for studying microbial ecology as they allow simple visualization of marked microorganisms. The primary consideration in choosing a marker gene is, therefore, the assays available for its detection. These must be technically simple, highly sensitive and affordable, so that they can be used routinely in the field by ecologists rather than their use being restricted to highly specialized laboratory environments. Additionally, there should be little or no detectable background activity in either the microbes being studied or in other components of the environment being assayed, such as soil or plant roots.

The *lacZ* gene, encoding β-galactosidase, has been used to study nodule infection by *Rhizobium* and for studies of root colonization in *Azospirillum*. The *E.coli gusA* gene encodes the enzyme β-glucuronidase (GUS) and it cleaves many glucuronide substrates such as X-glcA (5bromo-4-chloro-3-indolyl-β-D-glucuronide) or Magenta-glcA (5bromo-6-chloro-3-indolyl-β-D-glucuronide), releasing an indigo or magenta coloured precipitate by which marked strain can be identified. The cell β gene from the thermophilic *Pyrococcus furiosus* encodes a thermostable β-glucosidase with a high β-galactosidase activity.

Another critical component of the marker gene system is the choice of delivery system. The marker gene can be introduced into the recipient bacterium as a direct insertion in the genome of the recipient through the use of transposable elements.

Similar technique can be used to track the fate of free-living rhizobacteria like *Azospirillium, Azotobacter, Bacillus, Pseudomonas* etc. A suitable media containing kanamycin and amended with X- gal (5 bromo-4chloro-3indolyl-β-D-galactotyranoside) and IPTG (Isopropyl thiogalactoside) can be used to distinguish the blue colonies of the introduced strain from majority of the other microorganisms.

## 15.3 AMPLIFICATION OF 16S rDNA GENES BY THE POLYMERASE CHAIN REACTION (PCR)

PCR is an *in vitro* method for the enzymatic synthesis of specific pieces of DNA. The process was developed by Kary Mullis (K.B. Mullis, US patent 4,683,195; K.B. Mullis, US patent 4,683,202) for which he was awarded 1993 Nobel Prize in Chemistry. PCR has revolutionised research in the area of life sciences.

PCR process requires two oligonucleotide primers that hybridize to opposite strands of the target DNA and flank the region to be amplified. A suitable DNA polymerase, the four deoxynucleoside triphosphates (dNTPs), and $Mg^{2+}$ are also required. Amplification results from successive cycles of denaturation and oligonucleotide priming of the target DNA followed by the enzymatic synthesis of the strand of DNA that is complementary to the target DNA. To perform a PCR amplification, a mixture containing target DNA, oligonucleotide primers, dNTPs, and a heat stable DNA polymerase is heated to 90–95°C to denature the strands of the target DNA (Mullis 1990). The solution is cooled to a temperature that allows the primers to anneal to their complementary sequence on the target DNA and provide the 3′-OH required for DNA synthesis. The DNA polymerase then synthesizes a new DNA strand complementary to the target. The polymerase used (Taq polymerase) is isolated from the thermophilic bacterium *Thermus aquaticus*. This thermophilic enzyme can tolerate the high temperature denaturation step and is stable throughout all the reaction cycles. In each cycle, the template concentration doubles with respect to the previous cycle. Therefore, for initial double-stranded DNA molecule, there will be $2^c$ molecules, where c = number of cycles. The newly synthesized strand initially extends a variable distance primer-binding sites on the newly synthesized strand. This thermal cycling scheme is repeated numerous times with the DNA synthesized during the previous cycles serving as a template for each subsequent cycle. With each cycle, the strands are denatured at high temperature, the primers annealed, and then extended by the polymerase. The result is the doubling of the target DNA present with each cycle. This exponential accumulation of DNA or DNA sequences can theoretically produce over a millionfold amplification of the target in 20 cycles. In practice, the amplification efficiency is less than 100%, and 30 to 40 cycles are commonly done.

PCR-amplified products can be purified and examined by agarose gel electrophoresis. In addition, PCR-amplified products can be easily cloned and used for transformation or sequencing.

There are many variations of the PCR technique. In some cases it is helpful to amplify first a larger piece of the target by using primers that are not organism-specific and then re-amplify these products with primers specific for an organism of interest, a process called nested PCR.

PCR is an excellent technique for examining mixed microbial communities. However, care should be taken in interpreting the results. Minor contamination can lead to false positive signals, and false negative amplifications are often seen. It must be emphasized that standard PCR reactions are not quantitative.

One method to quantify PCR-amplified products for comparison is to include an internal control in the PCR reaction. Here, a known amount of target DNA is added to a PCR reaction containing DNA from the mixed microbial population. The known target DNA is complementary to the same primers and thus competes with the target sequences in the extracted DNA sample. By running a dilution series of the known and unknown DNA species, it is possible to quantify the amount of product produced from the complementary gene in the extracted DNA. Known targets can be generated by cloning the gene of interest or purifying the PCR-amplified product. Then an internal change (generally a deletion) is made that will give a differently sized PCR product. Alternatively, commercially available kits add primer-complementary sequences to 'benign' DNA to create known target DNA.

There are several different polymorphism-based procedures that have been applied to microbial ecological systems. The separation or detection of small differences in specific DNA sequences can give important information about community structure and the diversity of microbes containing a critical gene. Generally, these techniques are coupled to a PCR reaction to amplify sequences that might be in low abundance. PCR-amplified products can be examined by using techniques that detect single substitutions in nucleotide sequence. These techniques are important in separating and identifying PCR-amplified products that might have the same size but slightly different nucleotide sequences. For example, the amplified portions of 16S rDNA genes from a mixed microbial population might be of similar size with a particular set of oligonucleotide primers, but have small differences within the PCR-amplified product at the nucleotide level. One way of detecting these differences is to restrict the PCR-amplified product with endonucleases and examine the pattern of the restriction fragments. The PCR-amplified product can be either end-labelled or uniformly labelled for this technique and then cloned into a vector suitable for sequencing.

Two other protocols use a single primer to amplify fragments with PCR before examination or restriction. PCR amplification of repetitive genomic sequences (REP-PCR) takes advantage of repetitive sequences found in the microbial genome. Randomly amplified polymorphic DNA (RAPD) or arbitrarily primed PCR use primers that are not specific for a particular gene to amplify fragments. The fingerprints obtained by these techniques are organism specific. It is possible with modern computer software to examine complex patterns of DNA fragments and identify patterns representative of different known organisms. Both REP-PCR and amplified ribosomal restriction analyses have been used to monitor the microbial communities in a field test site.

There are also techniques for separating PCR-amplified products with such precision that single-base substitutions can be detected. In density-gradient gel electrophoresis (DGGE), a vertical polyacrylamide gel is poured with an increasing concentration of denaturant (formamide and/or urea). As duplex DNA fragments are subjected to electrophoresis, partial melting occurs at a denaturant concentration specific for the nucleotide sequence of the DNA. Once the proper DGGE conditions have been determined, DNA fragments that differ at a single nucleotide position can be resolved.

## 15.4 RESTRICTION FRAGMENT LENGTH POLYMORPHISM (RFLP) ANALYSIS OF PCR AMPLIFIED 16S rDNA GENE SEQUENCE

During the past decade, ribosomal RNA-based identification and detection methods have become techniques routinely used in all disciplines of microbiology. Highly conserved portions carry the information on early evolutionary events and more recent changes are documented within less conserved positions or stretches. The degree of divergence of present-day rRNA sequences gives an estimate of their polygenetic distances.

Comparison of sequences of 16S rRNA genes have permitted differentiation between bacterial genera, species and strains. Most of the validly described species have been phylogenetically characterized by comparative analyses of almost complete 16S rRNA sequences. Many strains and isolates can be assigned to phylogenetic groups by, full and partial rRNA sequences.

Variations in 16S rRNA genes can also be estimated by restriction fragment length polymorphism (RFLP) and analysis of 16S rDNA sequences amplified by polymerase chain reaction (PCR). The PCR amplified product is subjected to digestion with different restriction endonucleases, and the ones that are most discriminating are selected for identification purpose or assignment of phylogenetic group.

## 15.5 REP-PCR GENOMIC FINGERPRINTING

Both prokaryotic and eukaryotic genomes contain dispersed repetitive sequences separating longer single-copy DNA sequences. Various classes of repeated DNA sequences are present in diverse prokaryotic genomes. Interspread repetitive DNA sequences are typically between 15 and several hundred base pairs (bp) in length and are widely dispersed in bacterial chromosomes. These elements are non-coding and present in high copy number relative to the larger repeated elements, which contain coding sequences.

It has been shown that DNA primers corresponding to repetitive extragenic palindromic (REP, 38bp) enterobacterial repetitive intergenic consensus (ERIC 124 bp) and Box (154 bp) element sequences, coupled with PCR technique can be used to fingerprint the genomes of a variety of gram-negative soil bacteria.

Genomic fingerprinting with interspread repetitive sequence-based probe (BOXAIR) distinguishes unrelated organisms because individual bacterial strains vary with respect to the distances between the repetitive sequences. The PCR amplification of genomic regions between BOX copies produces a collection of distinct fragments (or typical fingerprint pattern) on an agarose gel.

## 15.6 IDENTIFICATION BY RANDOM AMPLIFIED POLYMORPHIC DNA (RAPD)

A single 10-base oligonucleotide primer is used to amplify genomic DNA. A DNA amplification product is generated for each genomic region that happens to be flanked by a pair of 10-base priming sites (in the appropriate orientation), which are within 5,000 base pairs of each other. Under the annealing conditions used, the single oligonucleotide acts as both a forward and a reverse primer.

Amplification products are analysed by electrophoresis in 1.4% agarose gels. They can be examined after staining with EtBr.

## 15.7 NUCLEIC ACID HYBRIDIZATION TECHNIQUES

The most straightforward way of detecting specific nucleic acid sequences or genes of interest is through direct hybridization of a DNA or RNA probe to microbial nucleic acid extracts. Whole-cell DNA or RNA is extracted from the microbes of an experimental sample such as soil or water and fixed to a positively charged membrane (nitrocellulose, nylon). Clonal bacterial colonies can also be replica-plated from agar plates to membranes and their nucleic acids exposed *in situ* for subsequent hybridization. The nucleic acids on the membrane are probed with a small piece, often an oligonucleotide of DNA or RNA containing a sequences complementary to the target sequence. The probe is labelled before hybridization with $^{32}P$, a chemiluminescent compound or digoxigenin (for later antibody-peroxidase detection). DNA probes can be used to detect genes in the bacterial genome or on plasmids (Southern blots). The signal from hybridization is quantitative; position bands or spots on a membrane can be directly compared through densitometry.

## 15.8 16S rDNA SEQUENCING

Another way of examining microbial communities is through the direct sequencing of 16S rDNA, although 23S rDNA has been used. This is a laborious task made somewhat easier through the use of PCR. The entire compliment of 16S rDNA species from a mixed population can be amplified by PCR and then cloned into a suitable vector for sequencing. In this way, organisms can be identified in a mixed population even if they are non-culturable. Individual clones can be examined, perhaps after sorting by restriction mapping.

## 15.9 FLUORESCENT *IN SITU* HYBRIDIZATION (FISH)

While examining an ecological system, it is often important to obtain information about how the functional components of the system relate to the organisation of the system. In communities that have an inherent architecture, such as biofilms and flocs, the question where various organisms (and their activities) are located is of paramount interest. These determinations are difficult, if not impossible, to make with standards techniques of microscopy. However, the distribution of microbes within an architectural framework can be addressed by using FISH, in which tagged DNA oligomers are used to probe intact microbial communities. Undistributed biofilms or flocs on gelatin-coated glass slides are treated with paraformaldehyde. The fixed and permeabilized cells are then exposed to a DNA probe, often an oligonucleotide, which has been tagged with a fluorescent dye such as fluorescein. The DNA probe binds to complementary DNA or RNA sequences within the cells. Microbes that contain sequences complementary to the DNA probe glow when excited in an epifluorescence microscopy. Coupling FISH with confocal scanning laser microscopy allows the labelled microbes to be placed in a three-dimensional reconstruction of the intact microbial community.

## 15.10 GENETICALLY ENGINEERED MICROORGANISMS AS MARKERS

The recent advances in molecular biology, in particular recombinant DNA (rDNA) technology, provides a remarkable understanding and control over biological processes. By these techniques it is increasingly possible to directly manipulate the heritable material of cells between different types of organisms, creating new combinations of characters and functions not previously achievable by traditional techniques. This rDNA technology or genetic engineering is the most revolutionary technology in the 21st century. Genetic engineering will have profound impacts on medicine, contributing to the diagnosis and cure of hereditary defects and serious diseases, the development of new biopharmaceutical drugs and vaccines for human and animal use, the modification of microorganisms in agriculture for improved and tailored food production and increased opportunities for environmental remediation and protection. The safety and impact of genetically modified organisms (GMOs) continues to be addressed by scientific research.

One way of monitoring the fate of a microorganism introduced into the environment is to engineer a marker into the genome of the microbe. For example, PCR has been used to detect a genetically modified organisms introduced into a freshwater microcosm. In this study, a specific synthetic PCR-amplifiable DNA fragment was introduced into the *Pseudomonas* chromosome to allow the genetically engineered microorganism to be identified easily.

# Immunology

## 16.1 INTRODUCTION

A healthy individual is one who is normally capable of carrying on all the functions necessary for its existence. These functions are guided by the metabolic and other processes of the various cells, which constitute the plant or animal. Once the rather indefinite limits of the healthy state have been passed, the individual is said to be diseased.

A number of hypotheses were advanced during early times to explain the nature and cause of disease. A number of workers attempted to produce infections in experimental animals by the transfer of such microbe, and in many experiments they were successful. However, much of this early work was of sloppy character, could not be readily confirmed, and as a result the "germ" theory of disease was in general dispute. We must remember that experimental methods in microbiology were very crude at that time, and the lack of pure cultures was a serious handicap. Jacob Henle (1840) protested against the poor experiments and loose reasoning of the time and demanded that evidence be submitted, which would show that the microorganism suspected of being the infectious agent of the contagion (1) could be constantly found in the infected tissue, (2) could be isolated from it, and (3) on injection could induce the disease. Almost thirty years lapsed before these conditions were met, and it was definitely established that microorganisms could produce infections in animals and plants.

A number of workers around 1850 observed the presence of relatively long bacteria in great numbers in the blood of animals which died due to anthrax. This led them to suspect that these rods might be the etiological agent of the disease. The studies of Davaine in the period of 1863 to 1868 gave considerable support to this belief, particularly when he was able to demonstrate that anthrax can result from the inoculation of a healthy animal with as little as one-millionth of a drop of blood from an infected animal.

In 1865 Pasteur was commissioned by the French government to investigate a disease, pebrine of silkworms, which was seriously crippling the French silk industry. He presented considerable evidence that the disease was caused by a protozoan and that a second disease, flacherie, was induced by bacteria. He was also able to introduce methods for the control of these infections.

Robert Koch (1876) was able to demonstrate that the bacterium observed in such large numbers in animals dying of anthrax was actually the etiological agent of this disease. He succeeded in isolating the organism in pure culture in sterile vitreous fluids from the eyeballs of cattle, passing it a number of times in the laboratory, and reproducing the infection in experimental animals inoculated with pure cultures of the anthrax *Bacillus.* This study (confirmed by Pasteur in 1877), together with one on the microbial causes of wound infections reported in 1878, Koch's discovery in 1882 of the tubercle *Bacillus*, and the demonstration that it is the causative agent of tuberculosis, firmly established the theory that a particular infection is induced by a specific organism. In the study of tuberculosis, Koch set forth his famous postulates, which should be fulfilled if an organism is to be definitely considered the causative agent of a disease. These, in essence is an amplification of the reasoning of Henle, one of Koch's teachers, are stated as follows:

1. The organism must be present in all cases of the disease.
2. The organism must be isolated and cultivated in pure culture.
3. The specific disease must be induced in susceptible animals following the injection of the pure culture.
4. The organism must be present in typical distribution in, and be recoverable from, the experimental host.

During this time and after a few years, Koch and other workers introduced a variety of technical methods for the study of microorganisms, and these, together with the impetus provided by the earlier observations of Koch and Pasteur, were responsible for rapid progress in the isolation and identification of various microorganisms as the causative agents of a wide variety of infectious diseases of both plants and animals.

It was also recognized that individuals who had recovered from certain diseases were generally not again susceptible to the same ailment. This was particularly true of infections such as measles, scarlet fever, and smallpox. Centuries ago, Chinese observed that individual inoculated with material obtained from smallpox lesions generally developed a mild localized infection and became resistant to the more generalized type of infection. Jenner, in 1796, observed that English dairymaids who had contracted cowpox (vaccinia), a mild, small pox-like disease, from cattle were subsequently immune to true smallpox (variola). He inoculated fluid from cowpox blisters into children and noted that they developed a mild localized form of cowpox, at the same time becoming resistant to smallpox. The introduction of vaccination as a preventive measure against a specific disease occurred long before the germ theory of disease had been established, but this early work did suggest that a specific material might be the cause of a particular disease and that the body can develop an increased resistance to the development of the disease.

By 1880, it was recognized that chicken cholera is caused by a bacterium now known as *Pasteurella multocida*. Pasteur observed that this organism did not multiply in sterile urine or yeast water, media favourable for the growth of the anthrax bacillus, while it did multiply in chicken broth. Reasoning by analogy, he concluded that this behaviour might be representative of what happens when a harmless microbe enters the body, it being harmless because conditions are not suitable for its growth. He also observed that the smallest drop of a culture of this organism on bits of food was sufficient to produce the disease in the chicken. This was an early observation of native species immunity and an important contribution to the science of public health, leading to later observations that one species may harbour a parasite with little or no harm to itself and serve as a possible reservoir for the dissemination of the organism to susceptible hosts.

It was also observed that chickens inoculated with an old culture of *P. avicidia* accidentally developed a mild form of chicken cholera from which they speedily recovered, and at the same time became highly resistant to subsequent infection with young, virulent cultures of the same organism. Subsequent observations showed that the virulence of this organism gradually decreased as the age of the culture increased, oxygen apparently being involved in the attenuation process. Pasteur then proceeded to develop "vaccines" for the immunization of chicken cholera. He recognized the similarities in this process of immunization and the earlier introduced against smallpox by Jenner and developed the hope that it would be possible to reduce artificially the virulence of all infectious agents to such an extent that they could be employed for immunization purposes. Pasteur was successful in attenuating both *Bacillus anthracis* by repeated transfers of the organism at 42°C and the virus of rabies by passage of the virus through rabbits and desiccation of their infected spinal cords for different periods of time, and was able to develop methods of employing these attenuated agents for the purpose of immunizing susceptible animals against the virulent form. In time, other methods for attenuating pathogenic forms were developed, and it was also noted that suspensions of killed bacteria were of value as immunizing agents. In 1890, Behring and Kitasato observed that animals may be artificially immunized against diphtheria toxin by repeated injections of sublethal doses of the toxin. Immune sera from these animals neutralized diphtheria toxin in the test tube or, when injected into a susceptible animal, would protect it against the subsequent injection of lethal doses of the toxin. This suggested a possible therapeutic use for diphtheria antiserum, and Behring first employed the antiserum (a serum containing neutralizing substances, antibodies, against a specific substance or cells) as a therapeutic measure.

Agents that can be employed for the artificial production of the immune state are popularly called *vaccines*, although strictly speaking, the terms "vaccine" and "vaccination" refer only to the material and procedure employed in immunization against smallpox. In immunology the term vaccine is commonly employed for suspensions of attenuated or killed organisms employed as immunizing agents. It must be borne in mind that the vaccine or immunizing agent does not of itself produce increased resistance on the part of the host. The material inoculated into the individual animal stimulates the animal to produce for and by itself substances that will neutralize the complex material, ordinarily proteinaceous, of which the immunizing agent is composed. The body itself builds up specific defence substances, antibodies, against the pattern presented by the particular immunizing agent. The production of increased resistance, either as a result of having the actual infection or of being inoculated with attenuated material, is spoken of as *active immunization*. Temporary immunization against a number of infectious agents may be accomplished by the injection of antibodies produced by and present in the serum of another animal, which had been actively immunized or has had the infection. The injected animal has not itself produced the antibodies involved in this protection, and therefore this means of establishing increased resistance is spoken of as *passive immunization*. Such immunization is temporary in character, as the injected antibodies disappear in a few days or weeks. The antibody-containing serum employed in passive immunization is known as an antiserum and is generally named after the material against which it was produced, e.g., diphtheria antiserum or diphtheria antitoxin. Antisera contain neither microorganisms nor their toxins, but instead antibodies preformed against these agents.

Passive immunization is of value in the attempted prevention of the development of certain infectious diseases after exposure to the infectious disease. Antisera have been employed with success as therapeutic agents, particularly in the treatment of diphtheria, pneumococcal pneumonia, and epidemic meningitis, and have proved to be of some value in the prevention or treatment of tetanus, gas gangrene, botulism, measles, and certain streptococcal infections. Antibodies are present in gamma globulin fraction of serum, and gamma globulins are often employed instead of whole serum.

Active immunization, vaccination, is highly successful against those agents responsible for infections, which leave a lasting immunity. It is of less value for general use against infectious agents, which do not elicit the establishment of a lasting immunity in the host, although it may be worthwhile against infections such as influenza when these appear in epidemic form.

The immunological response is a general system for the neutralization of foreign macromolecules through the formation of specific antibodies. Since the invading microorganisms contain a variety of macromolecules foreign to the host, antibodies develop against it.

(1) Antibodies are formed against a variety of foreign macromolecules, but ordinarily not against macromolecules of the animal's own tissues. Thus the animal is able to distinguish between its own and foreign macromolecules.
(2) In virtually every case, antibody against a foreign macromolecule is formed only if the animal is challenged with the foreign substance.
(3) Many but not all foreign macromolecules elicit the immunological response; those that do are called antigens.
(4) There is a high specificity in the interaction between antibody and antigen; antibodies will react against closely related antigens but usually with reduced efficiency.
(5) Antigen-antibody reactions are manifested in a wide variety of ways, depending on the nature of the antigen, the antibody, and the environment in which the reaction takes place.
(6) Not all antigen-antibody reactions are beneficial; some, such as those involving allergy and hypersensitivity are more or less harmful.
(7) The high specificity of antigen-antibody reactions makes them useful in many research and diagnostic procedures.
(8) The immunological response can be made the basic of specific immunization procedures for the prevention and control of specific diseases.

Antigens are defined as substances that under appropriate conditions induce the formation of specific antibodies; antigens then react specifically with these antibodies. Antigens are distinguished from haptens by the fact that although the latter react specifically with antibodies, they do not induce their formation. To establish whether or not a substance is an antigen must be operational; the material must be injected into an animal, which is then shown to form antibodies. However, the antibody response is not always an automatic reaction whenever an antigen is injected. It depends very much on the species (or even strain) of the animal, its previous history, the route of injection, the frequency of injection, the concentration of antigen injected, and the presence or absence of various nonspecific antibody-stimulating agents administered with the antigen, which are known as adjuvants.

An enormous variety of macromolecules can act as antigens under appropriate conditions. These include virtually all proteins and lipoproteins, many polysaccharides, some nucleic acids, and certain types of teichoic acids. One important requirement is that the molecules must be of fairly large molecular weight, usually greater than 10,000. However, the antibody is directed not against the antigenic macromolecule as a whole, but only against restricted portions of the molecule that are called antigenic determinants. Chemically, antigenic determinants include sugars, amino acid side chains, organic acids and bases, hydrocarbons, and aromatic groups. Antibodies are formed most readily to determinants that project from the molecular or terminal residues of a polymer chain. In general, the specificity of antibodies is comparable to that of enzymes, which are able to distinguish between closely related substrates. For instance, antibodies can distinguish between the sugars glucose and galactose, which differ only in the position of the hydroxyl group on carbon 3. However, specificity is not absolute, and an antibody will react at least to some extent with determinants related to the one that induced its formation. The antigen that induced the antibody is called the homologous antigen and others that react with the antibody are called heterologous antigens. Antibodies also are not uniform substances but exist as a spectrum of molecules having varying degrees of affinity for the antigenic determinant. The chemical substances such as sugars and amino acids, which function as antigenic determinants, do not themselves induce antibody formation, but will, if added, combine with antibody molecules. Such substances that combine with antibody molecules but do not induce antibody formation are called haptens. By combining with antibody and blocking the active site, haptens have the ability to inhibit antibody-antigen reactions.

Antibodies are found predominantly in the serum fraction of the blood, although they may also be found in other body fluids, as well as in milk. Serum is the fluid portion of the blood that is left when the blood cells and the materials responsible for clotting (fibrin and cofactors) are removed. Serum containing antibody is often called antiserum. When serum proteins are separated by movement in an electric field (electrophoresis), four predominant fractions are seen: serum albumin and alpha, beta, and gamma globulins. Antibody activity occurs predominantly in the gamma globulins. Antibody activity occurs predominantly in the gamma globulin fraction, which is composed of many distinct proteins. Those gamma globulin molecules with antibody properties are called immunoglobulins (abbreviated Ig) and can be separated into five major subclasses on the basis of their physical, chemical, and immunological properties; IgG, IgA, IgM, IgD, and IgE. Antibody molecules specific for a given antigenic determinant are found in each of the several classes, even in a single immunized individual. Upon initial immunization, it is IgM (the so-called macroglobulin), a protein with a molecular weight of about 900,000, which appears first; IgG appears later. In most individuals, about 80 per cent of the immunoglobulins are IgG proteins, and these have therefore been studied most extensively.

## 16.2 SEROLOGICAL TESTS

Antigen-antibody reactions are most easily studied *in vitro* using preparation of antigens and antisera. The study of antigen-antibody reactions *in vitro* is called serology, and is especially important in clinical diagnostic microbiology.

When an animal body is invaded by a pathogen, substances called *antibodies* usually appear in the blood. These antibodies react with the pathogen to inactivate it. Non-pathogens, dead cells, and complex cell products can also cause the formation of antibodies when injected into an animal. Such substances are called *antigens.* When blood is taken from an animal that has been inoculated with an antigen, it is the serum fraction that contains antibodies. Such antibody-containing serum is termed *antiserum* or immune serum. The reactions *in vitro* between antigens and antisera form the basis of serological techniques.

The antigens are crude mixtures of complex molecules, and are, in reality, mixtures of antigens. The antigenic components of such mixtures are highly proteins, although occasionally an antigenic polysaccharide is found. To behave as an antigen in a given animal, a protein must differ sufficiently from the proteins naturally present in the animal, otherwise antibodies will not be produced against it. There are vast numbers of different proteins present in different organisms. Many are antigens that produce differing antibodies, and because of this their parent organisms can be distinguished by serological tests. Serology may be used in detection, identification, and classification of microorganisms. It has been used much for the study of viruses, but less for fungi and bacteria.

Visible serological reactions are of two main types, agglutination, in which a particulate antigen, e.g., a suspension of bacterial cells, is caused to gather in clumps or agglutinate, and precipitation, in which a soluble antigen is precipitated from solution. Other reactions, not directly visible, may be demonstrated by the complement fixation test. This test has special importance in medicine.

### Preparation of Antisera

Antisera are produced by injecting an animal intravenously, intramuscularly, intraperitoneally, or subcutaneously at regular intervals with increasing doses of a suitable antigen preparation. Rabbits are most commonly used. Injection into the marginal vein of the ear is convenient and should cause no pain. Particulate antigen preparations are frequently made from fungi or bacteria by killing the cells with formalin or heat and suspending them in saline. Soluble antigen preparations are made from culture filtrates, or from cells disrupted by mechanical grinding, sonic treatment, or autolysis. Viruses may be partially purified and separated from host material by differential centrifugation, and some of the more stable ones may be further purified by methods such as precipitation by ammonium sulphate. The resultant soluble antigen preparations are often sterilized by filtration before injection. Courses of injection vary, but injection 2 or 3 times a week for 3 to 6 weeks of up to 2 ml of saline containing up to 100 mg soluble protein, or $10^8$ to $10^9$ cells or particles is usual. A week after the final injection, the animal is kept without food for 24 hr and then bled from the ear vein. The blood is allowed to clot and the serum decanted. Merthiolate (final concentration 0–01%), glycerol (50%), or phenol (0–5%) may be added as preservatives. Serum may also be freeze-dried or deep frozen, but it loses some activity in the freezing and thawing processes.

### 16.2.1 Precipitation

When an antibody has two combining sites (i.e., bivalent), it is possible for each site to combine with a separate antigen molecule, and if the antigen also has more than one combining site, a precipitate may develop consisting of aggregates of antibody and antigen molecules. Precipitation reactions for a wide variety of soluble polysaccharide and protein antigens are known. Since they are easily observed *in vitro,* precipitation reactions are very useful serological tests, especially in the quantitative measurement of antibody concentrations. Precipitation occurs maximally only when there are optimal proportions of the two reacting substances, since when the either reactant occurs in excess, the formation of large antigen-antibody aggregates is not possible.

Precipitation reactions carried out in agar gels have proved of considerable utility in the study of the specificity of antigen-antibody reactions. Precipitation bands form in the gel in the region where antibody and antigen meet in equivalent proportions. The shapes of the precipitation bands and the distance from the wells are characteristic for the reacting substances, and it is possible to determine whether two antigens reacting against antibodies in an antiserum are identical or not by observing the bands formed when the two antigens are placed in adjacent wells near the antiserum well.

### 16.2.2 Agglutination

If the antigen is not in solution, but is present on the surface of a cell or other particle, an antigen-antibody reaction can lead to clumping of the particles, called agglutination. When foreign cells are injected into an animal, antibodies are formed against a wide variety of their macromolecular constituents, including those in the cytoplasm as well as on the surface of the cells. Only the surface antigens are involved in agglutination, however, since only these are exposed in the intact cell. Antibodies against both the flagella (H antigens) and the lipopolysaccharide layer (somatic or O antigens) will agglutinate cells, but the nature of the clump differs from that of flagellar agglutination, which is much looser and more flocculent.

Agglutination of red blood cells can also occur, a phenomenon called hemagglutination. Antibodies can react with antigens of the red cells, or other antigens can be adsorbed to red cells and hemagglutination observed when antibody specific to the adsorbed antigen is used (Fig. 16.1).

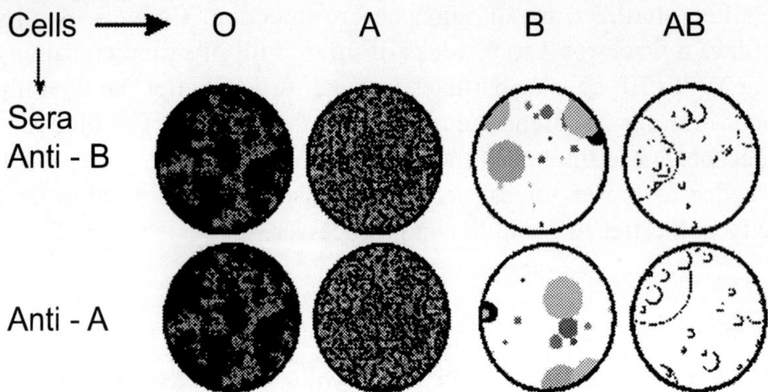


**Fig. 16.1.** Haemagglutination in blood grouping. Anti-B serum and anti-A serum are mixed on glass slides with erythrocytes of groups O, A, B and AB. Anti-B serum agglutinates erythrocytes of group B and AB; Anti-A serum agglutinates erythrocytes of groups A and AB.

### 16.2.3 Complement Fixation Test

A thermolabile substance called *complement* occurs in the serum of higher animals. It is able to combine with many antigen-antibody complexes, and in this way it can take part in serological reactions even though it is itself non-specific and not produced in response to immunization. The reaction of complement with antigen-antibody complexes is termed 'complement fixation'.

There are two important properties of complement that are made use of in the complement fixation test. Firstly, a mixture of antigen and antibody that does not react will not fix complement. Only when they react to form a complex does fixation occur. Secondly, haemolysin, an antibody specific for red blood cells, will only cause lysis of red blood cells in the presence of complement. Sheep red blood cells are used in the test. They are 'sensitized' before the test by mixing with haemolysin in the absence of complement.

To perform the test the unknown antigen and the antibody with which it is being tested are mixed together and a carefully measured amount of complement (usually guinea pig serum) is added. After allowing time for reaction, 'sensitized' sheep red blood cells are added. If the antigen and antibody have combined they will fix the complement, which will then not be available to take part in lysis of the red cells. The red cells will therefore not lyse, but will settle out of suspension leaving a clear liquid. If, on the other hand the antigen and antibody did not combine, free complement will be available to take part in the lysis. The red cells will then sediment out giving a red coloured supernatant fluid. The test is very sensitive.

### 16.2.4 Antibody Absorption

The interpretation of results of agglutination and precipitation tests needs a little care, especially as more than one antigen-antibody reaction may be going on. A positive result, when the test is properly conducted, shows that the organism has at least one antigen in common with

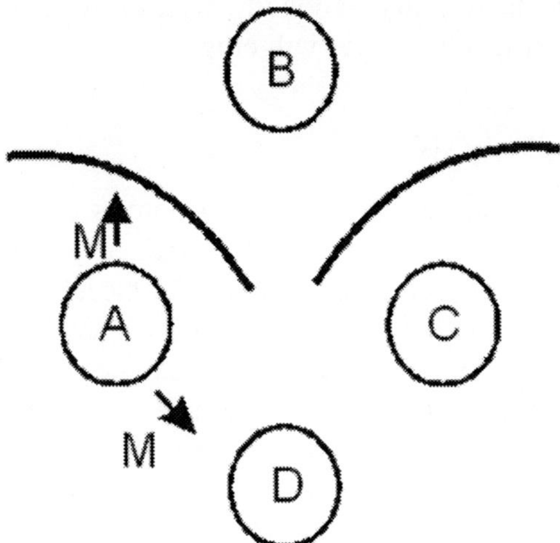


**Fig. 16.2.** Immunodiffusion—Precipitin reaction in agar gel.

the organism used to produce the antiserum, and that antibodies for it are present in the antiserum. A positive result with a second organism indicates that antibodies for it are also present in the antiserum. The two organisms may, however, have different antigens with two different antibodies. The possible serological relationship has not, therefore, been proved.

One technique that has been used with much success to prove and disprove such possible serological relationships is *cross absorption* or mirror absorption of agglutinin. In this procedure, washed cells of organism 'A' are added to a portion of antiserum 'b' (i.e., antiserum prepared using organism 'B'). After several hours at 37°C to allow the reaction to reach completion, the mixture is centrifuged to remove all solids. The supernatant may require a second absorption with 'A' to remove all antibodies that react with this organism, but one absorption is usually enough. Antiserum 'a' (prepared with organism 'A') is cross absorbed with organism 'B' in a similar way. Tube agglutination tests are then carried out on the four possible combinations of absorbed antisera and suspensions.

### 16.2.5 Immunodiffusion

An interesting extension of the precipitation test is to allow the reactants to diffuse through agar gel before coming together. The various antigens and antibodies frequently have different rates of diffusion and, because of this, may precipitate in different positions. In the double diffusion technique, the antigen preparation is separated from the antiserum by agar containing reactant. If more than one antigen-antibody system is present, it is likely that they will precipitate in different 'zones'. This test may be done in a tube, or, as is more frequent in plant virology, in a Petri dish with wells cut into the agar (Fig. 16.2). The latter method has the advantage that several antigen preparations can be tested at one time against one antiserum placed in a central well. Zones produced by identical antigen-antibody systems will then link up if the antigens are in neighbouring wells; non-identical ones will cross.

### 16.2.6 Immunoelectrophoresis

Antigens may be further separated in gels by electrophoresis before the antiserum is added. This is the technique of immunoelectrophoresis. The antigen preparation is placed in a well in a strip of agar about 1 mm thick on a glass slide. An electric current is passed along the length of the strip and the various component antigens will move according to their charges and their molecular sizes. When sufficient separation has occurred (a preliminary experiment may be required to determine this), the current is stopped and antiserum added, usually in a small trough prepared along one side of the strip. Precipitin lines then appear as arcs around the positions reached by the antigens. If desired, the agar strip may be cut in half longitudinally after the electrophoresis, and one half treated as above with antiserum. The other half may be stained with a protein stain to show the positions of the bands of antigens after electrophoresis, or the bands may be cut out and the pure antigens eluted for further study.

### 16.2.7 Fluorescent Antibodies

The conversion of antibody molecules into fluorescent substances can be effected by attaching them chemically to fluorescent organic compounds such as rhodamine B or fluorescein isothiocynate. This does not alter the specificity of the antibody significantly, but makes it possible to detect the antibody adsorbed to cells or tissues by use of the fluorescence microscope. Fluorescent antibodies have been of considerable utility in diagnostic microbiology since they permit the study of immunological reactions on single cells. The fluorescent-antibody technique is very useful in microbial ecology as one of the few methods for identifying microbial cells directly in the natural environments in which they live. Two distinct procedures, the direct and the indirect staining methods, are used. In the direct method, the antibody against the organism itself is fluorescent. In the indirect method, the presence of an antibody on the surface of the cell is detected by the use of another fluorescent antibody directed against the specific antibody itself. This is possible because immunoglobulins, as all proteins, are antigenic, and the immunoglobulins of one animal species can reduce antibody formation in those of another. For this reason, fluorescent goat anti-rabbit immunoglobulin can be used to detect the presence of rabbit immunoglobin adsorbed to cells. One of the advantages of the indirect staining method is that it eliminates the need to make the fluorescent antibody for each organism of interest.

# Environmental Microbiology

## 17.1 INTRODUCTION

The increase in population with concomitant industrial developments is resulting in the discharge of several synthetic chemicals into the environment. Majority of these xenobiotics after release enter soil and water. Some of the chemicals are broken down into simpler molecules by the microbial activity while others remain unaltered in the environment. The long term presence of these chemicals in biosphere poses threat to ecosystem. Many xenobiotics undergo bioaccumulation and biomagnification by primary food producers and thus cause more danger to the ecosystem. The discharge of industrial effluents into water reservoirs and land also poses similar threat to the environment. The oxygen depletion in aquatic ecosystem due to effluent discharge results in loss of aquatic life and makes it unproductive.

Hence, microbial degradation is usually the only means for complete mineralization of xenobiotics in water and soil. The relatively enormous catalytic power, large surface volume ratio and rapid multiplication rate of reproduction of microorganisms contribute to the major role they play in the chemical transformations occurring on the earth's surface. Microbes are also responsible for the maintenance of nitrogen, carbon, phosphorus and sulphur cycles in the environment.

There are certain metabolic processes (e.g., fermentation, anaerobic and chemolithotrophic metabolism, metabolism through exoenzymes and adaptability of microbes to adverse environments through mutation and induction), which take place only in microbes, making them unique tools for biodegradation or detoxification purposes.

## 17.2 PESTICIDE TRANSFORMATIONS

The increasing number and widespread use of herbicides, fungicides, and insecticides has aroused concern from the standpoint of their accumulation in soils and their possible injurious effects on the soil microflora.

Many different kinds of microorganisms, including bacteria, actinomycetes, and fungi, attack stable pesticides in order to obtain energy and carbon for growth. In this respect, the persistence of these compounds in soil will depend to a large extent on conditions, which influence the rate of microbial activity.

The contamination problem of pesticides is generally related to their persistent nature, susceptibility to biomagnification and toxicity to non-targets including higher animals. The major factors deciding the degree of persistence are chemical and physical characteristics of pesticides.

The microorganisms biodegrade/detoxify several chemicals by enzymatic or non-enzymatic processes. In these reactions, the substrate may be metabolized. In some cases, conversion may take place without energy gain. This is analogue induced metabolism or co-metabolism wherein the microorganisms are induced by chemicals, which structurally resemble the molecule of chemical meant for degradation. In this process, the microorganisms can grow on the given chemical, but not on the biodegraded chemical. The initial resistance of xenobiotics to microbial attacks is often overcome during this process. Hence, co-metabolism can offer possible alteration to biodegradation characteristics of many xenobiotics. Microorganisms usually transform compounds by certain reactions which are oxidation, reduction, hydrolysis and condensation.

## 17.3 CHLORINATED HYDROCARBONS

The pesticides of this group (e.g., DDT, lindane, heptachlor, chlordane) pose problem of environmental persistence. They remain unaltered for many years after their application. The important microorganisms that metabolize organochlorines are *Alcaligenes, Pseudomonas putida, Acinetobacter, Aspergillus flavus, Fusarium, Saccharomyes* etc. Sometimes the microbial transformation leads to formation of more stable/toxic compounds, e.g., isomerization of benzene hexachloride (to *a*-isomer) and epoxidation of heptachlor and chlordane.

## 17.4 ORGANOPHOSPHATES

Unlike organochlorines, the organophosphates are short lived in soil and aquatic ecosystems. Microorganisms detoxify this group of compounds through hydrolysis of P-O-alkyl and P-O-aryl bonds. Under low anoxic conditions, some organophosphates undergo reductive transformations to form intermediates, which are easily degraded by microorganisms in pure culture and nature. Diazinon, an organophosphate is utilized by synergistic cultures of *Arthrobacter* sp. and *Streptomyces*, and $^{14}CO_2$ is released from pyrimidinyl ring. None of the species could utilize diazinon individually. A strain of *Flavobacterium* in mineral solution hydrolysed diazinon to IMHP, which under anerobic conditions did not breakdown but under aerobic conditions was converted to pyrimidinol. By ring cleavage it was further broken down to $CO_2$. This strain of *Flavobacterium* hydrolyzed parathion, chloropyrifan and paraxon. The enzyme, phosphotriesterase, responsible for this reaction has been identified and purified. Another organophosphate, fenitrothion was aerobically metabolized to $CO_2$ via 3-methyl, 4-nitrophenol. Parathion, malathion, methyl parathion, paraxon, etc. undergo hydrolysis, reductive demethylation by microbial action. The important microorganisms that bring about these changes are *Pseudomonas, Arthrobacter, Nocardia, Corynebacterium, Trichoderma viride, Flavobacterium* etc.

Carbamates (ester derivatives of N-substituted carbamic acid), which include carbaryl and carbofuran, do not pose serious residue problem in nature and can be easily degraded by microbial action. Hydrolases have been isolated that could degrade different carbamates. Enzymes isolated from *Pseudomonas striata* and *Achromabacter* strain WM III can hydrolyse N-phenylcarbamates and N-methyl carbamates respectively.

Phenoxy alkanoic acid herbicides are metabolized by microbes by an oxidative attack on the ether bond that links the aliphatic acid portion of organic acids. Substituted phenol metabolites are degraded and used as carbon and energy source by organisms.

## 17.5 PETROL AND PETROLEUM PRODUCTS

The spills of crude petroleum and oil, which cause pollution, differ from other environmental pollutants because they are largely insoluble in water and form many coherent masses, which float on the water surfaces or become stranded on shores and thus cause damage at a considerable distance from the point of release. The waste water rising from various units contain oil, grease, phenols, pyridines, and carcinogenic and recalcitrant compounds like paraffins.

Bacteria and yeasts can grow on several fractions of hydrocarbons, e.g., heptane, decane, hexadecane etc. The mode of microbial uptake of these fractions has been well documented. The contact of microbial cell with large oil drops and utilization of submicron droplets play important roles in hydrocarbon uptake by cells. Addition of surfactant enhances the rate of oxidation. Usually, the hydrocarbon surface area is the rate controlling factor in petroleum degradation. The filamentous fungi have usually constitutive metabolism enabling them to grow on hydrocarbons. The utilization of hydrocarbons by these fungi is fortuitous and non-specialized, while yeasts and bacteria have highly modified characteristics induced during growth on hydrocarbons, which enable them to degrade hydrocarbons at a faster pace.

## 17.6 BIOFUELS FROM INDUSTRIAL WASTE

Biomethanation is an economically and environmentally acceptable solution for controlling pollution potential of high strength industrial effluents e.g., distillery stillage, tannery effluents etc. In this process, methane is produced as an end product, which is an alternate source of energy and fuel.

Biomethanation is a multistep reaction mediated through three distinct groups of microorganisms: (i) the heterogeneous fermentative microbes, (ii) the obligate proton reducing or hydrogenic bacteria, and (iii) the methane producing bacteria.

Waste products from thermal power stations, metal and refining industries, electroplating units, sewage sludge, tanneries, etc., often contain substantial amounts of toxic heavy metals. Environmental protection, restriction and environmental awareness has led to the establishment of desired waste water treatment plants.

*Pseudomonas putida*, *Arthrobacter viscous* and *Citrobacter* sp. are employed to remove several toxic metals from industrial effluents. This system is inexpensive, easy to operate and large volumes of aqueous solutions can be treated efficiently.

Removal of radioactive metals by microbial biomass is also feasible. The nonviable biomass of *Rhizopus arrhizus* can accumulate about 180 mg/g dry weight of uranium and thorium. The fungal biomass of *Penicillium chrysogenum* can adsorb 5 × 10 uCi/g of radium at an equilibrium concentration of 1000 pCi/l, which is fourteen times greater than the capacity of conventional adsorbents (Hutehing et al, 1986). The common yeast *Saccharomyces cerevisiae* can accumulate uranium from dilute metal solution (10–100 mg/l) with the capacity upto 20% of their weight.

## 17.7 MICROBES AND WATER-BORNE DISEASES

The human population worldwide suffers from several water-borne disease like typhoid, dysentery, jaundice, cholera, polio etc., caused by microbes present in water. In India, principal bacteria and viruses causing water-borne diseases are *Salmonella typhi, Shigella dysentriae*, Hepatitis virus, enteropathogenic *Escherichia coli* and polio virus. Thus the most important feature of water supply is sanitation or prevention of infection. This needs the detection of pathogenic microorganisms present in water supply, which also helps in designing proper control measure strategies for the prevention of infection. But 'coliform groups' of organisms are considered as the most reliable and widely used indicator organisms of faecal pollution. *Escherichia coli*, one of the coliforms, may be detected by two standard methods: (1) multiple tube fermentation technique (presumptive, confirmed and completed test), (2) membrane filter technique. These tests give the most probable number (MPN) of coliforms. Thus, by knowing the number of coliforms per 100 ml of water sample, one can predict the extent of microbial pollution. Indian Council of Medical Research (ICMR) published a manual in 1975 on Standard of Quality of Drinking Water Supply, giving guidelines for desirable and maximum permissible limits for various organic and inorganic compounds in drinking water. The WHO guidelines for drinking water quality (1984) also enlisted permissible limits for organic and inorganic compounds in drinking water. According to this manual, potable water should be totally free from faecal coliform and contain not more than 10 coliform per hundred ml of water sample.

## 17.8 BIOREMEDIATION

Bioremediation is a process of using living organisms or their principles for decontaminating the unwanted and toxic chemicals from the environment. Microorganisms are known to be versatile and live in extreme habitats, hence are used in the process of bioremediation. Bacteria, actinomycetes, blue green algae, fungi and others are known to co-exist in the soil with metallic ores and take part in natural cycling of mineral elements leading to the formation of deposits and in turn take part in their solubilization. Two broad mechanisms are involved in decontaminating the toxic elements, which are referred to as biosorption and bioaccumulation. Biosorption is a rapid process of metal ion binding to surface cell wall and is independent of cell metabolism. Bioaccumulation is a process, which involves cell membrane transporters and is dependent on energy.

Bioremediation has become important as our living habitats are getting contaminated with toxic elements. Soil, water, air are getting polluted due to over modernization, rapid industrialization, excess cultivation and unlimited exploitation of natural resources. Over utilization of synthetic chemicals in the form of pesticides, herbicides, dyes, detergents, drugs, petroleum products and waste materials etc., and their utilization into environment resulted in an alarming situation. In view of public health and environment concern, these chemicals and their transformation products are reported to be potentially toxic, teretogenic, mutagenic and carcinogenic to man and animals besides disturbing ecological balance of habitat. The biodegradation of such chemicals is influenced by microbial communities of the biosphere. Indigenous bacteria and fungi present in soil and water have versatile mechanism to degrade them. In view of their abilities to degrade and transform complex chemicals into simple compounds, microbial technology can find critical application in treatment of indigenous effluents, sewage and toxic wastes. Toxic metal ions unlike organic pollutants cannot be degraded. Hence, they accumulate in the environment leading to the toxic effects on all living systems. The classic examples of *minimata* disease due to mercury poisoning and *itai itai* due to cadmium toxicity in Japan are well known.

Large quantities of lead have been reported to be let into Thane creek in Mumbai each year. Due to inadequate treatment of such metallic effluents and lack of implementation of regulations to curb pollution by the responsible authorities, they are let out into open drains and sewers, which increase the toxic metal concentrations in water bodies and soil habitats. Heavy metal body poisoning has been reported in West Bengal, Bangladesh and Nepal due to consumption of ground water contaminated with arsenic metal. The bioremediation technology using microorganisms was discovered by G.M. Robinson in 1960, California, USA.

Bioremediation Technology is classified as *in situ* or *ex situ*. *In situ* bioremediation involves treating the contaminated material at the site, while *ex situ* involves the removal of the contaminated material to be treated elsewhere. Some examples of bioremediation technologies are bioventing, landfarming, bioreactor, composting, bioaugmentation, rhizofiltration, and biostimulation.

Not all contaminants are easily treated by bioremediation using microorganisms. For example, heavy metals such as cadmium and lead are not readily absorbed or captured by microorganisms. The assimilation of metals such as mercury into the food chain may worsen matters. Phytoremediation is useful in these circumstances, as many plants are able to bioaccumulate these toxins in their above-ground parts, which are then harvested for removal. The heavy metals in the harvested biomass may be further concentrated by incineration.

### 17.8.1 Genetic Engineering Approaches

The use of genetic engineering to create organisms specifically designed for bioremediation has great potential. The bacterium *Deinococcus radiodurans* (the most radioresistant organism known) has been modified to consume and digest toluene and ionic mercury from highly radioactive nuclear waste.

### Advantages

There are a number of cost/efficiency advantages to bioremediation, which can be employed in areas that are inaccessible without excavation. For example, hydrocarbon spills (specifically, petrol spills) or certain chlorinated solvents may contaminate groundwater, and introducing the appropriate electron acceptor or electron donor amendment, as appropriate, may significantly reduce contaminant concentrations after a lag time allowing for acclimation. This is typically much less expensive than excavation followed by disposal elsewhere, incineration or other *ex situ* treatment strategies, and reduces or eliminates the need for "pump and treat", a common practice at sites where hydrocarbons have contaminated groundwater.

### 17.8.2 Monitoring Bioremediation

The process of bioremediation can be monitored indirectly by measuring the Oxidation Reduction Potential or redox in soil and groundwater, together with pH, temperature, oxygen content, electron acceptor/donor concentrations, and concentration of breakdown products (e.g., carbon dioxide).

Bioremediation of a contaminated site typically works in one of two ways. In the case described above, ways are found to enhance the growth of whatever pollution-eating microbes might already be living at the contaminated site. In the second, less common case, specialized microbes are added to degrade the contaminants. Nonetheless, bioremediation provides a technique for cleaning up pollution by enhancing the same biodegradation processes that occur in nature.

### 17.8.3 Biotechnology and Bioremediation

Various microorganisms are being studied to see if they can remediate various chemicals often present at contaminated industrial sites. Also, scientists are currently looking into genetically engineering certain microorganisms to increase their ability to metabolize specific chemicals, such as hydrocarbons, in contaminated sites. More research needs to be done in order to completely understand the complex microbial processes which make bioremediation possible, especially the bioremediation of metals. Also, researchers are trying to understand why some microorganisms are better at degrading one kind of chemical than another. Bioremediation is frequently used in natural resource industries, like forestry, mining, and energy. The oil industry uses bacteria to clean up pollution created by spills and underground leaks, and to clean up waste products from oil production.

Wastewaters containing dissolved hydrocarbons make up the largest volume of waste material generated by the oil and gas industry. Some parts of hydrocarbons are very toxic and do not break down naturally. This wastewater is sometimes treated in bioremediation lagoons—enclosed ponds containing oil-degrading bacteria—before it goes out into the environment. However, this process is very time-consuming.

To remediate wastewater faster and more efficiently, Saskatchewan researchers developed a reactor that uses oil-degrading bacteria. The reactor, which uses bacteria from the *Pseudomonas* family, consists of a tubular structure with a bacterial film through which the wastewater is filtered.

### 17.8.4 Bioremediating Contaminated Sediment

Organic waste from sewage, pulp and paper mills, and steel/petroleum industries have contaminated many of the world's harbours, lakes, rivers and canals. They affect water quality and aquatic life, sometimes combining with other naturally existing compounds in water to produce foul-smelling toxic gases.

The most common way to deal with sediment contamination is to scoop it up and move it away for long-term storage or treatment. However, these procedures are expensive. As well, excavation disrupts the natural habitat of the aquatic plants and animals, and is not viable as a long-term solution.

### 17.8.5 Sustainable Development and Bioremediation

Bioremediation uses the resources available in nature to clean up contamination. Using biological processes, as in the case of bioremediation, usually means lowered costs compared to chemical treatment processes for various contaminated sites. It is also less disturbing to the environment. However, since it is a natural process, it requires time.

### 17.8.6 Limitations of Bioremediation

There are several limitations to bioremediation. One major limitation has to do with the nature of the organisms. The removal of pollutants by organisms is not a benevolent gesture. Rather, it is a strategy for survival. Most bioremediating organisms do their job under environmental conditions that suit their needs. Consequently, some type of environmental modification is needed to encourage the organisms to degrade or take up the pollutant at an acceptable rate. In many instances, the organism must be presented with low levels of the pollutant over a period of time. This induces the organism to produce the metabolic pathways needed to digest the pollutant. When using bacteria and fungi, it is usually necessary to add fertilizer or oxygen to the material containing the pollutant. This can be disruptive to other organisms when done *in situ.* In situations where simple compounds and metals are being taken up, it is likely that these pollutants are at toxic levels for the organisms. The organisms do not always live, as well grow on the pollutant diet as on other nutrients found more commonly in their environment. This is problematical when doing *in situ* remediation.

Two other limitations concern cost/benefit ratios: cost versus overall environmental impact. Neither the government nor industry wants to spend large amounts of money to clean up pollution. The popularity of bioremediation is further enhanced because it is perceived as being more "green" than other remediation technologies. Companies and individuals are investing in biotechnology futures in spite of the high risks. As a result, bioremediation companies have a viable future regardless of its long-term effectiveness.

Biostimulation refers to the addition of oxygen and/or inorganic nutrients to indigenous microbial populations in soils and groundwater. *In situ* or *ex situ* methods can be employed to stimulate biodegradation of contaminants.

Bioventing is a process of stimulating the natural *in situ* biodegradation of contaminants in soil by providing air or oxygen to existing soil microorganisms. Bioventing uses low air flow rates to provide only enough oxygen to sustain microbial activity in the vadose zone. Oxygen is most commonly supplied through direct air injection into residual contamination in soil.

In addition to degradation of adsorbed fuel residuals, volatile compounds are biodegraded as vapours move slowly through biologically active soil.

Bioventing is applicable to any chemical that can be aerobically biodegraded. Techniques have been successfully used to remediate soils contaminated by petroleum hydrocarbons, non-chlorinated solvents, some pesticides, wood preservatives, and other organic chemicals.

Factors that may limit the applicability and effectiveness of the process include: (1) low permeability soils (reduce bioventing performance); (2) air near the structure of concern has to be extracted in order to avoid vapour build up in basements within the radius of influence of air injection wells; (3) monitoring of off-gases at the soil surface may be required; (4) aerobic biodegradation of many chlorinated compounds may not be effective; and (5) low soil moisture content, which may be caused by bioventing, limits biodegradation.

## 17.9 TREATMENT OF WASTE

Huge amounts of sewage and waste water from industry and domestic habitats are released into atmosphere as effluents. Treatment of such wastes is done by filtration using up flow filters, biofilters and others besides using mechanical aeration process such as oxidation etc. Waste can be used for the production of energy on treatment and recycling. Microorganisms play an important role in fermenting such waste or transforming waste into aliphatic hydrocarbons, methane etc. Treatment of waste waters employing algae and microbes, results in production of a biomass having high proteins and trace elements.

There are different types of waste treatment systems of which biological treatment results in the production of useful materials. Waste from cattle houses on anaerobic fermentation results in methane production. Methanogenic bacteria are found useful in such process. Waste from paper industry, oil industry, oil refineries and plastics pose much problem for biological treatment though some bacteria, fungi, cyanobacteria and others help in their biodegradation.

### 17.9.1 Waste Treatment Under Anaerobic Conditions

Biogas is the resultant product of the anaerobic decomposition of organic material. The important product is methane which is not soluble but gets separated through fermentation. Methane content is around 60–75% with $CO_2$ ranging 25–30% and the remainder being $H_2$ and $N_2$.

The methane producing bacteria namely *Methanococcus, Methanobacterium, Methanospirillum* etc. have been isolated from anaerobic treatment plants.

### 17.9.2 Aerobic Waste Treatment

Treatment of effluents under aerobic system requires a large population of activity metabolizing microorganisms that are able to degrade both colloidal and soluble organisms along with a high rate of conversion to $CO_2$ and water. In most of the systems trickling filters and activated sludge ferments work under aerobic system. The flock of latter system depends on bacterial number including *Zoogloea*. The bacteria belong to *Pseudomonas, Alcaligenes, Achromobactor, Brevibacterium* and others.

Activated sludge treatments are the largest application of continuous or semi continuous fermentation with fresh effluent being added and sludge removed in order to maintain a constant population of microorganisms. The sludge is removed and is passed through settlement tank 1 where it is subjected to anaerobic digestion. Later it is released into aeration tank 2. Excess sludge undergoes anaerobic digestion. The treated effluent is released into river or other habitats.

The activated sludge process is expensive and requires more power for agitation and aeration. The low solubility of oxygen is usually considered to limit the size of microbial population and hence the overall rate of oxidation. The degradation of nitrogen containing compounds gives rise to ammonia as principal product. This ammonia is oxidised to nitrate by chaemolithotrophic nitrifying bacteria is an aerobic reaction. Denitrifying bacteria convert nitrate and nitrite into nitrogen gas. The removal of phosphorus release into liquid waste is an anaerobic process involving *Acinetobacter*.

## 17.10 BIODEGRADATION

It is a process by which the waste materials are degraded by the action of microbial systems. Alternatively the mechanism involves the use of enzymes in the metabolism of waste products. Co-metabolism is the transformation of anion growth substrate in the presence of a growth substrate in the presence of a growth substrate or another transformable compound. Genes coding for some enzymes essential for the biodegradation of a number of organic compounds are plasmid-borne and organisms have been constructed to degrade complex wastes. Genetically modified strain of *Pseudomonas putida* has been found useful to degrade camphor, metaxylene are besides being involved in the cleanup of oil discharges and oil spills (Dr. Chakravarthy's bug). It is also known that several microbes which include bacteria, actinomycetes, fungi etc. are known to decompose plastics, hydrocarbons, biocides, detergents, pesticides etc. Most of the non-biodegradable products which are known as xenobiotics get degraded by several microbes.

The following list gives some microorganisms which have reported to be metabolically able to carry out the biodegradation of some of these compounds (Table 17.1).

### 17.10.1 Microorganisms Able to Degrade Some Toxic Chemicals

A further mechanism which involves the cautious use of existing enzyme systems is co-metabolism. A cometabolite does not support the growth of an organism concerned and the products of transformation are accumulated. The transformation does not require energy consumption of the organism and this is usually measured as an increased uptake of oxygen by the culture. A number of examples of co-metabolism are enlisted in Table 17.2.

**Table 17.1.** Degradation of toxic chemicals by microorganism

| Organisms | Toxic chemicals |
|---|---|
| *Pseudomonas* sp. | anthracene, benzene, hydrocarbons, melathion, naphthalene, organophosphates, PCBs,parathion, phenanthrene, rubber, toluene, phenolics and pulp by products, halogenated hydrocarbons, alkyl benzene, polycyclic aromatics, benzene, hydrocarbons, pentachlorophenol, phenoxyacetates, polycyclic aromatic PCBs. |
| *Alcaligenes* sp. | aromatics, long chain alkenes, phenylureas |
| *Arthrobacter* sp. | halogenated hydrocarbons, phenoxyacetatesaromatics, branched hydrocarbons, benzene, cycloparafins, |
| *Bacillus* sp. | hydrocarbons, alkyl benzenes, naphthalene, phenoxy acetates, polycyclic aromatics |
| *Corynebacterium* sp. | diazinon, phenoxyacetates, halogenated-hydrocarbons |
| *Mycobacterium* sp. | hydrocarbons, polycyclic hydrocarbons |
| *Nocardia* sp. | PCBs |
| *Streptomyces* sp. | PCBs, polycyclic aromaticspropanil |
| *Xanthomonas* sp. | |
| *Candida tropicalis* | |
| *Cunninghamella elegans* | |
| *Fusarium solani* | |

## Examples of Co-Metabolism

The degradation of a wide range of compounds however, has been shown to proceed more readily with a mixed culture of organisms and the utilization of some compounds may not proceed at all in monocultures. A list of examples is given in Table 17.2.

**Table 17.2.** Examples of co-metabolism

| Organism | Growth substrate | Cometabolic substrate | Products |
|---|---|---|---|
| *Methylomonas* | methane | ethane | ethanol, acetaldehyde, acetic acid |
| *Nocardia* | hexadecane | toluene, ethyl benzene | 2,3dihydroxybenzoic acid, phenylacetic acid |
| *Achromobactor* | benzoic acid | 3-chlorobenzoate | 4-chlorocatechol, salicylic acid |
| *Corynebacterium* | hexadecane, glucose | naphthalene, anthracene | 2-hydroxy 3-napthoic acid, 3-napthoic acid |

Some compounds degraded by synergistic microbial communities are dalapon, chlorobenzoatecarathionisopropyl thiocarbamate, benzoatenitrosamines 2-(2-methoxy, 4-chloro) phenoxyalkyl phenol thioxylates, 3, 4-dichloropropionanilide, styreneisopropyl phenyl carbamate, etc.

While most of the functions essential for the growth of a bacterial cell are encoded in chromosomal genes, the presence of plasmids gives the host cell a survival or growth advantage under some conditions. Gene coding for some enzymes essential for the biodegradation of a number of organic compounds are plasmid-borne and organisms have been constructed to degrade difficult waste. For example, a patented process has been developed with a strain of *Pseudomonas putida* constructed to contain plasmids coding for breakdown of octane.

### 17.10.2 Other Technologies

Biodegradation is a process by which the waste materials are degraded due to the action of microbes. Microbes elaborate different enzymes such as cellulases, amylases, proteases, xylanases, tannases, lignases, lipases and others. Biotechnological application may laso help in controlling environmental pollution. It also includes biotreatment of effluents and toxic wastes. The cleaner technologies were patented in 1972 by Provesta Corporation along with Phillips petroleum in USA. Single cell protein forms dietary supplement for animals and human beings.

The present methodologies of pulp bleaching are being replaced by better procedures. Different chemical and mechanical methods are used to release cellulose from lignin. In this connection a fungus, *Phanerochaete chrysosporium* was found to degrade lignin by elaborating an enzyme lignase. Bioreactors have been devised and system has been developed to treat effluents using *Phanerochaete chrysosporium*. The lignin degrading enzyme was found useful to reduce energy costs and corrosion, which will enable in increasing life system and reducing environmental hazards. In paper industry, huge amounts of chlorine get converted into chlorinated organic compounds which are discharged into environment causing enormous damage to the environment. Therefore, biological methods using microbes and ultra filtration combined with biological treatment are adopted. Chlorine dioxide is substituted with lignin degrading fungi.

Alkenes like ethylene and propylene are released into environment from oil based raw material. There is a danger of their escape into atmosphere causing pollution. *Methylococcus capsulatus* have been used for converting alkenes into alkene oxide.

The distillery effluents have huge demand for oxygen. The effluents contain 9–10% of solids and fermentation sludge has 25–30% solids. The disposal of these effluents into environment causes pollution. Biomethanation and secondary treatments are recommended as cleaner technology.

**Table 17.3.** Role of plasmids in transformation of chemicals

| Chemical | Plasmid | Chemicals transformed |
|---|---|---|
| alkanes | OCT | octane, decane |
| aromatic hydrocarbons | TOL (pWWO) | oxylenes, toluene |
| polyaromatic hydrocarbons | NAH | napthalene |
| metabolic products | SAL | salicylate, benzoate |
| terpenes | CAM | camphor |
| alkaloids | NIC | nicotine, nicotinate |
| chlorinated hydrocarbons | pP1 pAC21 | 2,4 D p-chlorobiphenyl<br>3-chlorobenzoate |

It is also recommended to grow some bacteria and algae in such ponds/water bodies which receive distillery factory effluents. These bacteria and algae will extract the heavy metals like mercury, lead and cadmium and purify the distillery effluents. Microbes like *Pseudomonas* strains are isolated from different environments to degrade toxic chemicals and pollutants.

A number of microbial strains could be constructed using recombinant DNA technologies to control oil pollution and to remove oil from water. *Bacillus thuringiensis* is a novel bacterium which helps in genetic engineering. Biological insecticides are prepared by transferring *Bt* gene into naturally occurring strain of *Pseudomonas* so as to control insect pests.

Viral pesticides are also successfully developed and used for the control of pest. These pesticides are environmentally safe and can be used against the pest causing strains which have become resistant to chemical pesticides. Biofertilizers play an important role in the improvement of yield and can reduce the environmental hazards caused by chemical fertilizers. *Azolla, Azospirillum,* plant growth promoting rhizobacteria, mycorrhizae, phosphorus solubilizers, rhizobia and other microbes have been found to be beneficial and non-polluting agents. Recombinant DNA and gene cloning technologies were used to produce good and useful strains from the above bio-inoculants.

Biosensors are the biophysical devises employed to defect the presence and also to quantify specific substance present in a variety of environments. Microbial fertilization of mine soils, waste lands, unproductive and polluted soils resulted in their reclamations.

# Microorganisms of the Soil and Geochemical Cycles

# 18

## 18.1 INTRODUCTION

Soil is a three-dimensional, dynamic, natural body occurring on the surface of the earth, which is a medium for plant growth and whose characteristics have resulted from the forces of climate and living organisms acting upon parent material, as modified by relief, over a period of time.

Using modern methods of science, it is possible to study soil characteristics and the response to soil management practices in the field, greenhouse, and laboratory. Before the general availability of commercial fertilizers, farmers were mostly dependent on the natural nutrients supplying power to soils, supplemented by a small amount, farmyard manure to grow crops. In most parts of Tropical Asia, this condition still exists.

Nitrogen, phosphorus and sulphur in soils are constituents of organic matter. Approximately 95 per cent of the total soil nitrogen, 5 to 60 per cent of the total soil phosphorus, and 10 to 80 per cent of the total soil sulphur are present in the organic matter. These three nutrients become available to plants only after biological decomposition. The nutrient supplying power to the soil for nearly all of the nitrogen and for variable amounts of phosphorus and sulphur is therefore dependent upon the total amount of organic matter and its rate of decomposition in relation to the time that plants require these nutrients. Factors that favour organic decomposition are high temperatures, a favourable oxygen and moisture supply and an abundant supply of nutrients that are required by bacteria.

In addition to organic matter decomposition as a source of sulphur to plants, rain water returns to earth some amounts of sulphur.

The nutrient supplying power of sand and silt is very low because they are composed mostly of relatively undecomposed primary minerals. By contrast, clay is composed of secondary weathered minerals (kaolinite, montmorillonite, and illite) and hydrous oxides of iron and aluminium, and therefore has a greater power to supply nutrients to plants.

Phosphorus is taken into the plant in the water-soluble form from the soil solution. The concentration of phosphorus in the soil solution is then replenished from slowly soluble forms such as calcium phosphate, hydroxyapatite, fluorapatite, iron phosphate, and aluminium phosphate; and from the decomposition of organic matter.

Potassium occurs in the soil mostly as a constituent of primary minerals that exist in sand and silt particles. Minerals such as orthoclase and microcline release potassium slowly, but others such as muscovite and biotite are important sources of potassium, especially in tropical areas such as in Tropical Asia where temperatures and moisture are high. Some of the potassium that is released by the decomposition of minerals stays in the soil solution and is available to plants, and some is adsorbed on the surface of clay and humus particles in a readily exchangeable and available form. Of the total potassium in a soil, approximately 1 per cent will be in the exchangeable form, and of this amount, from 1 to 5 percent will be in the soil solution.

Calcium and magnesium in soils occur in both slowly as well as readily soluble minerals and rocks. Slowly soluble sources of calcium include oligoclase (a sodium and calcium silicate) and anorhite (a calcium aluminiosilicate), whereas readily soluble sources include calcite and limestone. Slowly soluble sources of magnesium include augite, biotite, and hornblende. Readily soluble sources of both calcium and magnesium include dolomitic limestone. Calcium and magnesium are supplied to plants in ionic form in a manner similar to that of potassium, i.e., from rocks and minerals to the soil solution and to an exchangeable form on the surface of clay and humus particles.

The soil is not a dead mass. There is life in the soil. Soil life consists of plant roots, earthworms, microscopic protozoa and nematodes and microorganisms such as bacteria, fungi, actinomycetes and algae. Besides, certain types of mammals (mice), arthropods (mites, millipedes, centipedes), and gastropods (snails) inhabit the soil. The total population of all soil life is numbered in billion per gram of soil, and the live weight per acre may be as much as five tons.

Due to the production of enzymes, carbon dioxide, and organic matter, the life in the soil is responsible for making numerous transformations, which changes plants nutrients to more readily available forms and make and stabilize desirable soil structure for more luxuriant plant growth.

The microscopic plant and animal world of soil is represented by the bacteria, actinomycetes, fungi, algae, and protozoa. While numerous methods have been devised to estimate total numbers of these groups in soil, all have certain inherent limitations. Such methods are, nevertheless, satisfactory for comparative purposes, and there exists a remarkable similarity in the qualitative distribution of microorganisms in different soils, along with distinct differences in their quantitative relationships.

Bacteria are the most abundant as well as the most diversified group of microorganisms. Numerically they are the largest group, ranging from as few as two million per gram in a poor sandy soil to hundreds of millions per gram in a soil where there is active decomposition of organic matter. The major bulk of bacteria is confined to the top 6 inches of soil, and in terms of biomass they have been estimated anywhere from 1 to 2 tonnes per acre. Numerous factors influence the number of bacteria, which are present in a soil at any one time. Thus, the total relative numbers fluctuate widely, depending on the soil type and environmental variables such as soil treatment, temperature, moisture content, aeration, pH, and the organic and inorganic nutrient supplies.

Since bacterial activity in soil is not necessarily related to the taxonomy of the groups involved, it is preferable to classify them from a physiological or functional standpoint. As a major division, bacteria can be divided into *autotrophic* and *heterotrophic* forms on the basis of their sources of energy and carbon. The obligate autotrophs and the less-restricted facultative autotrophs are the most self-sufficient of the bacteria. They are able to fulfil their carbon and energy requirements from carbon dioxide and oxidation of inorganic substances or simple compounds of carbon, respectively. Although autotrophy is confined to a relatively few species of bacteria, the transformations carried out are of immeasurable economic importance, particularly with respect to the oxidation of nitrogen and sulphur compounds to assimilable forms. Other forms are concerned with the oxidation of iron, manganese, hydrogen, and carbon monoxide.

The actinomycetes appear universally in soils, and are second in abundance only to the bacteria. Plating estimates give values of $10^5$ to $10^8$ colonies per gram of arable soils, and these forms make-up from one-tenth to one-half of the total soil population. Despite their lower numbers, the actinomycete population probably equals the bacteria in weight and thus adds appreciably to the biomass. The actinomycetes are not as subject to seasonal variations as the bacteria, but they are influenced markedly by the soil-organic-matter content, moisture level, pH, and temperature. Like the majority of bacteria, the actinomycetes are heterotrophic and thus dependent on organic substrates. In this respect they are very adaptable, and different species are able to utilize a wide range of food materials. They are, however, extremely slow growers, and in view of this characteristic, it is difficult to view them as active competitors with bacteria in the early stages of organic-matter decomposition. The fact that they are found in larger numbers during the later stages of decomposition has suggested that their main role in soil is the breakdown of the more-resistant components of plant and animal tissues remaining after the initial attack. In the same vein, they have also been implicated in the breakdown of the humus fraction of soil with the liberation and recirculation of nitrogen and other essential nutrients in soil.

The fungi, like the majority of the bacteria, are heterotrophic, and therefore their existence in soil is closely connected with the decomposition of organic matter. Standard plate counts place the total fungi between $10^4$ and $10^6$ viable units per gram of soil. Despite their relatively low numbers, the fungi, because of their larger cell size, probably equal the bacteria and actinomycetes in terms of microbial protoplasm. Unlike the bacteria and actinomycetes, the fungi have a high acid tolerance, and under the conditions of extreme acidity, these organisms may become the predominant flora. At less-acid reactions of pH 4.5 through the alkalinity, the fungi are less competitive and their number becomes more dependent on the source of organic matter present. As heterotrophs, the fungi are exceptionally well-equipped to undertake the rapid decomposition of virtually all of the major plant constituents—cellulose, hemicelluloses, pectin, starch, and even the lignins. They also play an active role in transforming the nitrogen constituents of soil. Although noted particularly for their degradative abilities, the fungi also participate in the synthesis of a wide range of complex compounds, a number of which have been implicated in the formation of soil humus.

Algae are universally distributed in soils, but are relatively few in numbers as compared to the bacteria, actinomycetes, and fungi. Numbers range from as few as hundreds to few thousands per gram of soil. The algae are characterized by their photosynthetic pigments and predominantly photoautotrophic nutrition. As photosynthetic forms, they are dependent on sunlight and carbon dioxide, although the later rarely becomes limiting. Their need for sunlight is reflected in the fact that the algae are concentrated in the surface layers, although they have been known to survive at considerable depths. Blue-green algae are also of considerable economic importance in the rice-growing areas. Here their role is the supply of nitrogen through nitrogen fixation and by supplying oxygen to the submerged roots of the plants.

Whereas the bacteria, actinomyctes, fungi, and algae constitute the microflora of the soil, the protozoa are the predominant microfauna present. The protozoa are the simplest forms of animal life and, with few exceptions, are microscopic and acellular. They are universal in their distribution in soils where the population ranges from a few hundreds to half a million per gram. Protozoa obtain their nutritional requirements either from preformed organic and inorganic substrates or by direct feeding upon microbial cells. These latter forms, which predominate in soil, are quite selective as to the organisms they will ingest.

## 18.2 ORGANIC MATTER TRANSFORMATIONS

The organic fraction of the soil is a complex of substances whose composition is determined in part by the plant and animal residues added to the soil, and to a greater extent, by the transformations of these substrates through biological, physical, and chemical means. As a result, a vast array of compounds is present: those representing the components of organic residues undergoing decomposition and the metabolic byproducts of the microorganisms utilizing such compounds as a source of energy. Included in this extremely diverse group one finds carbohydrates and related compounds (mono- and disaccharides, cellulose, hemicelluloses, pectins, pentosans, mannans, polyuronides, uronic acids, organic acids, alcohols hydrocarbons, aromatics); proteins and their derivatives (amino acids, amides, amino sugars, nucleoproteins, purine and pyrimidine bases); and lignins, fats, tannins, and their various decomposition products. In addition to the intermediary products of organic matter breakdown and secondary synthetic products of microbial action, there exists in soil an accumulation of the more-resistant products of decomposition—the soil humus. Microorganisms are implicated in all phases of humus formation. They effect the decomposition of the original plant and animal residues to simpler compounds and, at the same time, are responsible for the alteration of lignin and tannin molecules, which will become part of the structural units of humus.

The presence of humus in soil exerts a pronounced influence on the physical, chemical, and biological activities taking place. Humus improves soil structure, drainage, and aeration, increases the water-holding, buffer, and exchange capacity, influences the solubility of soil minerals, and serves as a source of energy for the development of microorganisms. Humus can also be considered a storehouse of nutrients essential to plant growth. During the slow microbial decomposition of the soil humus, there is a gradual release, with subsequent mineralization, of carbon, nitrogen, sulphur, phosphorus, and other elements immobilized during the formative processes.

About 90 per cent of the dry matter of plants is composed of carbohydrates of varying complexities. Of these, cellulose, hemicelluloses, lignins, and starches form the largest part, with lesser amounts of pectins, inulin, gums, waxes, and related substances. While considering the decomposition of these varied polysaccharides, it is important to realize that the critical sequence in all cases is the initial conversion of the long, complex molecules to simple easily metabolized derivatives. In most instances, these conversions are catalyzed by relatively specific extracellular enzymes.

## 18.3 CARBON CYCLE

Carbon, the major element in organic structures, is metabolized initially in the form of carbon dioxide. Since the carbon dioxide supply in the atmosphere is limited, it is essential that the carbonaceous materials produced by plants and animals eventually become completely decomposed with the formation of this gas. The carbon cycle is concerned, therefore, with carbon dioxide, its fixation, and regeneration (Fig. 18.1). Carbon dioxide is converted to organic carbon mainly through the photosynthetic activities of the higher plants and, to a lesser extent, by microorganisms.

Animals in turn feed upon these preformed carbon compounds, converting them to animal tissue and respiring carbon dioxide in the process. Eventually both plant and animal remains are deposited in the soil where the microbial population plays the dominant role in the continuing cycle. In the soil the residues undergo decomposition through the activities of successive groups of microorganisms. Initially, activity is high as the readily decomposable residues are oxidized to carbon dioxide and water. As the available material decreases, evolution of carbon dioxide lessens until the products of microbial metabolism and the more resistant fractions of the organic residues and carbon is made available with the final, slow decomposition of the soil humus.

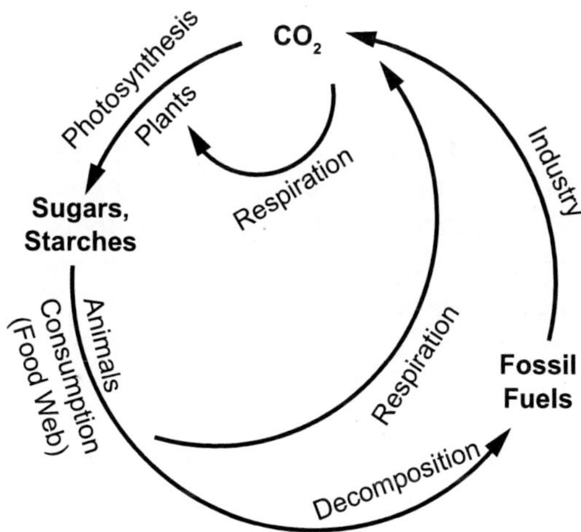


**Fig. 18.1.** Carbon cycle.

Cellulose occurs abundantly in nature, primarily as the principal constituent of the cell walls of most plants. It is often found in close association with hemicelluloses, lignin, pectin, and other polysaccharides. Cellulose is a long-chain polymer of glucose joined by β-1,4-glucosidic linkages and containing between 1500 and 9500 sugar residues.

Cellulose is broken down by the extracellular enzymes of microorganisms, which cleave the β-1,4-glucoside linkages in a nonspecific manner. It has been shown that a single cellulose is capable of hydrolyzing native cellulose to glucose. On the other hand, the dimer cellobiose frequently occurs as the end product of cellulase activity, and it is only through the action of a second enzyme, cellobiase (β-glucosidase), that cellobiose is split into glucose units.

### 18.3.1 Hemicelluloses

Apart from cellulose, the hemicelluloses form one of the major constituents of the native plant cell. Hemicelluloses refer collectively to less water-soluble cell wall polysaccharides, excepting cellulose and the pectins. Two distinct types of hemicellulose are recognized: (a) the polyuronides, composed of repeating units of sugars and uronic acids, and (b) the cellulosans, which are characteristically polymers of hexose or pentose sugars, although occasionally uronic acids are present. In nature, xylose polymers occur most frequently as polyuronides in association with glucuronic acid or as the cellulosan (xylan), made up of repeating units of xylose. While many hemicelluloses are of these types, others have been found to contain arabinose, galactose, mannose, and galacturonic acid.

Enzymatic hydrolysis results in the eventual liberation of the simple sugars or uronic acids of the polymers, all of which are actively metabolized by the soil flora. Frequently, the synthetic activities of these microorganisms give rise to new hemicelluloses of microbial origin in the form of excretion products, capsular material, or other cellular constituents, which are recycled as part of the soil organic fraction.

### 18.3.2 Lignins

Lignin represents the third most abundant constituent of plant tissues, where it is found in the middle lamella in close association with the cell wall carbohydrates. It is presumed that lignins do not occur in the plant in the free state but rather as chemically combined lignin-carbohydrate complexes. Lignins do not represent a uniform family of compounds, as individual lignins differ chemically, depending on their source.

The lignin molecule contains carbon, hydrogen, and oxygen, but the structure is aromatic as opposed to the carbohydrate nature of cellulose and hemicelluloses. Basically, lignin is a polymer built up from substituted derivatives of phenylpropane. Hydrolysis of many plant lignins frequently yields structures such as vanillin, syringaldehyde, coniferaldehyde, and p-hydroxybenzaldehyde. It is generally accepted that the basic units of lignin are typified by a phenylpropionic structure containing hydroxyl and methyoxyl groups. Polymerization of these basic units results in the formation of chain-like molecules, which, in turn, are condensed to form the complex branched structures found in native lignin.

It has been pointed out that in the formation of humus, the lignin molecule undergoes modification. It has been suggested that the lignin molecule is first split oxidatively with the formation of its primary structural groups (vanillin, syringaldehyde, etc.). The compounds then lose their methoxyl groups and are oxidized to quinines, which are then polymerized and condensed with amino acids to form humic acids. Many soil fungi are capable of oxidizing vanillin, ferulic acid, syringaldehyde, etc. Similarly, many soil bacteria are capable of decomposing a wide range of aromatics with an ultimate splitting of the ring structure resulting in the formation of easily metabolized, low-molecular-weight organic acids.

### 18.3.3 Starches

Starches are the most common reserve food materials of plants and are found principally in the seeds, fruits, tubers, roots, and stems. They vary considerably in their physical properties when isolated from different species, although chemically they have definite similarities. Starches are not generally single substances, but rather mixtures of two structurally different polymers of glucose. The amyloses are linear, thread-like molecules in which glucose units are joined by α-1,4 glucosidic linkages, while in the amylopectins, the glucose is joined by the same linkage, but the molecules are branched with side chains attached through α-1-6-glucosidic bonds.

Starches are rapidly decomposed by many soil microorganisms through the elaboration of extracellular enzymes known collectively as amylases. The enzyme β-amylase completely hydrolyzes amylose to maltose, but stops at the 6-1 branch points of amylopectin, leaving a residual unhydralyzed dextrin. The α-amylases are characterized by their reduction of the starch molecule to large units containing approximately six glucose residues. Neither the α- nor β-amylases can hydrolyze the α-1, 6 linkages—a process which is carried out by special debranching enzymes. The final product of starch hydrolysis, maltose, is converted by α-glucosidase to glucose which is rapidly metabolized.

### 18.3.4 Pectins

Pectins are found most abundantly in the intercellular layers and in the primary and secondary cell walls of all higher plants. The molecules of these substances are polymers of galacturonic acid units in which some of the carboxyl groups are esterified with methyl groups. Pectins are commonly broken down by two distinct enzymes: pectinesterase, which hydrolyzes the methyl ester linkage to yield pectic acid and methanol; and polygalacturonidase, which destroys the linkage between the galacturonic acid units and either pectin or pectic acid to release galacturonic acid.

Many bacteria, fungi, and actinomycetes are capable of bringing about the rapid hydrolysis of pectic substances in soil. The metabolic activities of these organisms are important in the retting of flax and in the decay of fruits and vegetables. In addition, they have been implicated in the wilt disease of plants.

### 18.3.5 Chitin

Chitin differs from the majority of polysaccharides found in soil in that its basic unit consists of an amino sugar. Chitin is found in the cell walls of lower plants and fungi, but the major source in soil is in the form of insect remains. The structure of chitin is similar to that of cellulose (β1,4-linked chain of hexose molecules), but in place of glucose, N-acetylglucosamine serves as the repeating unit. Decomposition of this material is through the elaboration of extracellular chitinase. Some evidence suggests that only one enzyme is implicated in the formation of N-acetylglucosamine. On the other hand, the frequent formation of N, N'diacetylchitobiose necessitates the action of a second enzyme, chitobiase, for the final hydrolysis. N-acetylglucosamine is readily metabolized by the soil flora.

## 18.4 NITROGEN CYCLE

The availability of nitrogen is of prime importance to growing plants, since they are dependent on an adequate supply of nitrate and ammonia for the synthesis of their nitrogenous constituents. Animals, in turn, depend on plants for their energy and synthetic processes. When plant and animal remains are added to soils, their nitrogenous constituents undergo numerous transformations (Fig. 18.2).

Mineralization of nitrogen is concerned with the degradation of the organic nitrogenous compounds of freshly added residues, as well as those in organic complexes of the soil humus. In mineralization, the initial reduction to $NH_4$ constitutes ammonification, while the

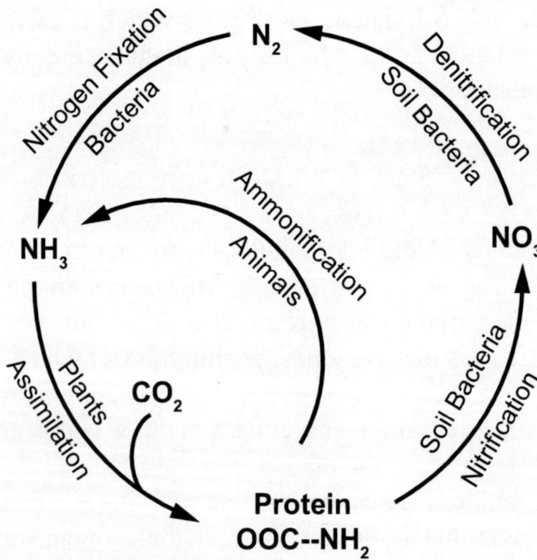


**Fig. 18.2.** Nitrogen cycle.

oxidation of this compound to nitrate is termed nitrification. In opposition to mineralization, a large proportion of the ammonia and nitrate may be immobilized through the biological activities of microorganisms utilizing these compounds in the synthesis of protoplasm. Eventually, much of this protoplasm and the nitrogenous by-products of metabolism enter the stable soil humus fraction. In view of the fact that active mineralization implies active growth of the soil flora, it is clear that the two processes—mineralization and immobilization—occur simultaneously.

Nitrate may be lost from the soil through biological denitrification, whereby nitrates are reduced to free nitrogen, which is unavailable to plants and most microorganisms. Certain groups of microorganisms have the ability to fix atmospheric nitrogen. The first group, or symbiotic nitrogen-fixers, are able to utilize free nitrogen when growing in symbiosis with the roots of certain plants, while the nonsymbiotic organisms are able to carry out this process independently. In the later instance, nitrogen gas is transformed and incorporated as microbial tissue, which is eventually mineralized.

Since the bulk of nitrogenous compounds in soil are protein in nature, their decomposition constitutes a major phase of the mineralization process. In soil, the degradation of proteins is accomplished by innumerable heterotrophic microorganisms. Very few organisms fail to utilize some complex form of nitrogen, and it is uncommon to find that nitrogen is not mineralized even under the most extreme conditions. As with most macromolecules, proteins are broken down through the elaboration of extracellular enzymes. Proteolytic enzymes attack the protein molecule by hydrolyzing the peptide linkages. The initial products of hydrolysis are long, amino acid chain fragments or peptones, which, on continued hydrolysis, yield peptides and eventually free amino acids. Amino acids may then be: (a) metabolized by microorganisms (immobilization), (b) transformed by microbial enzymes with the formation of ammonia, (c) adsorbed by clay minerals or incorporated in the humus fraction, and (d) utilized by plants. The amount of free amino acids that can be detected in soils is very small, attesting to the fact that these compounds are rapidly immobilized or transformed. Ammonia, in turn, can be: (a) oxidized to nitrates (nitrification), (b) utilized by higher plants, (c) immobilized by non-nitrifying organisms, and (d) adsorbed by clay minerals or organic matter.

In addition to proteins, nucleic acids provide a major portion of the nitrogenous substances available to microorganisms. While small amounts of these compounds are derived from plant and animals tissues, they are to a large extent of microbial origin. Both ribonucleic acid (RNA) and deoxyribonucleic acid (DNA) are polymers of mononucleotides, which consist of a purine or pyrimidine base, a pentose sugar, and phosphate. Many bacteria and fungi elaborate extracellular deoxyribonuclease and ribonuclease, which cleave the large polymers of nucleic acids into their constituent mononucleotides. Dephosphorylation of mononucleotides with subsequent hydrolysis of the ribosides or deoxyribosides yields sugar and free purines or pyrimidine bases. These nitrogenous bases are readily metabolized by a variety of soil organisms to form ammonia, amino acids, organic acids, and urea. Urea may also enter the soil as fertilizer or as an excretory product of animal metabolism. Again a wide range of microorganisms is capable of elaborating the enzyme urease, which hydrolyzes urea with the eventual liberation of ammonia.

### 18.4.1 Nitrification

The capacity of soil microorganisms to convert ammonia to nitrate is of utmost significance in providing growing plants with this assimilable form of nitrogen. The oxidative sequence involved in the conversion of ammonia to nitrate is termed nitrification. In soil this is generally brought about by two highly specialized groups of aerobic autotrophic bacteria, which derive their energy from the oxidation of their specific inorganic nitrogen compounds. The first group, the *Nitrosomonas*, oxidizes ammonia to nitrite; the second group, the *Nitrobacter,* oxidizes nitrite to nitrate.

*Nitrosomonas:* $NH_4^+ + 1\ \frac{1}{2}\ O_2 \rightarrow NO_2^- + 2H^+ + H_2O$

*Nitrobacter:* $NO_2^- + \frac{1}{2}\ O_2 \rightarrow NO_3$

### 18.4.2 Denitrification

Denitrification, the most important of these transformations, is of considerable economic importance and, under adverse conditions, may account for significant losses of nitrogen from the soil. The processes are: (a) nitrate assimilation or assimilatory nitrate reduction, and (b) nitrate respiration or dissimilatory nitrate reduction. Nitrate assimilation denotes the biological reduction of nitrate to the ammonia or amino level, with the products being used for the biosynthesis of nitrogen containing cellular constituents. In nitrate respiration, nitrate is used as the terminal electron acceptor in place of oxygen by a variety of microorganisms under anaerobic conditions. Nitrate respiration is carried out by many microorganisms in the absence of oxygen, and two distinct groups have been noted. The first group includes those organisms, which reduce nitrate to nitrite and allow the product to accumulate; the second group, or true denitrifiers completely reduce nitrate to nitrogen and nitric or nitrous oxides. Thus, the term denitrification has been reserved to describe microbial processes whereby nitrate is reduced to gaseous products. Both nitrate assimilation and nitrate respiration involve reductive pathways. The end products of nitrate respiration (denitrification) are, however, volatilized and lost from the soil, whereas the process of nitrate assimilation results in the nitrogen remaining in the soil as microbial protoplasm.

The organisms responsible for denitrification in soil are all facultative species, which utilize nitrate as an alternate electron acceptor in the absence of oxygen. In most instances, these organisms will oxidize the same organic substrates under either condition. The capacity for denitrification appears limited to certain bacteria, and so far no fungi or actinomycetes have been implicated.

Included among the denitrifiers are a number of autotrophic organisms. *Micrococcus denitrificans,* a facultative autotroph, develops anaerobically while oxidizing hydrogen at the expense of nitrate. *Thiobacillus denitrificans* oxidizes sulphur while reducing nitrate in the absence of oxygen.

### 18.4.3 Nitrogen Fixation

Plant growth, leaching, and the activities of denitrifying organisms all tend to remove the soil nitrogen necessary for continued existence. Although the nitrogenous constituents of plants are decomposed and again made available, denitrification would eventually deplete the soil of this element if nitrogen gas were biologically inert. Fortunately, atmospheric nitrogen can be used by a number of free-living microorganisms and in symbiotic associations with higher plants. In this way large amounts of free nitrogen are initially fixed into microbial protoplasm to be recycled eventually in the soil.

The free-living microorganisms capable of utilizing molecular nitrogen include, in addition to the classical *Azotobacter* and *Clostridium* species, a wide range of heterotrophic bacteria, autotrophic organisms, various photosynthetic bacteria, and a number of blue-green algae.

The classical example of symbiotic nitrogen fixation is the association between bacteria of the genus *Rhizobium* and plants of the leguminosae. Most legumes differ from other plants in that they use, in addition to the soluble nitrogen of the soil, the atmospheric nitrogen made available by *Rhizobium* species living in symbiosis with them in root nodules. Mature nodules consist of a central core containing non-dividing resting cells of *Rhizobium* (bacteroids) surrounded by host tissue. It is this central area of bacterial tissue that is the site of nitrogen fixation.

Symbiotic fixation is not confined to plants of the legume family. Eight genera of the angiosperms, comprising some 190 species of widely distributed trees and shrubs, also produce nodules. Investigation of a number of these species revealed that they do fix appreciable quantities of atmospheric nitrogen and that this fixation is confined to the nodules.

## 18.5 PHOSPHORUS TRANSFORMATIONS

Phosphorus is an essential constituent of every living cell. It is concerned with structural compounds, nucleic acids for reproductive purposes, and the conservation and transfer of energy in the metabolic reactions taking place. Phosphorus ranks next to nitrogen in importance, and its availability to the growing crop is of prime importance in soil fertility (Fig. 18.3).

Inorganic phosphorus compounds found in soils are, to a large extent, insoluble in the soil solution. These compounds range from the relatively soluble mono- and dicalcium phosphates, which are present in low concentrations, to the relatively insoluble fluorapatite and hydroxyapatite. Iron and aluminium phosphates predominate in neutral soils.

The result of many laboratory and greenhouse studies have implicated the activities of soil microorganisms in the solubilization of inorganic soil phosphates. In general, solubilization is due to microbial production of organic or inorganic acids, which effectively dissolve inorganic phosphates to the soil flora and growing plants. In the case of heterotrophic organisms, the amount of phosphorus brought into solution varies with the carbon substrate used. Autotrophic organisms, primarily concerned with the oxidation of sulphur and nitrogen, are also able to solubilize inorganic phosphates through the formation of sulphuric and nitric acids.

Many soil microorganisms produce lactic, glycolic, citric, formic, and acetic acids, which can solubilize compounds such as tricalcium phosphate, synthetic apatite, and natural apatites.

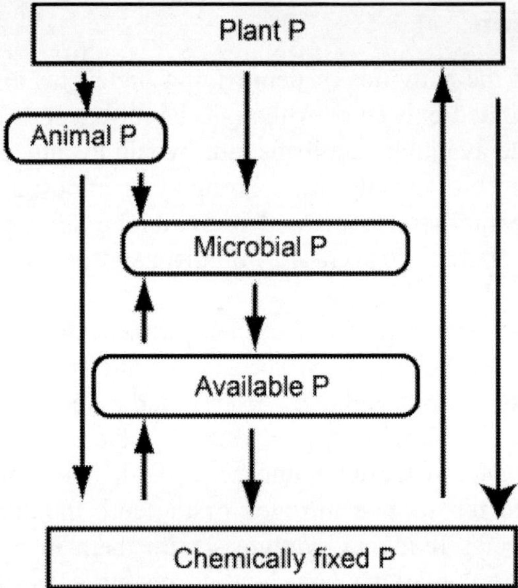


**Fig. 18.3.** Phosphorus cycle.

A number of other organic acids, including 2-ketogluconic acid, are effective chelating agents capable of complexing with calcium, copper, nickel, manganese, iron, and aluminium salts of phosphates with resultant solubilization of phosphorus.

The large proportion of phosphate dissolving organisms present in soils (10 to 15 per cent) and in the rhizosphere of plants implies that these organisms may play a significant role in the phosphorus economy of the soil.

## 18.6 SULPHUR TRANSFORMATIONS

The cycle of sulphur in nature is very similar to that of nitrogen. Sulphur compounds undergo many transformations as a result of activities of plants. Plants build up their sulphur-containing components primarily through the assimilation of sulphates from soil, although small amounts may be obtained through the absorption of S-amino acids or the direct assimilation of sulphur dioxide from the atmosphere. Animals in turn utilize plants or their residues, converting the reduced sulphur compounds to sulphates for incorporation into organic molecules. When plant and animal remains are returned to the soil, they are decomposed by microorganisms with the liberation of sulphur in inorganic forms. These inorganic sulphur compounds may be oxidized to sulphates or reduced to hydrogen sulphide by many physiologically diverse organisms.

The cycle of sulphur in soil can be divided into a number of distinct phases: (a) mineralization—the decomposition of large organic molecules into simple inorganic sulphur compounds; (b) immobilization—the assimilation of these compounds into microbial tissue; (c) oxidations—the conversion of inorganic forms of sulphur (sulphides, thiosulphates, polythionates, and elemental sulphur) to sulphate; and (d) reductions—the conversion of sulphates and other oxidized forms of sulphur to hydrogen sulphide (Fig. 18.4).

### 18.6.1 Inorganic Sulphur Oxidation

The formation of sulphate and other sulphur compounds from more reduced forms of sulphur, both organic and inorganic, is a common characteristic of numerous microorganisms. Four groups have been implicated in these oxidations: (a) the chemoautotrophic bacteria of the genus *Thiobacillus;* (b) a poorly defined group of heterotrophic bacteria, fungi, and actinomycetes; (c) the colourless filamentous bacteria of the genera *Beggiatoa, Thiotrix,* and *Thioplaca;* and (d) photosynthetic sulphur-bacteria belonging to the families Thiorhodaceae and Chlorobacteriaceae.

### 18.6.2 Reduction of Inorganic Sulphur

The reduction of sulphate and incompletely oxidized inorganic sulphur compounds to hydrogen sulphide is relatively unimportant in well-aerated arable soils. Under these conditions, many heterotrophic organisms produce sulphides during the decomposition of organic matter or through the reduction of inorganic sulphur compounds. Sulphides in turn are assimilated into

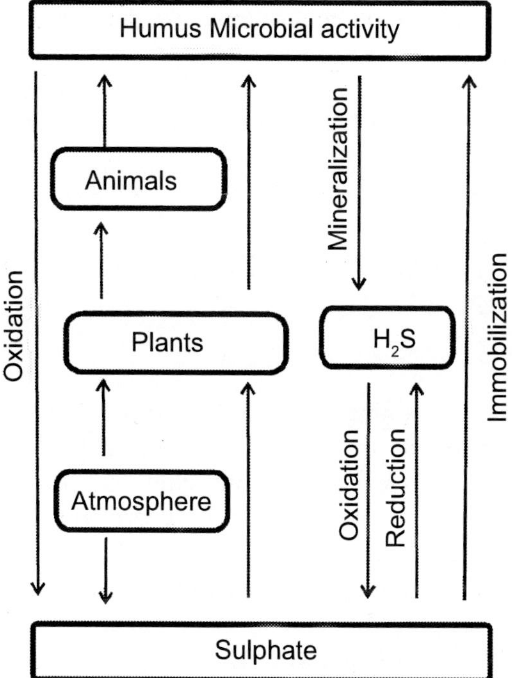


**Fig. 18.4.** Sulphur cycle.

microbial tissue or oxidized. Under anaerobic conditions, such as in waterlogged soils, hydrogen sulphide will accumulate at the expense of oxidized forms of sulphur. The most active agent in sulphate reduction is the bacterium *Desulfovibrio desulfuricans.*

## 18.7 TRANSFORMATION OF OTHER ELEMENTS

In addition to the numerous transformations of N, S, and P achieved by microorganisms, many of the minor nutrients necessary for plant nutrition are also affected by the metabolic activities of soil organisms. In many instances, the availability of iron, manganese, potassium, selenium, tellurium, arsenic, zinc, copper, boron, calcium, magnesium, aluminum, and molybdenum is also affected, to varying extents, by microbial action. Transformations of these compounds are brought about by the same general types of reactions as for the major elements, i.e., mineralization as a result of decomposition of organic plant and animal residues; immobilization due to the reduction of oxidized forms of the elements, or assimilation and immobilization in microbial protoplasm; increased availability resulting from the oxidation of reduced forms or vice versa; and chemical transformations occurring indirectly through the production of organic acids or chelating agents. Iron receives preferential treatment since a number of microorganisms are specifically involved in its oxidation. The iron bacteria that deposit insoluble ferric hydroxide through the oxidation of the ferrous form are primarily aquatic organisms and range from autotrophic bacteria, which depend on this oxidation for energy, to miscellaneous heterotrophs in which immobilization of iron is a secondary process. Under anaerobic conditions, sulphate-reducing organisms may also bring about a concomitant reduction of iron to ferrous sulphate.

On the other hand, the production of organic acids by certain bacteria often results in the formation of soluble iron complexes. Similarly, ferric iron may be reduced to the available ferrous form by certain bacteria, which utilize the ferric ion as an electron acceptor at suboptimal oxygen levels.

## 18.8 MICROORGANISMS AS FERTILIZERS AND PLANT PROTECTION AGENTS

### 18.8.1 Biofertilizers

Biofertilizers are the products containing living cells of different types of microorganisms that have an ability to mobilize nutrients from non-usable to usable form through biological processes. These broadly include nitrogen fixers (both symbiotic as well as non-symbiotic bacteria), phosphate solubilizing bacteria and fungi and mycorrhizal fungi, capable of mobilizing non-liable nutrients and transporting metals to and across the plant roots. On global basis, biological nitrogen fixation (BNF) potential has been estimated at 175 Mt/ha/year as against 70 Mt N fixed chemically. The figures of some of the commonly used biofertilizers and biocontrol agents are given at the end of this chapter.

Conjoint application of biofertilizers, chemical fertilizers and organic manures, in addition to inclusion of legumes in cropping systems and incorporation of on- and off-farm generated crop residues constitutes an efficient nutrient management strategy. The use of efficiency of biofertilizers is highly crop-location and management-specific and requires a reliable system of quality control and efficient system of storage, transportation and management system in the field. Use of *Rhizobium* cultures in legumes and blue-green algae in rice is picking up and is expected to supplement substantially for the fertilizer-N.

## 18.8.2 Nitrogen Fixers

### *Rhizobium*

Nitrogen fixation by *Rhizobium* in root nodules of legumes is of the order of 14 million tonnes on a global scale and almost half that of industrial N-fixation, which has been estimated to be 30 million tonnes per year (Fig. 18.5).

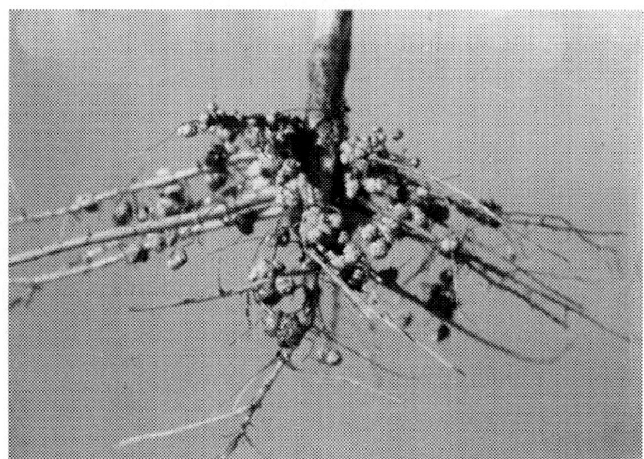

**Fig. 18.5.** Root nodulation by rhizobia in legumes forming tumour-like growths called nodules which acts as factories of ammonia production.

It is now known that the yield of pulses and oilseed crops can be stepped up substantially by the use of rhizobial culture as shown by the extensive field studies in different parts of the country.

Legume inoculation sometimes proves unsuccessful in acid and alkaline soils. In order to protect rhizobia from effects of acidity and alkalinity, seed pelleting is usually recommended. Among the various pelleting materials, lime seems to be quite promising under acid soil conditions. In alkaline soils, use of gypsum as a seed pelleting agent has been advocated. Other pelleting agents like rock phosphate, talc, bentonite clay, charcoal, etc., have also been reported in literature.

Legumes are known to leave behind some residual nitrogen in soil. The extent of benefit is normally judged by the yield of subsequent non-leguminous crop. A well inoculated soybean crop leaves as much as 40 kg N/ha for subsequent crop (wheat) under Indian soil conditions.

The medium in which rhizobia are allowed to multiply is an important factor in rhizobial culture production. The term 'carrier' is generally used for this purpose. The carrier-based cultures have almost replaced liquid cultures and agar slope cultures, which were commonly used a few years ago.

Among various carrier materials, peat was found to be superior. Peat-based cultures offer some outstanding advantages such as increased protection for rhizobia when in contact with acid fertilizers because of high organic matter content, which acts as a buffering agent. In India, high quality peat is not available although peat-like material of medium quality, designated as peat soil, is located in pockets of Nilgiri valley in Tamil Nadu and is being used a career for *Rhizobium*. Lignite, a low grade coal available in Neyveli, Tamil Nadu has also been reported to be a suitable carrier for *Rhizobium* and is currently being used commercially. Finely powdered charcoal is also used in some places. The availability of suitable carrier in sufficient quantity is often a pressing problem for practical adoption than the superiority of an individual career or a combination of carrier, which has been adjudged as an ideal material.

There exists a great variation among strains of rhizobia with regards to nitrogen fixing capacity, competitive ability and ability to tolerate environmental stresses like temperature, moisture, soil N, acidity, salinity and alkalinity. Selection of strain for being exploited as commercial inoculant is therefore of utmost importance. The selection process should be rigorous and should include large pool of genetic variants. Lack of inoculation responses when large populations of rhizobia are present is largely due to the very limited strain selection, which has been done. Even in the case of some major legume crops, a 'recommended' strain may simply represent a strain selected as the best nodulating and $N_2$-fixing strain under axenic, glasshouse conditions compared with a limited number of isolates collected from different regions or countries. Only very few strains ever reach the field testing stage in rhizobial assessment. The selected rhizobial strains should be evaluated in field soils, which had large indigenous populations of rhizobia in order to give a response in plant nitrogen yield. An inoculant strain had to be a good competitor, be well adapted to hostile soil environment and of course be a good $N_2$ fixer.

### 18.8.3 Free-Living Nitrogen-Fixers

#### *Azotobacter*

In India field experiments with various crops like wheat, rice and vegetables are in progress since 1963. Some of the results of field trials with various crops using *Azotobacter chroococcum* cultures indicate positive responses to inoculation resulting in 10–15 % increase in yields of wheat, mustard, maize, cotton etc. These beneficial effects of *Azotobacter* inoculation appear to be qualitatively real and are attributable to the multiple action of *Azotobacter*, namely, the fixation of atmospheric nitrogen, synthesis of growth promoting substances and antifungal substances.

#### Am Fungi

Arbuscular mycorrhizae or Vesicular arbuscular mycorrhiza are omnipresent and are being studied extensively throughout the world. Mycorrhizal associations are known to improve the growth and yield of crops in nutrient-deficient conditions. They have an important role in economy and efficient use of phosphatic fertilizers and in improving nitrogen fixation. Many of the AM fungi inhabit the roots of majority of plant species resulting in a dual beneficial association from nitrogen fixers as well as AM fungi. Lack of suitable inoculation production technology is the major limitation for commercial exploitation of the system. Therefore, AM inoculation is not congenial for field crops since the requirements of inoculum is very high. These cultures hold good for tree-legumes, vegetable crops and fruit crops (Fig. 18.6).

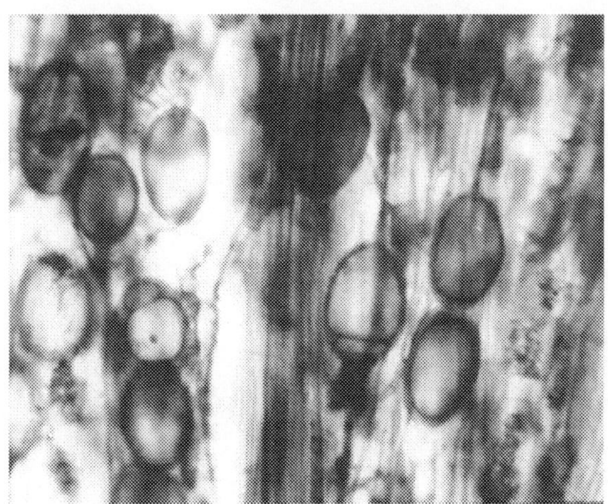

**Fig. 18.6.** Root infection by arbuscular mycorrhizal fungi.

### *Azospirillum*

The crops that respond the *Azospirillum* inoculation are maize, barley, sorghum, pearl millet, forage and other crops. *Azospirillum* inoculations increase grain productivity of cereals by 5–20 per cent, millets by 30 per cent, and fodder by over 50 per cent. Apart from nitrogen-fixing ability, *Azospirillum* is known to produce auxins, cytokinins and gibberellins (Fig. 18.7).

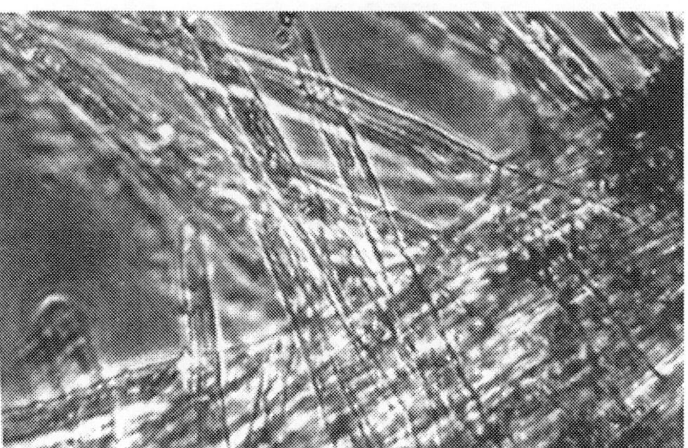

**Fig. 18.7.** Presence of *Azospirillum* in root hairs as endorhizosphere bacteria.

## Blue-Green Algal Biofertilizer for Rice

A group of free-living microorganisms, commonly known as blue-green algae like *Nostoc, Plectonema, Oscillatoria, Gleocapsa* etc. has been demonstrated to be an ideal candidate for biological nitrogen input in rice ecosystems. They have the ability to carry out both photosynthesis and nitrogen fixation, the former process meeting the energy bill of the later. A farm-orientated algal biofertilizer technology for rice has been developed in our country, which involves introduction of specific strains of blue-green algae and their decentralized production by the small and marginal farmers (Fig. 18.8).

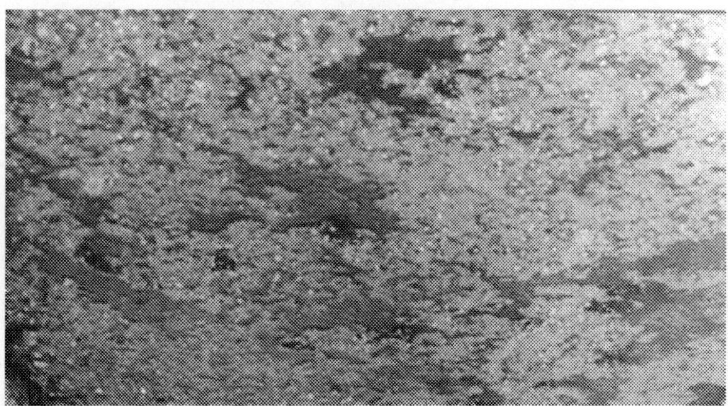

**Fig. 18.8.** Cyanobacterial (BGA) blooms in ponds.

### *Anabaena-Azolla symbiosis*

*Azolla*, a water fern is commonly seen in low land rice fields and in shallow freshwater bodies. This fern harbours a blue-green alga, *Anabaena azollae*. The *Anabaena-Azolla* association

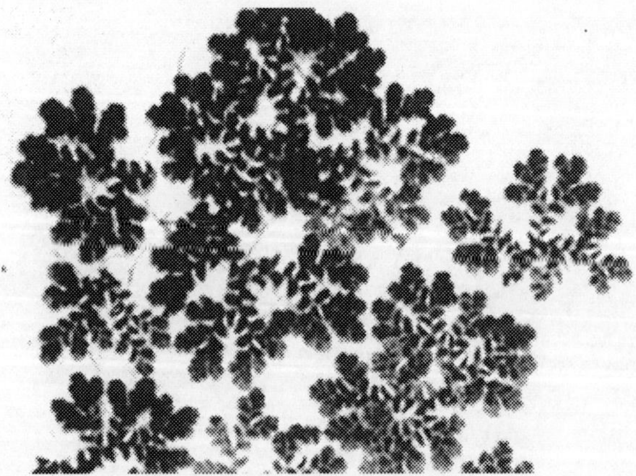

**Fig. 18.9.** *Azolla*, a water fern in waterlogged rice fields.

is a live floating nitrogen factory using energy from photosynthesis to fix atmospheric nitrogen amounting to 100-150 kg per hectare per year from about 40-60 tonnes of biomass. Its use as a bioinoculant for rice is reported. An integrated system of rice-*Azolla*-fish has also been advocated. Greater thrust need to be given on this system (Fig. 18.9).

### Phosphate Solubilizers

Field experiments conducted with P-solubilizers like *Aspergillus awamori*, *Pseudomonas striata* and *Bacillus polymyxa* significantly increased the yields of various crops like wheat, rice, cowpea etc. In the presence of rock phosphate, a saving of 30 kg $P_2O_5$/ha with the use of phosphate solubilizing microorganisms has been reported.

## 18.9 MICROORGANISMS IN PLANT PROTECTION

The microorganisms that are exploited for insect control purposes are relatively few in our country. The *Bacillus thuringiensis* research and use have picked up only after the ban was lifted due to their pathogenicity to silkworms. It is now being marketed in our country and pests like *Plutella xylostella* and *Heliothis armigera*, that have developed resistance to many pesticides could be controlled by *B. thuringiensis* products.

Over 500 insect pests belonging to the order Coleoptera, Diptera and Lepidoptera against which various strains of *B. thuringiensis* could be used. There is a possibility of isolation and development of novel indigenous strains of *B. thuringiensis* against the crop pests. Since it is a facultative pathogen, the technique of production through fermentation technology is standardized. Now this bacterium is commercially available as dust, wettable powder and flowable formulations under different trade names. The *B. thuringiensis* products are registered for use in all parts of the world and it amounts to more than 80 per cent of microbial insecticides sold and the industry is keen on developing *B. thuringiensis* products.

The baculoviruses are currently developed for controlling specific insect pests of agriculture and forestry. These are obligate pathogens and are produced using laboratory reared host insects. There are nearly 40 insects pests on which viral pathogens are recorded in our country. Out of these, the baculovirus of two notorious pests *H. armigera* and *Spodoptera litura* are being tried on large scale in many parts of our country. The production techniques of these viruses and standardized and few biotechnology firms and private entrepreneurs are manufacturing and selling to a limited extent. The field trials on various crops have resulted in adequate control comparable to synthetic chemical insecticides.

The other group of microorganisms that is of importance are fungi. Besides, they cause natural epizootics, some of these could be exploited for pest control purposes. The fungal pathogens like *Beauveria bassiana, Metarrhizium anisopliae, Nomuraea rileyi* and *Verticillium lecani* are already commercially available in countries like USA, UK, Europe etc. These fungal pathogens are produced with minimum investment. Fungi like *Trichoderma viride* and *T. Harzianum* are the efficient biocontrol agents of many bacterial and fungal diseases which are soil-borne and root-borne (wilts, root rots etc.). Further, commercialised products of *Trichoderma* such as *Tina-T* is available in European market.

*B. thuringiensis* and baculoviruses molecular biology have led to either development of transgenic crops or recombinant strains with increased efficacy.

### *Pseudomonas fluorescens*—a biocontrol agent

*Pseudomonas fluorescens* is well known for biocontrol, plant growth promotion, and recognized as an important plant growth promoting rhizobacteria (PGPR) throughout the world. It has been roughly estimated that about 2.5% of the bacteria adopted to rhizosphere environment are PGPR and *P. fluorescens* dominates the group. The widely recognized mechanisms of biocontrol mediated by these bacteria are competition for an ecological niche, food, production of inhibitory allelo-chemicals and induction of systemic acquired resistance in host plants against a broad spectrum of pathogens and also in increasing the endurance of plants to various abiotic stresses (Fig. 18.10).

**Fig. 18.10.** Control of cotton wilt by *Pseudomonas fluorescens*—a biocontrol agent.

## Bacteria Parasitizing Nematodes

So far, only three species of *Pasteuria* (*P. penetrans, P. thornei* and *P. nishizawae*) have been identified to parasitize important group of nematodes. They are obligate and host specific parasites. Of the three species, *P. penetrans* alone has been the subject of intensive studies because of its wide prevalence in more than 50 countries including India. In India *P. penetrans* has been observed parasitizing juveniles of root-knot and cyst forming nematodes. The isolates of *P. penetrans* have been found to be very effective in inhibiting the egg production, reducing nematode penetration and improving the plant growth. Maximum reduction in multiplication and penetration of *Meloidogyne incognita* on tomato was recorded at a level of 3.5 g parasite per kg. Similarly, studies conducted on *Heterodera cajani* revealed that spore load of 510 spores/juvenile cause of 86.7% in cyst population and 96.5% in juvenile population of *H. cajani* in soil at a bacterial load of 40 spores/juvenile.

## *Bacillus cereus* – as probiotic agent

*Bacillus cereus* is an endemic, soil-dwelling, Gram-positive, rod-shaped, beta hemolytic bacterium. Some strains are harmful to humans and cause foodborne illness, while other strains can be beneficial as probiotics for animals. *B. cereus* bacteria are facultative anaerobes, and like other members of the genus *Bacillus* can produce protective endospores (Fig. 18.11).

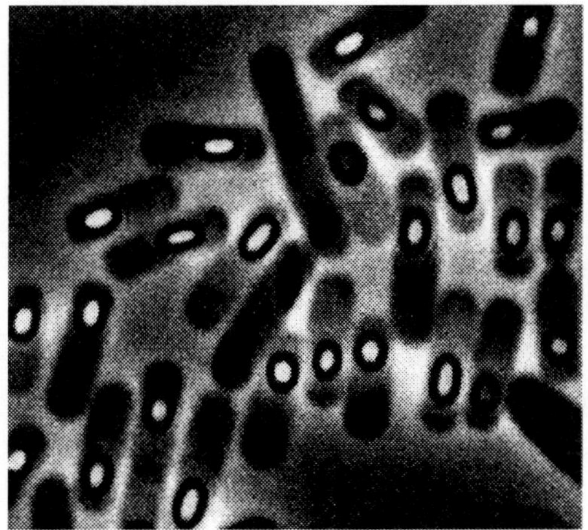

**Fig. 18.11.** Cells of *Bacillus cereus* with spores.

# Microbiology of Food and Industry

## 19.1 INTRODUCTION

Food is a rich nutrient source that is essential for the health and growth of human beings, infants and animals. Without food, oxygen, water, shelter, etc., life is miserable. Rice, wheat, pulses, fats, fruits, vegetables, oils, milk, medicines, cloth, meat, fish etc. are derived either from plants or animals. For the healthy growth of an individual, vitamins, proteins, carbohydrates, macro- and micro-elements etc. are essential. India is going to be most highly populated country by 2100. In view of the growing population, the demand for food supply has also grown. Green revolution, though has solved our food problem, the growing population has created problems including for food.

It is a known fact that milk is converted into curd by *Lactobacillus*. The milk protein, casein curdles due to lactic acid bacteria. Rennin, the enzyme also plays an important role. Cheese is produced due to the action of lactic acid bacteria. Fungi and bacteria help in ripening of cheese. *Penicillium roqueforti* and *Penicillium camembert* are known to help in soft cheese production and ripening. Butter making process, the traditional method is because of lactic acid bacteria i.e. *Lactobacillus*, Streptococci or yeast grow in low fat milk and help in the production of sour cream, yoghurt, kefir etc. Yeast is used in bread making. Sugars in bread dough are fermented by yeast to produce ethanol and $CO_2$. Pickles, soy sauce are also the products of microbial fermentation. Yeast is used in fermenting starch containing food grains, potatoes and molasses to produce alcoholic beverages and vinegar. Lactic acid bacteria play an important role in converting malic acid into lactic acid during wine making. Vinegar is produced due to *Acetobacter* and *Gluconobacter*. Grape fruits are fermented by yeast to produce wine.

Microorganisms are also rich source of food such as protein. Microorganisms during growth on diversified substrates produce protenaceous cell material. Microbial cells are converted into single cell protein by a blue-green alga, *Spirulina.* Yeast, *Methylophilus* and *Fusarium graminearum* are used in the production of single cell protein. Amino acids used in foods and medicines are produced by bacteria. Source enzymes used in the preparation of foods, medicines etc. are also produced by microbes.

Traditional methods used in food preservation were drying, addition of salt, sugar and fermentation. The food was stored as canned food also. Nicholas was the first scientist to devise canning in the year 1810. Canning process involves keeping air out of sealed cans and heating food in the cans. Steam was also used in canning. As the science and technologies have progressed, commercial sterilization by heating is used in canning. Commercialized sterilization process uses sufficient heat to destroy bacteria like *Clostridium botulinum.*

Canned food stored at 45°C gets spoiled by thermophilic microbes and this is accompanied by gas production also. Mesophilic bacteria spoil food under improper heating procedures or leakages. Acidic foods can be preserved by heating at 100°C. *Bacillus coagulans, Aspergillus* etc. are the microbes that tolerate acidity and heat above 100°C. These are responsible for the spoilage of acidic foods.

## 19.2 PASTEURIZATION

Louis Pasteur was the first person to determine the spoilage of milk and other foods, which was due to certain microbes. He advocated that it was not necessary to kill such microorganisms. Instead he advocated the application of heat that selectively kills microorganisms or contaminants without altering the quality of food such as milk.

## 19.3 PRESERVATION

Materials that are presterilized are packed and aseptically filled with heat sterilized liquid foods. Many microbial pathogens also grow slowly at refrigeration (0°–7°C). *Clostridium botulinum* and *Yersinia enterocolitica* are some of the pathogens that grow at low temperatures. Freezing helps in preventing the growth of some bacteria. Gamma radiation helps in sterilizing the food and also kills insects. Microwaves kill bacteria by heating the food.

There are some chemical preservatives that are added to inhibit the growth of moulds. Such chemicals are also used as active ingredients. Sodium benzoate, sorbic acid, potassium sorbate, calcium propionate are added to foods to inhibit the growth of moulds and bacteria.

Food-borne infection is a serious health hazard as the pathogen grows in the infected host tissue. Gastroenteritis, typhoid and dysentery are food-borne infections. The ingestion of infected food also results from ingestion of a toxin produced by microbes. Examples of microbial intoxications are botulism, staphylococcal food poisons and mycotoxicoses. It is also reported that pasteurized dairy milk may contain gram-negative psychrophiles. The pasteurized milk has a standard count of fewer than 20,000 bacteria per million and not more than 10 coliforms per ml. Dry milk products contain a count of 30,000 bacteria per gram, coliform bacteria being 10 per gram.

In unpasteurized milk, the bacterial contamination is not controlled and raw milk transmits the disease. Natural fermentations are known since times immemorial to make alcoholic drinks, bread, dairy products, pickles, and no one understood that microbes were responsible for these processes. Fermentation gives food a good taste, texture and smell and it brings in changes that inhibit the growth of undesirable food microbes, improving its storage life and safety.

Fermentations are also used to make an amazingly wide range of foods and drinks. Microbes are involved in the production of foods. Microbes help as catalysts to break down complex molecules in food so as to provide the energy for their growth. As by-products they sometimes produce the fermented foods and drinks that we consume.

## 19.4 YEASTS

Yeasts are simple, unicellular, eukaryotic fungi that are widely found in the environment. They break down sugars into ethanol (alcohol) and carbon dioxide and this action is harnessed to make bread, beer, wine and many other alcoholic drinks. Yeasts also contribute to the development and flavour of some fermented milk, vegetable and meat products.

**Bread:** Yeast contributes to the texture and taste of bread. Dough is made by mixing yeast with flour, salt and water. The yeast ferments sugars in the mixture to make alcohol and carbon dioxide. The gas gets trapped in the sticky proteins of the dough and causes it to rise. The alcohol is converted into compounds that add to bread flavour when cooked.

**Alcoholic Drinks:** Fruit or vegetable containing fermentable carbohydrates can be used as substrate to produce alcoholic drinks with the help of yeast. Yeast carries out the fermentation and other microbes contribute to the flavour and aroma of alcoholic drinks as they mature. Most drinks are pasteurised before they are packed. The yeast residues may be recycled or used as food flavourings or supplements (yeast extract). Marmite is a by-product of beer manufacture.

**Beer:** It is made from cereals (usually barley, but sometimes wheat) by a complex process known as brewing.

**Wine:** It is made from fruit juice—usually crushed grapes.

**Cider:** It is made from apples and perry from pears.

**Spirits:** Like whisky and brandy are made by distilling alcoholic liquors.

## 19.5 LACTIC ACID FERMENTATIONS

Lactic acid bacteria convert sugars to lactic acid. Along with the production of flavours, increased acidity of the product helps to prevent the growth of unwanted spoilage bacteria and pathogens. Changes in texture may also occur.

**Dairy products:** Cheese, yoghurt and fermented milk drinks are made using cultures that can produce acid from the milk sugar lactose, break down milk proteins and add flavour compounds and sometimes gas.

Yoghurt is a thickened, fermented milk produced using a mixture of bacteria, which work together to produce desired acid and flavours. The microbes are killed only if the product is pasteurised to extend its shelf life. Many people believe that eating live yoghurt bacteria is good for health.

Cheese manufacture depends on microbial activities at several stages to produce the required flavour, texture and appearance. There are many different types of cheeses. Blue cheese is made by inoculating the whole cheeses with a fungus, which gives the typical taste and appearance.

**Fermented meats:** Salami and chorizo are the fermented meats produced through lactic acid fermentation with curing salts and drying for their keeping quality, safety and colour. Yeasts and moulds may also contribute to the flavour of these products.

**Fermented vegetables:** Sauerkraut, olives and gherkins are made by lactic fermentation of the brine-soaked food. Different lactic acid bacteria are involved, sometimes together with certain yeasts and moulds.

**Fermented doughs:** Lactic acid bacteria work alongside baker's yeast in some doughs to improve the structure, flavour and storage life of French bread and similar products.

**Wine:** Some of the flavours in wine are partly due to fermentations by bacteria, which break down malic acid in the fruit juice to lactic acid and carbon dioxide gas.

## 19.6 FOOD POISONING

Food-borne illness is the result of bacterial contamination. Some bacteria form endospores that are resistant to drying and heating. They can survive cooking and grow again in favourable conditions.

**Food poisoning bacteria cause illness in the following ways:** They grow in food and produce a toxin. When swallowed, this irritates the stomach lining, causing pain, nausea and sickness. The toxins may also enter the intestines and cause diarrhoea. Toxins can survive cooking, even though the bacteria that produced them are killed.

They are eaten with food and burrow into the intestine wall where they multiply and may produce toxins. Painful inflammation and diarrhoea follow. More severe symptoms such as kidney and liver damage may occur if the bacteria and toxins enter the blood stream.

They enter the body and do not produce symptoms in the gut but in other parts of the body.

## 19.7 FOOD BIOTECHNOLOGY

Natural fermentation of a variety of substrates due to the action of microbes is a process known since 8000 years. Bread, alcoholic beverages, vinegar, cheese and yogurt, and many other foods are due to the activity of enzymes produced by microorganisms. Biotechnology helps food industry by providing new products, lowering costs and improving the microbial processes on which food producers have long relied.

Biotechnology processes improve the quality, nutritional value and safety of the crop plants and animal products that are the basis of the food industry. It offers many ways to improve the processing of raw materials into final products: natural flavours and colours; new production aids, such as enzymes and emulsifiers; improved starter cultures etc.

### 19.7.1 Improving Raw Material

Genetically cloned plants or animals, which have genes incorporated from outside, are the genetically modified organisms. The first generation of transgenic crops primarily benefited farmers. Although there are consumer benefits in growing these crops, the benefits are largely invisible to consumers. Studies have shown that since insect-resistant corn (Bt corn) sustains relatively little insect damage, fungi and moulds cannot infect these plants as easily as non-insect-resistant crops. Therefore, the level of toxins, such as aflatoxin, produced by these pathogens, some of which are fatal to livestock, is much lower in Bt corn than non-Bt corn.

A variety of healthier cooking oils derived from biotechnology are already in the market. Using biotechnology, plant scientists have decreased the total amount of saturated fatty acids in certain vegetable oils. They have also increased the conversion of linoleic acid to the fatty acid found mainly in fish that is associated with lowering cholesterol levels.

Another nutritional concern related to edible oils is the negative health effects produced when vegetable oils are hydrogenated to increase their heat stability for cooking or to solidify oils used in making margarine. The hydrogenation process results in the formation of trans-fatty acids.

Biotechnology companies have given soybean oil with these same properties, not through hydrogenation, but by using biotechnology to increase the amount of the naturally occurring fatty acid—stearic acid.

Animal scientists are also using biotechnology to create healthier meat products, such as beef with lower fat content and pigs with a higher meat-to-fat ratio.

Biotechnology also promises to improve the health benefits of *functional foods*. Functional foods contain significant levels of biologically active components that impart health benefits beyond our basic needs for sufficient calories, essential amino acids, vitamins and minerals. Familiar examples of functional foods include compounds in garlic and onions that lower cholesterol and improve the immune response; antioxidants found in green tea; and the glucosinolates in broccoli and cabbage that stimulate anticancer enzymes.

Biotechnology is also used to change the characteristics of the raw material inputs so that they are more attractive to consumers and more amenable to processing. Biotechnology researchers are increasing the shelf life of fresh fruits and vegetables; improving the crispness of carrots, peppers and celery; creating seedless varieties of grapes and melons; extending the seasonal geographic availability of tomatoes, strawberries and raspberries; improving the flavour of tomatoes, lettuce, peppers, peas and potatoes; and creating caffeine-free coffee and tea.

Japanese scientists have now identified the enzyme that produces the chemical, which makes us cry when we slice an onion. Knowing the identity of the enzyme is the first step in finding a way to block the gene to create "tearless" onions.

Much of the work on improving how well crops endure food processing involves changing the ratio of water to starch. For example, potatoes with higher starch content are healthier because they absorb less oil when they are fried. Another important benefit is that starchier potatoes require less energy to process and therefore cost less to handle. Many tomato processors now use tomatoes derived from a biotechnological technique, somaclonal variant selection. The new tomatoes, used in soup, ketchup and tomato paste, contain 30 per cent less water and are processed with greater efficiency. Another food processing sector that will benefit economically from better quality raw materials is the dairy products industry. Scientists in New Zealand have now used biotechnology to increase the amount of the protein casein, which is essential to cheese making, in milk by 13 per cent.

Biotechnology also allows economically viable production of valuable, naturally occurring compounds that cannot be manufactured by other means. For example, commercial-scale production of the natural and highly marketable sweetener known as fructans has long eluded food-processing engineers. Fructans, which are short chains of the sugar molecule fructose, taste like sugar but have no calories. Scientists found a gene that converts 90 percent of the sugar found in beets to fructans. Forty percent of the transgenic beet dry weight is fructans and this crop can serve as a manufacturing facility for fructans.

### 19.7.2 Safety of the Raw Materials

The most significant food-safety issue that food producers face is microbial contamination, which can occur at any point from farm to table. Any biotechnological product that decreases microbes found on animal products and crop plants will significantly improve the safety of raw materials entering the food supply. Improved food safety through decreased microbial contamination begins on the farm. Transgenic disease-resistant and insect-resistant crops have less microbial contamination. New biotechnology diagnostics, detect microbial diseases more accurately, so farmers can identify and remove diseased plants and animals before others become contaminated.

Biotechnology improved the safety of raw materials by helping food scientists discover the exact identity of the allergenic protein in foods such as peanuts, soybeans and milk, so that they can then remove them. Although 95 per cent of food allergies can be traced to a group of eight foods, in most cases we do not know which of the thousands of proteins in a food triggered the reaction. With biotechnological techniques, we are making great progress in identifying these allergens. More importantly, scientists have succeeded in using biotechnology to block or remove allergenicity genes in peanuts, soybeans and shrimp.

Biotechnology is helping us to improve the safety of raw agricultural products by decreasing the amount of natural plant toxins found in foods such as potato and cassava.

## 19.8 FOOD PROCESSING

Microbes have been recognised essential to the food-processing industry for decades. They play a role in the production of fermented foods listed in Table 19.1. They also serve as a rich source of food additives, enzymes and other substances used in food processing.

### 19.8.1 Improving Food Fermenters

The bacterium responsible for many of our fermented dairy products, such as cheese and yogurt, is also susceptible to infection by a virus that causes substantial economic losses to the

**Table 19.1.** Production of fermented foods by microbial fermentation

| Fermented foods | |
|---|---|
| beer | pickles |
| bologna | salami |
| bread/baked goods | sauerkraut |
| buttermilk | sour cream |
| cheeses | soy sauce |
| cider | tamari |
| cocoa | tea |
| coffee | tempeh |
| cottage cheese | tofu |
| distilled liquors | vinegar |
| kefir | wine |
| miso | yogurt |
| olives | idli |

food industry. Through recombinant technology, researchers have produced some strains of this bacterium and other important fermenters resistant to viral infection.

We have known for years that some bacteria used in food fermentation produce compounds that kill other contaminating bacteria, which cause food poisoning and food spoilage. Using biotechnology we are equipping many of our microbial fermenters with this self-defence mechanism to decrease microbial contamination of fermented foods.

Microorganisms have been essential to the food industry not only because of their importance as fermenters, but also because they are the source of many of the additives and processing aids used in food processing (Table. 19.1). Biotechnology advances will enhance their value to the food industry even further.

Food additives are substances which increase nutritional value, retard spoilage, change consistency and enhance the flavour. The compounds of food processors used as food additives are substances usually of plant or microbial origin, such as xanthan, gum and guar gum, which are produced by microbes. Many of the amino acid supplements, flavours, flavour enhancers and vitamins added to breakfast cereals are produced by microbial fermentation. Through biotechnology, food processors will be able to produce many compounds that could serve as food additives but are now in scant supply or that are found in microorganisms or plants and difficult to maintain in fermentation systems.

Food processors use plant starch as a thickener and fat substitute in low-fat products. Currently, the starch is extracted from plants and modified using chemicals or energy-consuming mechanical processes. Scientists are using biotechnology to change the starch in crop plants so that it no longer requires special handling before it can be used.

Enzymes produced by microbial fermentation play essential roles as processing aids in the food industry. The first commercial food product produced by biotechnology was an enzyme used in cheese making. Prior to biotech techniques, this enzyme had to be extracted from the stomach of calves, lambs and baby goats, but it is now produced by microorganisms that were given the gene for this enzyme.

The production of high-fructose corn syrup from corn starch requires three enzymes, and these enzymes are important in making baked goods and beer. Other enzymes are essential for the production of fruit juices, candies with soft centres, and cheeses. The food industry uses more than 55 different enzyme products in food processing. This number will increase as we discover how to capitalize on the extraordinary diversity of the microbial world and obtain new enzymes that will prove important in food processing.

### 19.8.2 Food Safety Testing

Biotechnology offers much help in enhancing the safety of the food supply. Biotechnology tools help to detect microorganisms and the toxins they produce. Monoclonal antibody tests, biosensors, polymerase chain reaction (PCR) methods and DNA probes are being developed that will be used to determine the presence of harmful bacteria that cause food poisoning and food spoilage, such as *Listeria* and *Clostridium botulinum.*

It is now possible to distinguish *E. coli 0157:H7*, the strain of *E. coli* responsible for several deaths in recent years, from many other harmless *E. coli* strains. These tests are portable, quicker and more sensitive to low levels of microbial contamination. The new diagnostic tests for *Salmonella* yield results in 36 hours, compared with the three or four days the older detection methods required.

Biotechnology-based diagnostics and kits have also been developed that allow us to detect toxins, such as aflatoxin produced by fungi and moulds that grow on crops, and to determine whether food products have inadvertently been contaminated with peanuts, a potent allergen.

## 19.9 MICROBES IN FOOD

Food is rarely sterile when it reaches the consumer. It usually contains bacteria or other microbes, most of which are harmless. Sometimes it may also contain pathogenic microbes, which could potentially be a threat to food safety. The internal tissues of plants and animals have many defences to keep microbes at bay, with the result that healthy, freshly-harvested crop plants and fresh meat are usually sterile. However, changes after harvesting or slaughter, or during processing may allow microbes to enter the food. These may originate from the crop plant or animal itself, from the environment, the factory environment (via soil or animals such as insects, birds and rodents) or from human sources.

Much of the food spoilage is due to microbial activity. While spoilage does not necessarily make food unsafe to eat, it can make it unpalatable. Examples that pose little health risk are the moulds, which can ruin the appearance of fruits and bread; yeast and lactic acid bacteria, which can spoil sauces and beverages; and slime produced by microbes, which can make chilled meat look unappetising.

### 19.9.1 Food-Borne Illness

Even today food-borne diseases remain a major public health concern both in developed and developing countries. There are two main categories of food-borne illness:

(1) Infections where the microbes attack the intestines or other organs directly, causing symptoms such as nausea, vomiting, diarrhoea and fever. Here, there may be a gap of several days between eating the infected food and the appearance of symptoms due to the time taken for the microbe to multiply. Examples are the infections caused by the bacteria *Salmonella, Campylobacter* and *Listeria monocytogenes.*

(2) Infections where symptoms such as diarrhoea result from poisons or toxins produced by the microbes as they grow in the intestine. Here, the time taken for symptoms to appear can vary from several hours to several days after eating infected food. An example is the illness caused by a toxic strain of the bacterium, *Escherichia coli.*

### 19.9.2 Intoxications

These result from eating contaminated food, which contains toxins produced by microbes that do not need to grow in the human body to cause illness. Symptoms can begin only a few hours after eating contaminated food. An example is botulism, caused by a toxin from the bacterium *Clostridium botulinum.*

Foods of animal origin are the primary source of many food poisoning microbes, such as *Salmonella, Listeria, Campylobacter, E. coli* and *L. monocytogenes.* These may occur on the live animal, and remain in the meat after slaughter. Without appropriate treatment to kill the microbes, or if conditions of hygiene or temperature control are poor, microbes may still occasionally be present in the final food product.

### 19.9.3 Poultry

Microbial contamination is probably highest in poultry. The rapid growth in poultry production has resulted in a readily available source of meat. However, there has been increased infection with food poisoning microbes in poultry, meat and eggs. Bacteria such as *Salmonella* enteritidis can be present within the egg or on their shells. *Salmonella* infections are on the increase across Europe. An important precaution in preventing food-borne illness from poultry and eggs is thorough cooking.

### 19.9.4 Dairy Products

Raw milk may contain various pathogens from the dairy animal or its environment. Pasteurization destroys all pathogens, and sterilisation ensures that the product is free from all microbes. While pathogens are inactivated by many of the methods used to produce dairy products—including acidification and fermentation of milk—certain types may sometimes survive. Hard cheeses, yoghurt and butter are considered safe because of their acidity or lack of moisture, but mould-ripened soft cheeses can allow growth of *Listeria monocytogenes.*

### 19.9.5 Herbs and Spices

These also frequently carry large numbers of bacteria such as *Bacillus cereus, Clostridium perfringens* and *Salmonella.*

### 19.9.6 How Processing Provides Protection

Commercially manufactured foods are designed to be safe. A number of common preservation methods used to destroy microbes or stop them from growing, include:

- heat treatments such as pasteurisation and sterilisation;
- canning;
- low temperature storage, for example, refrigeration and freezing;
- the addition of chemical preservatives, such as organic acids and nitrites;
- natural antimicrobial products, such as nisin;
- fermentation;
- control of water content through drying, salting or smoking;
- modification of the atmosphere in which the food is packed, for example, vacuum packing and gas packing;
- less widely-used preservation methods include high pressure and irradiation, together with novel technologies such as treatment with electricity or pulsed light treatment; and
- there remains a slight possibility that foods can become contaminated after processing. This can largely be avoided if everyone involved in handling food follows a good hygienic practice.

### 19.9.7 Avoiding Food Contamination

Food safety is a shared responsibility of everyone involved in the food chain from farm to fork. This includes primary producers, food companies, establishments that serve food, and consumers. Measures currently used to help prevent contaminated food from reaching the consumer include:

- using good quality raw materials from assured suppliers;
- following good manufacturing practices. Using management systems that allow the identification, monitoring and control of hazards during production, processing and sale of food;
- providing training programmes for all food industry personnel. Carrying out research on pathogens and how best to control them; and
- exchanging information on food safety.

A number of simple rules are recommended by the World Health Organization to ensure the safe preparation of foods:

- Avoid contact between raw and cooked foods to reduce the risk of cross-contamination. Wash hands before handling and after handling raw foods to minimise possible contamination.
- Cook food thoroughly in order to kill any microbes present. All parts of the food should reach a temperature of at least 70°C.
- Cool cooked foods as quickly as possible and then refrigerate. This slows down or stops microbial growth, which occurs best at 10–60°C.
- Reheat cooked foods thoroughly to kill tiny microbes, which may have developed during storage.
- Keep all kitchen surfaces clean to prevent cross-contamination. Protect foods from insects, rodents and other animals that may carry pathogenic microbes.

***Microbes used in food fermentation:***

*Lactocooccus lactis*—used in dairy fermentation
*Streptococcus thermophilus*—used in dairy fermentation
*Leuconostoc* sp.—used in wine making, dairy fermentation
*Pediococcus* sp.—meat fermentation, vegetable fermentation, ripening of some cheeses
*Lactobacillus* sp.—meat fermentation, vegetable fermentation, dairy fermentation, sourdough bread
*Bifidobacterium* sp.—added to dairy products to promote intestinal health. *Propionibacterium* sp.—Swiss cheese
Yeasts—bread, beer, wine, liquors
Moulds—ripening cheeses, soy sauce
*Lactobacillus delbruekii*, subspecies *bulgaricus* and *Streptococcus thermophilus*—making of yogurt

## 19.10 SPOILAGE BY MICROBES

Although microbes are useful for the production of some foods, they can also be the culprits in the spoilage of foods (Table 19.2). These microbes must first be introduced into the food product. After introduction, environmental conditions must be appropriate for the microbes to grow. This includes time, temperature and water availability.

If they are allowed to grow, the spoilage occurs either by the use of the food by the microbes for growth, or by the release of by-products from the microbes during their lifespan or after they die. Bacteria usually cause the fastest spoilage of foods because of their shorter reproduction times. They are followed by yeasts and moulds. However, if conditions are not favourable for bacterial growth, then yeasts and moulds may dominate spoilage. Spoilage of foods is based on the numbers of microbes that are present on the food product. For most products, the level of microbes that constitutes spoilage is approximately $1 \times 10^7$ microbes per gram of solid material, and in some cases per square centimetre. For liquid products this level is counted per millilitre of product. This level of microbes will usually produce a visible colour change of the product, or there will be odours from the by-products of the microbes.

### 19.10.1 Important Spoilage Microbes

Since the spoilage level is a count of microbes, essentially any microbe that is on the food product and can reproduce in enough numbers can spoil the product. However, there are specific organisms that have been associated with spoilage in most foods produced today (Table 19.2).

**Table 19.2.** Spoilage of foods by microorganisms

| Food Product | Spoilage Organism(s) (includes species of listed genera as given below) |
|---|---|
| Raw Meats | *Pseudomonas* |
| | *Proteus* |
| | *Micrococcus* |
| | *Escherichia* |
| | *Lactobacillus* |
| | *Clostridium* |
| Eggs | *Pseudomonas* |
| | *Proteus* |
| | *Aeromonas* |
| | *Penicillium* (mould) |
| Fish | *Pseudomonas* |
| | *Alcaligenes* |
| | *Vibrio* |
| | *Flavobacterium* |
| Raw Milk | *Pseudomonas* |
| | *Alcaligenes* |
| | *Flavobacterium* |
| Pasteurized Milk | *Micrococcus* |
| | *Lactobacillus* |
| | *Bacillus* |
| | *Clostridium* |
| | *Streptococcus* |
| Butter | *Pseudomonas* |
| | *Candida* (yeast) |
| | *Geotrichum* (mould) |
| Vegetables | *Pseudomonas* |
| | *Bacillus* |
| | *Penicillium* (mould) |
| | *Aspergillus* (mould) |
| Fruits | *Acetobacter* |
| | *Penicillium* (mould) |
| | *Aspergillus* (mould) |
| | *Rhizopus* (mould) |
| Condiments | *Bacillus* |
| | *Clostridium* |
| | *Lactobacillus* |
| | *Saccharomyces* (yeast) |

### Psychrotrophs

Psychrotrophs are organisms that can grow at low temperatures. Since many foods today are stored under refrigeration conditions, spoilage organisms belonging to this category are very important. There are many different types of psychrotrophs, including those which are aerobic, anaerobic, and facultative in their use of oxygen.

### Thermophiles

These organisms are capable of growth at high temperatures. They can easily spoil foods that are kept warm for extended periods of time, such as soups.

### Acidophiles

These microbes are capable of growing in acidic conditions that would kill most other organisms. They are responsible for the spoilage of foods that are packaged in an acidic environment. These products include pickles, mayonnaise, and certain fruit juices.

## 19.11 MICROBES IN INDUSTRY

Microorganisms have been associated with man for their mutual benefit. For example, yeasts have been known to be involved with man since times immemorial and have served a steady and useful purpose. They are involved in the fermentation processes of fruit juices, wine making, brewing and bread making etc. Pasteur showed that the process of grape juice fermentation was the result of microbial activity and the organism involved in bringing this desirable change in grape juice was identified as yeast. Pasteur's work showed that fermentations characterized by the formation of particular products such as alcohol, acetic acid, lactic acid, butyric acid etc. were dependent on the activities of different kinds of microorganisms. This recognition of the ability of microorganisms to bring about desirable chemical changes in appreciable amounts led to the development of industrial microbiology and establishment of several industries.

Three important characteristics that have made microbes useful in industry are: (1) their ability to grow rapidly using easily available and cheap raw materials, (2) their ability to maintain physiological constancy, and (3) their ability to bring about biochemical transformations under simple cultural conditions. These properties have made the microorganisms highly preferable over synthetic processes. The wine, beer-making industries and the baking industry are mainly dependent on the activities of the microorganisms. Aceone, butyl alcohol, lactic acid, acetic acid and many other industrial products such as enzymes, vitamins, hormones etc. are manufactured with the help of microorganisms. Antibiotics are always manufactured with the help of microorganisms.

The applications of microbes in some of the processes are enlisted below:

A. In food products (e.g., fermented vegetable products and dairy products) for the purpose of producing certain flavours, consistencies and nutritive values in the products;

B. In media where they decompose (ferment) various substrates (usually carbohydrates), the products of fermentation (various alcohols, organic solvents, lactic, acetic and citric acids) being recovered, purified and sold;
C. In flavoured nutritive solutions, notably fruit juices and extracts of grains that are fermented, the entire culture fluid (after clarification and processing) then becoming beverages such as beer and wine;
D. In contact with a specific substances such as steroids, so that a single, specific enzyme of the microorganisms brings about a specific transformation of the substrate molecule into a molecule of another desired substance;
E. In media so that the enzymes or other substances that the organisms synthesize (amylase, protease, antibiotic) may be collected, purified and sold for commercial or medical use.

In addition to these uses, organisms themselves (principally yeasts, some algae) are sometimes cultivated for use as food. Such food products are generally used as feed for poultry and livestock. Microorganisms are sometimes used for the removal of certain sulphur compounds from petroleum, and for vitamin assay.

The uncontrolled actions of fortuitous mixtures of microorganisms in processes such as retting of flax, preparing hides for leather, and coffee-bean hulling are time-honoured practices. They are now being replaced in more advanced industries by processes in which purified cultures or enzymes of the effective species in the mixtures are used under carefully controlled conditions.

In developing an industrial process based upon the action of microorganisms, many details must be given consideration. Important among these are the type of culture needed, cultural condition and the adaptability of the organism to large-scale production.

### 19.11.1 Types of Fermentation Process

Industrial fermentation process may be divided into two main types: (1) batch fermentation and (2) continuous process. There are various combinations and modifications of these.

#### Submerged Aerobic Cultures

Many industrial processes, usually called "fermentations," are carried on by strictly aerobic microorganisms. For example, production of penicillin by *Penicillium notatum*, a strictly aerobic mold. Now it is a common commercial practice to carry on such "fermentations" in closed tanks with submerged cultures. Aerobic conditions are maintained by constant agitation of the contents of the tank with an impeller and constant aeration by forcing sterilized air through a porous diffuser. The flow of air through the tank removes gases such as ammonia and carbon dioxide. In each sort of process, very careful adjustment of O-R potentials, mechanical agitation, ratio of dissolved oxygen to other ingredients in the medium, and pH are necessary.

#### Industrial Ethyl Alcohol Production

The production of ethyl alcohol by fermentation using yeasts is an important industry. It serves to illustrate industrial fermentation processes, in general. Crude molasses is often used as mash.

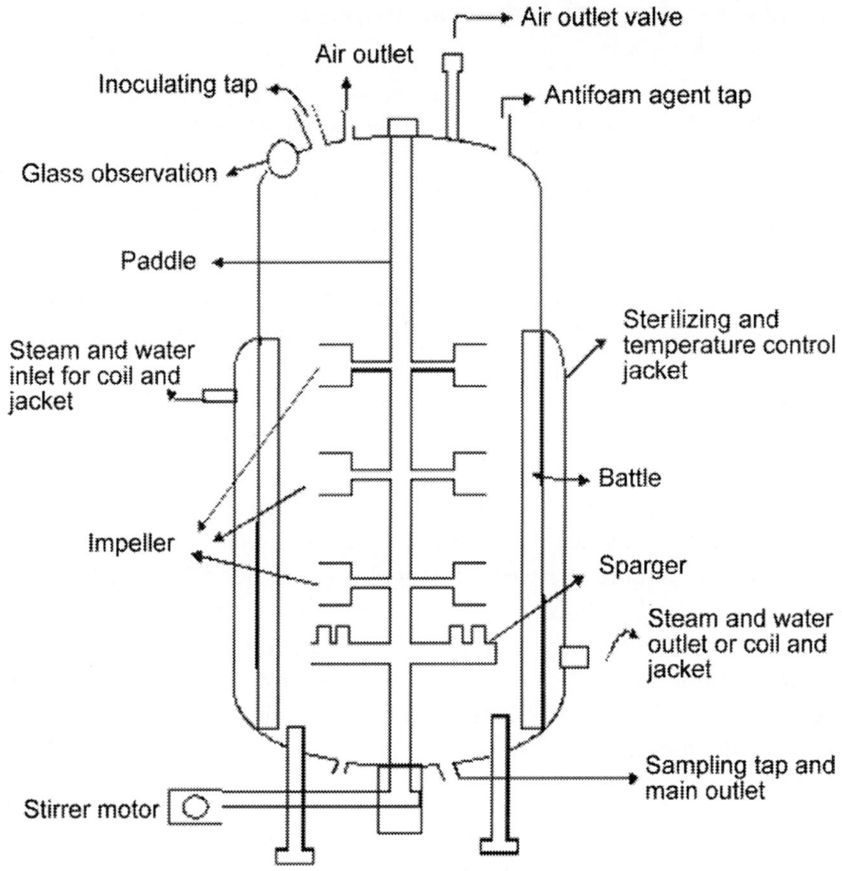


**Fig. 19.1.** Fermenter vessel.

It generally requires only to be diluted and the pH adjusted (usually with sulphuric acid). This pH is favourable to the yeast and unfavourable to many bacteria. A source of nitrogen such as ammonium sulphate or ammonium phosphate is usually added. The final solution is a rich nutrient carbohydrate culture medium or mash.

This is rather heavily inoculated with an aciduric and alcohol-resistant strain of yeast, the variety depending on the conditions under which the fermentation is to proceed and the exact end products desired. A good strain of *Saccharomyces cerevisiae* is commonly used. The inoculum comes from a large tank of carefully maintained pure culture, previously inoculated from a smaller seed tank, and the latter from a flask or tube of culture in the laboratory. At present, stainless steel continuous-culture apparatus is available for maintaining constantly large amounts (many gallons) of pure cultures of inoculums (Fig. 19.1).

The fermented mash contains the crude alcohol. This is usually a mixture of ethyl alcohol and a small amount of glycerol with fusel oil. It contains amyl, isoamyl, propyl, butyl and other alcohols with acetic, butyric and other acids, as well as various esters. The high wine is driven off from the mash or beer by heat, and further purified by fractional distillation.

## 19.11.2 Microbial Products of Industrial Importance

### Microbial Enzymes

Enzymes are biocatalysts produced within the living cells to bring about specific chemical changes. They are universally present in all the living organisms, and the metabolic reactions common to all cells are catalysed by enzymes.

In oriental countries, microorganisms are directly employed as enzyme source for the preparation of products such as shoyu, miso, natto and sake. Other important processes in which enzymes are primarily used are cheese making, leavening of bread, manufacture of vinegar, tanning of leather etc. It was Takamine who laid the foundation for the industrial production of microbial enzymes by developing a process for producing diastatic enzymes from fungi. Boidin and Effront of France were the first to produce industrial enzymes from bacteria. During the past 40–50 years, more than 1300 enzymes have been identified, out of which nearly 100 enzymes have been obtained in crystalline form.

Like chemical catalyst, enzymes alter the velocity of a reaction without appearing as part of the final product. However, enzymes do not behave as ideal catalysts since it has been established that enzymes form complexes with the reacting substrates. Enzymes are water-soluble, simple, globular or conjugated globular proteins. They vary widely in molecular weight. Simple globular enzymes having molecular weight of 11,000 and conjugated proteins having molecular weight of several million have been reported. Certain enzymes like pepsin and urease are found to be pure proteins, while some others have been found to consist of a non-protein component in addition to the protein component.

| Protein (Apo-enzyme inactive) | + | Non-protein Prosthetic group (Whole enzyme, inactive) | ⟶ | Holo enzyme (Co-enzyme active) |
|---|---|---|---|---|

The prosthetic groups are usually dialyzable, heat stable portion of the enzyme systems. If the dialyzable portion of the enzyme is an organic molecule, it is referred to as a co-enzyme.

Some enzymes are first produced by the cell in an active form. They are called Zymogens or pre-enzymes. These must then be activated by contact outside the cell with another agent called an activator or a co-factor. Co-factors may be metal ions, co-enzymes, vitamins or other enzymes.

Enzymes are remarkable not only for their very high effectiveness as catalysts, but also for their high degree of specificity with respect to their substrates.

Some enzymes attack certain specific sites on the substrate molecule. For example, some proteolytic enzymes attack peptides at sites adjacent to amino acids containing aromatic group.

In the process of enzyme manufacture by microbial sources, various factors have to be considered if we have to produce a desirable product at a reasonable cost.

## A) Selection of Microorganisms

The following conditions are essential for selecting a microorganism for enzyme production:

1. The organism should be nonpathogenic.
2. It should produce no products, which would be toxic when used.

**Table 19.3.** Some important microbial enzymes used in food industries

| S.No. | Enzyme | Microorganisms | Reaction Catalyzed | Food industry where used |
|---|---|---|---|---|
| 1. | Amylases | *Bacillus* sp.<br>*Aspergillus* sp. | Endohydrolysis of starch | Bakeries, breweries, sugar industries |
| 2. | Glucoamylase | *Aspergillus* sp. | Exohydrolysis of starch | Bakeries, breweries, sugar industries |
| 3. | Cellulases & hemicellulases | *Trichoderma.* sp.<br>*Aspergillus* sp. | Hydrolysis of cellulose and hemicellulose | Feed industries, fruit processing |
| 4. | Pectinases | *Apergillus* sp. | Hydrolysis of polygalacturonic acid | Fruit industries, wineries |
| 5. | Pullulanase | *Bacillus* sp. | Hydrolysis of α-1, 6-bond | Bakeries, breweries |
| 6. | Invertase | *Saccharomyces* sp. | Hydrolysis of sucrose into glucose and fructose | Confectioneries |
| 7. | β-Galactosidase | *Aspergillus* sp. | Hydrolysis of lactose | Dairies |
| 8. | Protease | *Bacillus* sp.<br>*Aspergillus* sp. | Endohydrolysis of proteins | Bakeries, dairies, meat processing |
| 9. | Glucose isomerase | *Streptomyces* sp. and other Actinomycetes sp.<br>*Bacillus* sp. | Isomerization of D-glucose to D-fructose | Sugar and starch industry, bakeries |
| 10. | Glucose oxidase | *Penicillium* sp.<br>*Micrococcus* sp. | Oxidation of glucose to gluconic acid | Fruit juice industry wineries |
| 11. | Catalase | *Aspergillus* sp. | Decomposition of $H_2O_2$ | Fruit Juice industry |

3. It should exhibit biological stability, and not be susceptible to degeneration.
4. The organism should be capable of growing on inexpensive substrates and yields should be high.

Microorganisms can be isolated in pure form from soil, decaying organic material and air, and tested individually for their capacity to produce the desired enzymes. A large number of isolates are to be tested and screened for enzyme production. It is also customary to grow the selected organisms on their substrates. In certain cases, as in the case of pectinase, the organism is induced to secrete the desired enzyme. The selected strain is maintained in pure form by lyophilization on agar slants or soil culture. The isolate must be periodically checked for purity and for the retention of its original activity.

Secondary screening is conducted in flasks or small fermenters. This will give information to evaluate the true potential of the microorganisms. Once the potential of the microorganism is established, investigations are undertaken to work out a suitable medium and other conditions like pH, temperature, aeration etc., for maximal enzyme production.

## B) Methods of Cultivating Microorganisms

The most important part of the industrial production of enzymes is the method of cultivating the microorganism in which high titres of enzyme are obtained. Several methods of culturing are being employed in the industry, which can be broadly classified as follows:

**Solid culture:**

a) Conventional koji culture
b) Bulk koji culture
c) Rotary drum culture

**Liquid culture:**

a) Stationary
b) Submerged

Enzymes manufactured by fermentation include bacterial amylases, proteinases, glucose oxidase, pectinases and amyloglucosidase, invertase, penicillinase, streptokinase, streptodornase, cellulose, pentosanases, hesperidinase and naringinase etc.

# Microbes and Diseases in Humans, Plants and Animals

## 20.1 INTRODUCTION

Bacteria, fungi, actinomycetes, viruses and protozoans are known to cause severe diseases in humans and animals. Any disease that gets transmitted from one host to the other host is called infectious disease. In some diseases there are carriers called vectors through which the disease causing microbe gets transmitted. There are a number of outbreaks of diseases in past few years. Smallpox, cholera, tuberculosis, yellow fever, plague, dengue, diphtheria, recent spread of cancer and HIV/AIDS are some of the examples, which caused havoc. Most of the diseases such as smallpox, cholera, tuberculosis etc. are now under control and the cases reported are negligible.

Disease is nothing but any change in the normal health or a state of abnormal physiology due to which part or the whole body of an individual gets affected or does not function. Disease is manifested externally in the form of symptoms. For example, high temperature is the indication of viral flu. The process of entry, establishment, multiplication of the disease causing agent in host tissues or within the host cells is known as infection. The infectious microbe, which causes the disease and derives all nutrients for growth from the susceptible host and lives at the expense of the living host, is called pathogen. If the microbe requires a living host and multiplies in it, it is called obligate parasite. This requires a specific host and the pathogen does not infect all types of hosts. However, it requires a specific host. Obligate parasites cannot grow artificially, while some pathogens can be cultivated under laboratory conditions. The ability of the pathogen to cause disease is called pathogenicity. The process of infection and sequential steps involved in infection and disease development is termed pathogenesis.

Disease development is a complicated process and involves many changes not only in morphology but also in biochemistry, physiology and other related aspects. All such complexities are considered disease syndrome. Three important components are involved in disease production, namely the virulent pathogen; susceptible host and congenial environment. For disease production, a single microbial pathogen such as spore or other infectious agent may be required. In some diseases, a load of infectious agent called inoculum is required. The amount of inoculum required for the causation of disease is called inoculum potential.

In other words, if sufficient infectious inoculum is not available, then the disease is not produced. Susceptible host not only provides space but also supplies all nutrients for the growth, multiplication and establishment of pathogen. Virulent pathogen is the potential infectious agent required for initiation of the infection. In case of resistant host, the host has all the genetic makeup, which resists the pathogen entry and spread. In case a large population is affected by the disease covering a large geographic area, the disease is called epidemic. Epidemiology is the branch of medical science, which deals with regular occurrence, distribution of pathogen, extent of disease spread and control of disease. The scientist dealing with above aspects is called epidemiologist. Any disease that occurs irregularly affecting human population and occurs occasionally is known as sporadic disease (typhoid). A disease that has low level frequency and occurs at a regular interval is called endemic disease (common cold). In case a disease increases within a large population covering a wide region, it is called pandemic. Repeated occurrence of disease as fraction in a given population is known as frequency. It is related to mortality i.e., death rate or morbidity.

Different diseases have different characteristic signs and symptoms, which are the objective changes in body (e.g., fever). Based on fever, a disease can be recognized and symptoms are the objective changes or external signs of disease development (e.g., pain). Diseases do not develop all of a sudden, but during certain phases. Incubation period is the time required after infection till the appearance of symptoms. In second phase, there is onset of signs and symptoms. In the third phase, the illness stage, the disease becomes fully established and becomes severe with characteristic signs and symptoms. The last stage is called convalescence, where the decline of disease starts and patient gets recovered gradually.

The in-born capacity or genetic ability of the individual in resisting infection and spread of pathogen is called immunity. This immunity forms the basis in resisting the pathogen. The infectious agent has different stages of development and chain of events that includes epidemiological aspects also. The events related to infection till the control of disease are considered under disease cycle. Disease cycle includes the study of sources of disease, reservoirs and carriers. The disease causing agents are transmitted from one host to other by means of air (flu, mumps, pneumonia etc.), contact (Aids, syphilis), inanimate materials (*Vibrio, Salmonella*), vector-borne transformation (flies carry *Shigella*, mosquito – *Plasmodium*) etc.

Control of infectious agent is possible by breaking the disease cycle. This includes pasteurization, destruction of vectors, chlorination etc. It is also possible to eliminate the source of infection by quarantine regulations, destruction of animal reservoir and treatment of sewage etc.

Immunization of individuals by vaccination etc. can eliminate the source of infection. Mass immunization of children against chickenpox and pulse polio immunization are some of the control measures adopted in controlling infectious diseases. Many bacteria are known to produce cytototoxins, enterotoxins and neurotoxins during their infectious phase. They act on membrane, Hageman factor, disrupting membrane transmit etc. Such toxins (aflatoxins, mycotoxins) are also elaborated by fungi, which are hazardous to health.

## 20.2 DISEASES CAUSED BY PROTOZOA

Protozoa are known to cause diseases in humans. The following is the brief account of some protozoan diseases:

### 20.2.1 Malaria

Malaria is a common disease and causes human death en masse. *Plasmodium vivax* is the disease causing agent. However, with the development of awareness, vaccines and chemotherapeutic methods, the disease is under control though not eradicated from India. About 2–4 million people die every year in the world. *Anoepheles*, a mosquito is the vector of malarial disease.

Mosquito carries the inocula of *Plasmodium*, multiplies and produces the sporozites. As soon as the mosquito bites human, the sporozites present in saliva are delivered into the blood of human. Thirty minutes later, sporozites enter liver cells and undergo schizogamy. A large number of merozites enter the blood stream and infect them. This cycle gets renewed thus infecting RBCs. RBCs are disintegrated, releasing toxic substances that cause chill and fever. As soon as the body reaches a temperature of 40°C, fever subsides. Due to breakdown of RBC, patient suffers from anaemia, enlargement of spleen and liver. Merozites enter divertive trait and 1% infected RBC contain 10,000 million parasites.

Control of malaria is possible through chemotherapy by using quinine, chloroquinine, primaquine etc. In recent times, efforts are in the way for the development of malarial vaccines. DDT has been extensively used to eradicate mosquitoes. It is learnt that mosquitoes have also developed resistance towards DDT. DDT usage has been banned in USA and other countries. In spite of many toxic effects, DDT is still in use in India.

### 20.2.2 Giardiasis

*Giardia lamblia* is a flagellated protozoan parasite in human intestine. It is a water-borne disease distributed all over the world and the inoculum comes from human faeces. The disease persists for a week. The symptoms are nausea, weakness, weight loss and chronic greasy diarrhoea. Seven per cent of total population in the world carries this disease. Filtration of water, use of metronidazole or quinocrine hydrochloride may help in removing contaminants.

### 20.2.3 Sleeping Sickness

*Trypanosoma* infects the blood and the disease is called trypanosomiasis or sleeping sickness. *Trypanosoma gambiense* is the commonest haemoflagellate. Trypanosomiasis has been reported from Africa, America and other parts of the world, however, species are different. Rodents, arthropods, tse-tse fly, and bugs are the vectors of the disease and trypanosomes grow inside the gut of bugs. The bug bites near the lips and transmits the disease. The pathogen infects the lymph and is carried by the blood to the other part of the body organs such as spleen, liver and heart. Damage to the macrophages, nervous system, loss of tissue and weakness are the symptoms of the disease. Patients may fall into coma and die.

### 20.2.4 Amoebic Dysentery

This disease, commonly called amoebiasis, is known to occur throughout the world. The parasite is *Entamoeba histolytica* and it spreads rapidly. The source of infection is contaminated food. Chlorination does not solve the problem as the amoebae cysts are resistant to many chemicals. The amoebic cells multiply on the wall of the large intestine. The amoeba lives on RBC and damages the tissue of the digestive system. In chronic and severe cases, amoeba may enter liver, lungs and intestinal walls. The amoebic cysts can be observed in faeces. All this results in dysentery and the faeces contain blood and mucus. Metronidazole plus Iodoquinol are the chemotherapeutic agents used.

## 20.3 BACTERIAL DISEASES

### 20.3.1 Tuberculosis

It is a serious disease caused by *Mycobacterium tuberculosis*. This disease affects three million people every year around the world. It is very common in slum areas, malnourished individuals and under unhygienic conditions (Fig. 20.1). The symptoms of the disease are cough, pain in chest, fever, sputum release and others. The sputum is red or rust coloured and bacteria spread into the lungs, chest and other body parts. The disease is diagnosed by tuberculin skin test, isolation of acid fast bacterium and chest X-ray. The bacterium is rod-shaped, acid fast and cells take red stain. The disease can be controlled by chemotherapeutic treatment. Treatment has to be done in consultation with doctor.

### 20.3.2 Diphtheria

Diphtheria is an air-borne infection. It is a serious disease and affects poor people living in crowded conditions. The organism that causes diphtheria is *Corynebacterium diphtheriae,* which is a gram positive bacterium (Fig. 20.2). Symptoms of the disease include mucus and pus formation (mucopurulent), nasal flow, vomiting, headache, cough, stiffness of neck and back fever. The bacterium releases exotoxin that causes inflammation and greyish

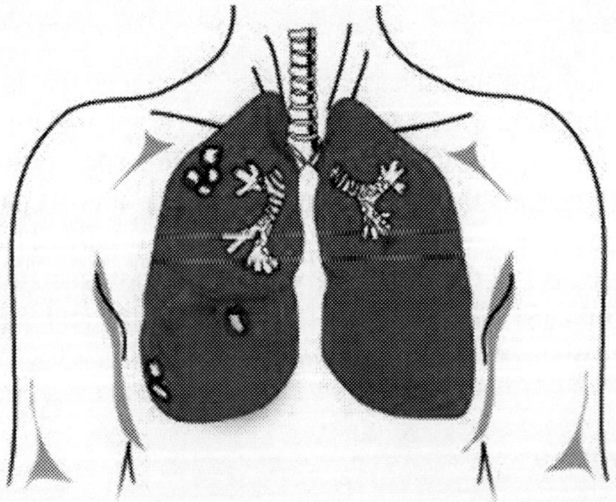

**Fig. 20.1.** Lungs infected with *Mycobacterium tuberculosis.*

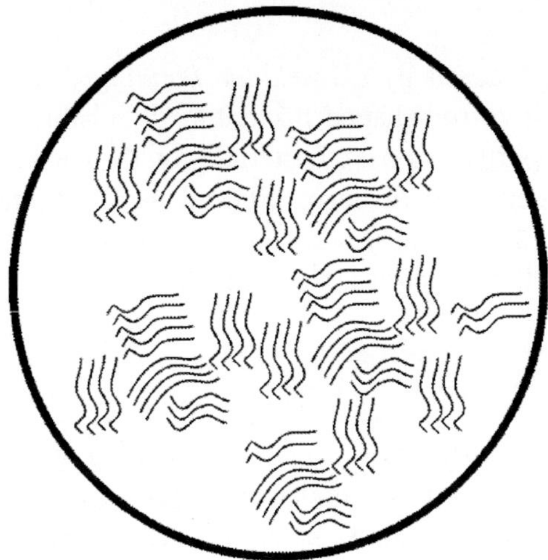

**Fig. 20.2.** Cells of *Corynebacterium diphtheria.*

pseudomembrane in respiratory mucosa. The exotoxin affects the kidney, nervous system and indirectly the heart. Skin lesions or wounds cause ulceration. Penicillin and Erythromycin are used in treatment. Worldwide immunization is launched with Diptheria-Pertussis-Tetanus (DPT) vaccine.

### 20.3.3 Whooping Cough

*Bordetella pertussis* is the disease causing agent. The microbe is a small, non-flagellated, gram negative *Coccobacillus,* which is capsulated. It produces an endotoxin besides being an obligate aerobe. It is a childhood disease. Initially it resembles common cold and later cough by mucus accumulations. Whooping sound is produced in between coughs. It is a respiratory infection and white blood cell count normally increases. Broken ribs are also seen due to cough. The disease is common among children. Whooping cough can be controlled by using erythromycin, tetracycline and chloramphenicol. Vaccination is also recommended.

### 20.3.4 Skin Infection

Streptococci are gram-positive, facultative anaerobic cocci growing in chains. These bacteria, mostly the nonpathogenic, inhabit gastrointestinal tract and upper respiratory system. Some streptococci cause skin infection and secrete many toxins, which damage red blood cells. Skin infection, impetigo is characterized by randomly distributed small, round elevations containing pus. It is spread by contact. Some strains cause skin infection in the form of reddish patches. The drugs needed for control are penicillin and erythromycin.

### 20.3.5 Botulism

It is a kind of food poisoning caused by *Clostridium botulinum*. This is an obligate, anaerobic endospore forming, gram-positive rod-shaped bacterium. It inhabits soil, fresh water and food. Food, which is not canned properly, will have bacterial growth. The bacterial toxin is a neurotoxin causing damage to the synergetic end of nerve. Flaccid paralysis may also occur, followed by respiratory and cardiac failure. Nausea may lead to neurological problem. Difficulties in swallowing, general weakness etc. are the symptoms. Scientific and hygienic storage of foods is essential.

### 20.3.6 Tetanus

*Clostridium tetani,* an obligate anaerobic, endospore forming gram positive rod-shaped bacterium causes tetanus. The bacterium is found in the soil—which contaminates animal faecal waste. The symptoms are caused by tetanospasmin a potent neurotoxin released upon the death of bacterium. The neurotoxin blocks the relaxation pathway so that the muscle contraction is affected, resulting in muscle spasms called spastic paralysis. Jaws and skeletal muscles are affected. Death results from spasms of the respiratory muscles. Immunization includes tetanus toxoid resulting in the formation of antibodies.

### 20.3.7 Leprosy

*Mycobacterium leprae* is the non-culturable bacterium that grows in the peripheral nervous system and skin cells. G.W. Hangen of Norway was the first to isolate and identify leprosy bacterium in 1870. It grows at 30°C on the outer cooler portions of the human body. The neural form is identified by regions of the skin that have lost sensation and are surrounded by nodules as a border. These symptoms are seen in persons having effective immunity. In the progressive form of leprosy, skin cells are infected and disfiguring root nodules are formed all over the body. Lion-faced appearance is associated with this kind of leprosy.

It is spread by transfer of bacterial discharge from lesions of the diseased individuals. Clothes of diseased persons may also transfer the disease. However, it is not that much contagious. This is one bacterium that is not cultured.

The disease is difficult to cure. The long treatment with sulfone drugs is effective in the control of leprosy. Refemycin, Colifazimine and other drugs are used along with dapsone. A vaccine is under trial. BCG vaccine might be effective.

### 20.3.8 Anthrax

Robert Koch isolated *Bacillus anthracis* in 1877. This bacterium is endospore bearing rod-shaped bacterium swarming in the blood of dead animals. It is aerobic and gram-positive bacterium. It is recorded that the endospores of anthrax bacterium survive up to 60 years. The toxins produced by the bacterium inhibit phagocytosis. The disease occurs in cattle and sheep. Endospores enter through grass into grazing animals. The endospores germinate in stomach, pass through intestinal mucus and invade the blood causing dangerous septicemia. Human anthrax cases are also reported. The bacterium enters the skin through cuts or abrasion and produce malignant pustules. Pulmonary anthrax is the most dangerous disease resulting in wool sorters disease (a kind of pneumonia). This causes high fever, difficulty in breathing and pain in the chest. Penicillin is the antibiotic used in initial stages and this proves useless in controlling advancement of disease. Recently, it caused fear psychosis as many people in different continents were afraid of bioterrorism.

### 20.3.9 Plague

A blackish area of skin caused by haemorrhages leading to Black Death is called plague. Around 10 million people have died due to this disease in India. It is caused by *Yersinia pestis*, a gram-negative, rod-shaped bacterium. This disease spreads from one rat to another by the rat flea *Xenopsilla cheopis*. As soon as the host dies, the rat flea searches for replacement.

The replacement may be a rodent or a human. The bacterium, after entry into flea, blocks the digestive tract and infects the blood. The dispersal of disease is also from infected animals, scratches of domestic cats and others. Due to flea bite, the bacterium enters the human blood and multiplies in the lymph and blood. The lymph nodes and other infected parts get enlarged and fever develops. Swellings are seen on the body (Buboes). The most dangerous situation is when bacterium is carried by the blood to the lungs resulting in pneumonial plague causing 100% mortality. Control is by improving sanitary conditions, garbage disposal, sources of food.

### 20.3.10 Cholera

It is an endemic disease in Asia and particularly in India. Lapses in sanitation is the root cause. The organism responsible is *Vibrio cholerae*, comma-shaped, gram-negative, rod-shaped bacterium with a single polar flagellum (Fig. 20.3). Quarantines can reduce the disease. It releases

**Fig. 20.3.** Cells of *Vibrio cholerae*.

enterotoxins that results in the secretion of chlorides, bicarbonates and water. Mineral electrolytes and excess water gets leaked out as an excretion. Rice water stools are excreted out from the mosses of intestinal mucus, epithelial cells and bacteria. This is due to growth of bacteria in the small intestine. Blood becomes viscous and makes vital organs weak. Vomiting is also present. The organism can be isolated from faeces. Good sanitation, vaccination, tetracyclines are the control measures.

### 20.3.11 Gonorrhea

It is a communicable disease caused by gram-negative diplococcus, *Neisseria gonorrhoeae.* Symptoms in infected males are painful urination and discharge of pus material from urethra. Sterility is also noticed when testes get infected. In females, the infection results in abdominal pain and uterine tube infection. Unlimited sexual activity with many sex partners attributes to severe gonorrhoeal infections and serious health problems. Penicillin has been an effective drug. Streptomycin is the drug of choice, since penicillin resistant strains were observed.

### 20.3.12 Gastroenteritis

*Salmonella* is a gram-negative facultatively anaerobic, non-endospore forming, motile rod bacterium causes salmonellosis. It takes 12–36 hours to produce symptoms. The bacterium first invades the intestine mucosa, re-enters cardiovascular system and spreads to other parts. Fever is associated with infection and endotoxins are released. The sources of inoculum are intestinal tracks of many animals and contaminated meat, poultry, eggs and egg products. Milk is also another source. Pasteurization of milk is compulsory. Proper refrigeration, proper cooking and observing good sanitation can control the disease.

### 20.3.13 Typhoid

*Salmonella typhi* causes the disease. The infected person suffers from fever and continuous headache. Diarrhoea appears after a week or two and then fever comes down. Disease is prevalent in areas of poor sanitation. Already infected but recovered patients become carriers of the disease. They harbour pathogen in the gall bladder. Chloramphenicol is the drug used, besides ampicillin and trimethoprim sulfamethaxazole. Immunization is not done normally but implemented for military personnel.

### 20.3.14 Syphilis

Syphilis is more common among the age group of 20–39 years. It is transmitted by sexual contact via syphilitize stage of genitals. It is caused by *Treponema pallidum,* which is a gram-negative spirochete. It is cultured *in vitro* and propagated in rabbits. Initially there is a small, hard based chancre or sore at the site of infection. The fluid is infectious. After the primary stage, the skin rash appears on the body. Other symptoms being loss of hair in patches, mild fever etc. Secondary syphilis lead to latent period. Many lesions are formed on the body. It has become important in the venereal diseases. Benzathine, penicillin are effective in the body for about two weeks. Penicillin therapy is effective during primary stage. Tetracyclines have been proved to be effective.

## 20.4 VIRAL DISEASES

### 20.4.1 Smallpox

Smallpox is caused by pox virus affecting a population with 1–20% mortality. The virus passes through the respiratory route, infects internal organs and moves into blood stream. The virus grows in the epidermal layer and produces lesions. Virus is prick-shaped having DNA packed in nucleocapsid. It is surrounded by a series of fibres. The symptoms are chill, fever and general prostration. At low temperature pinkish red spots develop and are converted into pink pustules. These pustules at a later stage cover the whole body. It spreads through contact with skin, body fluids such as urine and blood droplets. In the year 1798, Edward Jenner developed vaccination i.e., using the disease causing virus against disease.

### 20.4.2 Chickenpox

It is a disease common among children. The infection is through respiratory system and infection is on skin cells. It forms vesicles with pus, which rupture and form scab. Severe complication of chickenpox, influenza and other viral diseases result in complex Reye's syndrome. The agent that causes chickenpox is a herpes virus. The virus enters peripheral nerves and moves to a central nerve region. It is a DNA virus. The disease is best controlled by vaccination.

### 20.4.3 Measles (Rubeola)

It is a contagious disease and spreads through respiratory route. Humans are the reservoir of infection. Due to vaccination, there is a decline in the incidence of measles. The infection and symptoms are similar to that of smallpox and chickenpox. To start with symptoms are sore throat, headache, cough, rash appear on the skin etc. It is a dangerous disease among the young and elders.

### 20.4.4 Polio

Polio is a kind of paralysis. Hundred per cent of infected individuals develop identifiable symptoms. It is caused by poliomyelitis virus. The symptoms are headache, sore throat, fever, nausea, influenza etc. The reasons attributed are poor sanitary conditions. Humans are the only hosts for polio virus. The infection is due to ingestion of contaminated water. The virus also infects lymph nodes of the neck and terminal portion of small intestine and then enters blood. Nerve cells are affected. The motor nerve cells etc. die and respiratory failure results in death. Polio vaccine was introduced in 1963 and it has achieved immunity.

### 20.4.5 Rabies

It is an infectious disease that results in fatal encephalitis. Rhabdovirus is the bullet shaped virus causing the disease. Dog bite and bites from an infected animal including a lick is enough to cause the disease. Rabies virus enters the healthy animal through saliva also. The virus multiplies in skeletal muscle and connective tissue. Symptoms are spasms of muscles of mouth and pharynx that occur when liquids are being swallowed. Extensive damage is to the nerve cells of the brain and spinal cord, followed by death. HDCV vaccine is useful in treatment. Effective vaccines are prepared from animal nerve tissue. However, new vaccines are being developed.

### 20.4.6 Influenza

This disease is also called flu. It is characterised by chills, fever, headache and muscular pain. Recovery is possible in few days. The influenza virus has eight distinct RNA fragments of differing lengths enclosed by a layer of protein and lipid bilayer. There are numerous projections namely hemaegglutinin spikes and neuraminidase spikes. As on today, it has not been possible to make a vaccine for the control of influenza. An antiviral drug, Anamtadine is known to control flu.

### 20.4.7 AIDS

AIDS (Acquired Immunodeficiency Syndrome) is a sexually transmitted disease. AIDS results in disabling cognitive or motor dysfunction that interferes with the patient's ability to function and HIV wasting disease (Involvementory weight loss). AIDS was first identified in 1981 and cases doubled every year. It is a sexually transmitted disease among homosexual men, intervener drug users, haemophilia patients etc. AIDS virus was isolated both from semen and saliva. The virus gets transmitted by sexual contact or chiefly by contaminated blood but not by casual contacts. It is caused by Human immunodeficiency virus, a lenti virus of Retroviridiae.

The structure of HIV virus includes RNA as a central core. It destroys immune system. It is 100 nm in diameter containing a protein envelope to which spicules or glycoproteins are attached. The envelope has icosahedron, capsid core that has macromolecules of RNA as a genetic material. It is a sphere made up from assembles of 12 pentameres and 20 hexameres. HIV is of 100 nm and is covered by a protein envelope. On the envelope glycoprotein spicules are found. Protein envelope encloses Icosahedral capsid core, which has dentical macromolecules of RNA as genetic material. It is a sphere made up from assembles of 12 pentameres and 20 hexameres. HIV is of 100 nm and is covered by a protein envelope. On the envelope, glycoprotein spicules are found. Protein envelope encloses icosahedral capsid core, which has identical macromolecules of RNA as genetic material. Glycoprotein hexameres play very important role in binding HIV to its host. The glycoprotein 120 is attached to glycoprotein 41. Under the protein envelope, two helix of RNA molecules in the folded form gets attached to the reverse transcriptase. Reverse transcriptase controls the transformation of RNA into DNA. The spicules of the protein envelope of HIV get attached to the target cell. The glycoprotein 120 has affinities to T4 or Cd4 molecule. The molecules are secreted by the lymphocytes which play a major role in the immune system. AIDS symptoms are expressed after their destruction. The HIV infected cells produce antibodies against glycoprotein 120, which is recognised as an antigen. The infected individual spreads the infection to the sexual partner. T4 lymphocytes become incapable to secrete interleukins. The monophages that are infected become the reservoir of viruses. In HIV infected patients, lymphocytes, macrophages, and T4 cells are destroyed, thus leading to the failure of immune system. HIV replicates in host cell after coming out of capsid envelope.

The HIV is a sexually transmissible disease and is prevalent among homosexuals, prostitutes, drug users, blood transfusion patients, haemophilic and children born to HIV infected mothers. It is also spread by contaminated needles, and syringes. India too has the largest number of HIV patients. Initially the first HIV infected case was reported in 1986. There are 3.20 million suspected HIV infected persons.

HIV infection leads to the total breakdown of immune system. Government of India has launched the National AIDS programme, emphasizing the increased awareness of AIDS besides screening blood samples every year. 1st December is celebrated as world AIDS day. Prevention and control of AIDS is possible by educating the people.

## 20.5 DISEASES CAUSED BY OTHERS

### 20.5.1 Scabies

It is a skin disease caused by *Sarcoptes scabies*, an arthropod. It is common among schoolchildren. The fingers, wrists and elbows are infected. The mites enter the skin, lay eggs and multiply. It results in hypersensitivity. Symptoms are itching and red lesions appear. It is treated by the topical application of benzene hexachloride and cleaning of beds etc.

### 20.5.2 Candidiasis

*Candida albicans*, a fungus (yeast), causes the disease. Candidiasis is very common among newborn infants and is called thrush. It also causes vaginitis. The fungus produces white patches on the tongue and mouth parts. It can be controlled by topical application of micanozole, nystatin etc.

### 20.5.3 Trachoma

It is an infection caused by *Chlamydis trachomatis*, which causes blindness. It is transmitted by hand contact or by towels. Flies may also carry the bacteria. The disease is a conjunctivitis leading to blindness. It can be treated by tetracyclines.

### 20.5.4 Meningitis or Cryptococcosis

*Cryptococcus,* the yeast is the disease causing agent. *Cryptococcus neoformans* is the only pathogen for humans causing the disease (Fig. 20.4). The inoculum comes from soil, pigeon

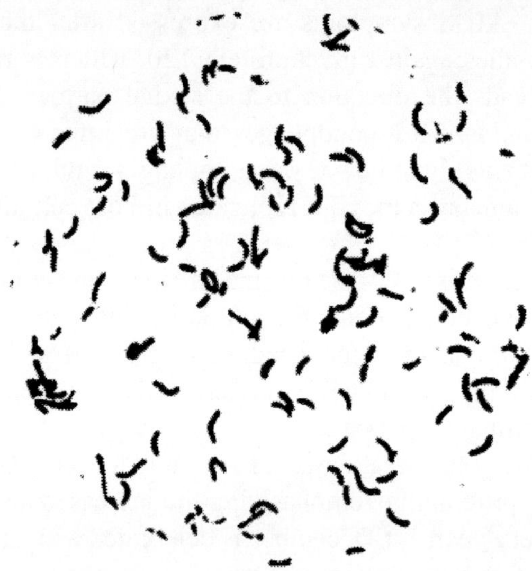

**Fig. 20.4.** Cells of *Cryptococcus neoformans.*

dropping etc. The inhalation of pigeon dropping causes this disease. It initially infects lungs and spreads through blood stream including the brain and meninges. It is finally expressed as meningitis, which is usually progressive and fatal if untreated. The drugs used for treatment are Amphotericin-B and Flucytosine.

## 20.6 DERMATOPHYTIC FUNGI CAUSING DISEASES IN HUMANS AND ANIMALS

Fungal infection is caused by moulds or yeasts that happen to find the right set of conditions for growth in the human body and virtually no animal is free from the possibility of incurring an invasion mast frequently by fungi that are found in soil. Many of the fungi initiating diseases in man and animals are closely related and in a few cases some species can infect both. In such cases, animals act as reservoir of the inoculum and increase the chances of exposure of humans to infections. Some fungal diseases are confined to specific geographical areas, others are worldwide in distribution while some are seen only in certain populations, relating to age, sex, race, occupation, or general physical conditions. Fungal infection of man and animals are usually categorized into four categories: (i) superficial, where only the outer surface of the skin is invaded; (ii) cutaneous, where the keratinized tissues of the body (the outer epidermal cells of the skins, hair and nails) are invaded, (iii) subcutaneous where invasion occurs in the deeper layers of the skin, extending into muscles, sometimes into the bone, and may involve the lymphatic system; and (iv) systemic, where the infection of lungs takes place by direct inhalation of fungal spores and from the lungs the infection may be disseminated to all organs of the body and may cause death in man and animals. The fungi that are able to invade human tissues are divided in two groups: (a) the primary pathogens which are capable of producing special kinds of fungal infection in healthy individuals and have specific clinical characteristics; and (b) the opportunists which will only invade the tissue of a compromised host. The importance of the study of fungi inciting human mycoses is becoming recognized in recent years. Medical mycology, which was considered to be the domain of the medical man is now taken up by the mycologists also. The subject is gaining importance rapidly. At present there are over 80 human pathogens among the approximately 1,00,000 species of fungi. Thirumalachar (1968) has discussed the saprophytic habitats of causative agents of human myoses in India and the gamut/ galaxy of the fungal infections in Tamil Nadu state by Kamalam (1984). Ecology of human pathogenic fungi has been discussed by Srivastava (1984). Some of the important mycoses and their causative agents are given the Table 20.1.

## 20.7 DISEASES CAUSED BY ACTINOMYCETES

These organisms have been shown to be higher bacteria, but they were thought to be fungi for many years because they have filamentous forms, 0.5 to 0.8 microns in diameter, which appear to branch. Some species form aerial mycelia in culture. The clinical manifestations of infection are similar to those of a systemic fungal infection. It is now clear that they are not fungi but are closely related to the mycobacteria. Some facts that one should know about these genera are that: *Actinomyces* are anaerobic, while *Nocardia* and *Streptomyces* are aerobic. *Nocardia* stain partially acid-fast, *Actinomyces* and *Streptomyces* are not acid-fast. *Actinomyces* produce granules. Most actinomycetes in tissue do not stain with the H & E stain commonly used for general histopathology. All genera may produce granules; *Actinomyces* almost always produce granules.

### 20.7.1 Actinomycosis

Actinomycosis is a chronic suppurative and granulomatous disease of the cervico-facial, thoracic or abdominal areas. The most common cause of actinomycosis is the organism *Actinomyces israelii* which infects both man and animals. In cattle, the disease is called "lumpy jaw" because of the huge abscess formed in the angle of the jaw. In man, *A. israelii* is an endogenous organism that can be isolated from the mouths of healthy people. Frequently, the infected patient has a tooth abscess or a tooth extraction and the endogenous organism becomes established in the traumatized tissue and causes a suppurative infection. These abscesses are not confined to the jaw and may also be found in the thoracic area and abdomen. The patient usually presents with a pus-draining lesion. By rotating the vial of pus, the yellow sulphur granules, characteristic of this organism, can be seen with the naked eye. One can also see these granules by running sterile water over the gauze used to cover the lesion. The water washes away the purulent material leaving the golden granules on the gauze. This organism, which

**Table 20.1.** Dermotophytic diseases (courtesy of Prof. R. Sin)

| Classification | Disease | Affected area | Organism |
|---|---|---|---|
| Superficial cutaneous mycoses | Black piedra | Scalp and beard | *Piedraia hortae* |
| | White piedra | Beard, scalp, | *Trichosporon beigelii* |
| | | Pubic hair | *Pityrosporum orbiculare* |
| | Tinea vesicular | Smooth body | *Cladosporium werneckii* |
| | Tinea nigra | Skin, thick stratum Corneum, palms and feet | *C. manonii* |
| | Dermatomycoses | Keratinized layers of body; skin, hair and nails | *Epidermophyton, Microsporum and Trichophyton spp.* |
| | Tinea capitis | Scalp | *Microsporum canis, M. audouinii*<br>*Trychophyton tonsurans* |
| | Tinea favosa | Scalp, skin, nails | *T. schoenlenii* |
| | Tinea unguium | Nails | *T. rubrum*<br>*T. mentagrophytes*<br>*Epidermophyton floccosum* |
| | Tinea barbae | Beard and coarse body hair | *T. verrucosum* |
| | | Smooth body hair | *T. mentagrophytes* |
| | Tinea corporis | Smooth body hair | *T. rubrum* |
| | Tinea imbricata | Feet, interdigital spaces and soles | *T. mentagrophytes*<br>*T. concentricum* |
| | Tinea pedis | Palms and fingers | *T. rubrum*<br>*T. mentagrophytes*<br>*E. floccosum* |
| | Tinea manuum | Groin | *T. rubrum*<br>*T. mentagrophytes*<br>*T. floccosum* |
| | Tinea cruris | Nonbearded area of adult males, face of adult females and children | *T. rubrum*<br>*T. mentagrophytes* |
| | Tinea facei | Groin and genital areas | *T. violaceum*<br>*T. rubrum*<br>*T. rubrum* |
| | Tinea axillaries | Moist skin areas; groin, glans penis scrotum, folds of buttocks, under the breast axilla, inter-digital spaces | *T. tonsurans*<br>*M. gypseum*<br>*T. mentagrophytes*<br>*T. tonsurans*<br>*E. floccosum* |
| | Intertriginous candidiasis | Hands, feet, face, scalp and nails | *C. albicans* |
| | Candidal granuloma | Mucocutaneous areas | *C. albicans* |

*(Continued)*

**Table 20.1.** *Continued*

| Classification | Disease | Affected area | Organism |
|---|---|---|---|
| | Onychomycosis | Mouth and tongue, corners of mouth, vagina | *Candida sp.* |
| | Mucocutaneous candidiasis | Anal area<br>Esophagus | Scopulariopsis sp.<br>*Aspergillus sp.* |
| | Thrush | Ear | *C. albicans* |
| | Perleche | | *C. albicans* |
| | Vaginal candidiasis | | *C. albicans* |
| | Perianal candidiasis | | *C. albicans* |
| | Esophageal candidiasis | | *C. albicans* other *Candida sp.*, *Aspergillus* sp., other common yeasts and molds |
| | Otomycosis | | *C. albicans* |
| Subcutaneous mycoses | Chromomycosis | Skin surface, mostly lower extremities | *Fonsecaea pedrosoi*<br>*F. compactum*<br>*Phialophora verrucosa*<br>*Wangiella dermatitidis*<br>*Cladosporium carrionii* |
| | Mycetoma | Skin surface, mostly lower extremities of foot and is of (maduramycosis, madura various types; white or foot) yellow or black | *Allescheria boydii*<br>*Aspergillus sp.*<br>*Madurella mycetomii* |
| | | Skin, primarily hands, arms and legs | *M. grisea, ASS, actinomycetes* |
| | Sporotrichosis | Nasal mucosa | *Sporothrix schenckii* |
| | Rhinosporidiosis | Smooth skin | *Rhinosporidium seebri* |
| | Lobomycosis (Lobo's disease) | | *Loboa loboi* |
| Systematic mycoses | Histoplasmosis | Lung, reticulo-endothelial system, bone, kidney and other organs | *Histoplasma capsulatum* |
| | Coccidiomycosis | Lungs | *Coccidiodes immitis* |
| | Blastomycosis | Lungs, may spread to other organs, skin lesions are common | *Blastomyces dermatitidis* |

*(Continued)*

**Table 20.1.** *Continued*

| Classification | Disease | Affected area | Organism |
|---|---|---|---|
| | Cryptomycosis | Lungs, central nervous system, skin<br>Subclinical infection in lung, mucous membranes and skin | *Cryptococcus neoformans* |
| | Paracocciciomycosis | Lung, skin, mucocutaneous tissue, other body organs, face, sinuses, gastrointestinal tract, lungs | *Paracoccidioides brasiliensis* |
| | | Aspergillosis | *Aspergillus fumigatus* |
| | Mucoromycosis | Brain or central nervous system | *Absidia corymbifera* and other members of Mucorales *e.g. Mucor, Rhizopus* |

occurs worldwide, can be seen histologically as "sulphur granules" surrounded by polymorphonuclear cells (PMN) forming the purulent tissue reaction. The organism is a gram-positive rod that frequently branches. The laboratory must specifically be instructed to culture this anaerobic organism. These lesions must be surgically drained prior to antibiotic therapy and the drug of choice is large doses of penicillin (18–20 million units/6 h).

## 20.7.2 Nocardiosis

Nocardiosis primarily presents as a pulmonary disease or brain abscess in the U.S. In Latin America, it is more frequently seen as the cause of a subcutaneous infection, with or without draining abscesses. It can even present as a lesion in the chest wall that drains onto the surface of the body similar to actinomycosis. Brain abscesses are frequent secondary lesions. The most common species of *Nocardia,* which cause disease in human beings, are *N. brasiliensis* and *N. asteroides*. These are soil organisms that can also be found endogenously in the sputum of apparently healthy people.

*N. asteroides* is usually the etiologic agent of pulmonary nocardiosis, while *N. brasiliensis* is frequently the cause of subcutaneous lesions. The material sent to the lab, depending on the presentation of the disease, is sputum, pus, or biopsy material. These organisms rarely form granules. The *Nocardia* are aerobic, gram-positive rods and stain partially acid-fast (i.e., the acid-fast staining is not uniform). There are no serological tests, and the drug of choice is Bactrim (Trimethoprim plus sulfamethoxazole). The *Nocardia* grows readily on most bacteriologic and TB media. The geographic distribution of these organisms is worldwide.

### 20.7.3 Streptomycosis

The *Streptomyces* species usually cause the disease entity known as mycetoma (fungus tumour). These infections are usually subcutaneous, but they can penetrate deeper and invade the bone. Some species produce a protease, which inhibits macrophages. Material sent to the lab is pus or skin biopsy. The streptomycetes are aerobic like *Nocardia,* and can grow on both bacterial and fungal (Sabouraud) media. They produce a chalky aerial mycelium with much branching. It is important to let the lab know the organism you suspect because most bacterial pathogens will grow out overnight, but the actinomycetes take longer to be visible on the culture plates (48–72 hours). The various species of *Streptomyces* produce granules of different size, texture and colour. These granules along with colonial growth and biochemical tests allow the bacteriologist or mycologist to identify each species. The organisms are found worldwide. There are no serological tests, and the drugs of choice are the combination of sulfamethoxazole/trimethoprim or amphotericin B. In the tropics, this disease may go undiagnosed or untreated for so long that surgical amputation may be the only effective treatment.

## 20.8 MYCOPLASMAL, BACTERIAL AND VIRAL DISEASES IN PLANTS AND ANIMALS

### 20.8.1 Mycoplasmal Diseases of Plants

In 1967, wall-less microbes resembling mycoplasmas were seen with electron microscope in the phloem of plants infected with one of several yellow type diseases. Such diseases were till then thought to be caused by viruses. Similar microbes in the same year were also seen in the insect vectors of these diseases. These microbes were shown susceptible to tetracycline and not penicillin. Since then about hundred plant diseases affecting several hundred genera of plants have been shown to be caused by mycoplasmas.

Morphologically, mycoplasmas associated with plant diseases resemble typical mycoplasmas found in animals and man and those living saprophytically. However, plant mycoplasmas could not be grown on artificial nutrient media. Also, so far no plant disease has been reproduced on healthy plants inoculated directly with mycoplasmas obtained from diseased plants. The pathogens of only two diseases, citrus stubborn and corn stunt could be grown on nutrient media and have also been reproduced on healthy plants. But both these are different from typical mycoplasmas in having helical structure, being motile and in some other respects. There are three genera, *Mycoplasma, Acholeplasma* and *Spiroplasma*.

True mycoplasmas (*Mycoplasma* and *Acholeplasma*) are the prokaryotes that lack a true cell wall and are bounded by a single trip layered "until" membrane. They are small, sometimes ultra-microscopic containing cytoplasm, randomly distributed ribosomes and strands of nuclear material. They measure 175–250 nm dia. during reproduction but later grown into various shapes and sizes. They reproduce by budding and binary fission of coccoid and filamentous cells. They lack flagella and spores and are Gram-negative.

Among the most important plant diseases caused by mycoplasmas are aster yellows, apple proliferation, coconut lethal yellowing, elm phloem necrosis, papaya bunchy top, peach X-diseases, peach yellows, pear decline, stolbur diseases of solanaceous plants, tomato big bud, clover phyllody, little peach, sugarcane white leaf, witch's broom of groundnut etc.

## Little Leaf of Brinjal

The disease, perhaps present only in our country, is found in all brinjal growing areas. It was first reported from Coimbatore (Tamil Nadu). The disease causes heavy losses and young plants fail to produce flowers and fruits due to infection by Mycoplasma.

### *Symptoms*

The chief symptoms are very short leaves produced on the attacked plants. The petioles become very much reduced and the leaves appear very close to the branches and stem. Attacked leaves are narrow, soft, smooth and yellowish in colour. Any new leaf coming out also becomes reduced in size. The internodes are much shortened and at the same time large number of axillary bunds are stimulated to grow into short branches with small leaves. This gives the plant a bushy habit. The plants do not form flowers. If flowers are formed at all, they remain green. Fruits are formed rarely (Fig. 20.5).

### *Causal Organism*

The disease is caused by a mycoplasma. Till 1969, the disease was thought to be caused by a virus. The structures from sap appear spherical or oval in shape, 40–300 μm in dia. and without a cell wall. In the field, the disease is transmitted by the vector, *Cestius phycitis.* This mycoplasma also attacks *Datura fastuosa* and *Vinca rosea.* It is possible to transmit the agent to tomato, potato and tobacco. Perhaps the agent survives and weed hosts also.

**Fig. 20.5.** Little leaf disease of brinjal.

**Control:**

1. Antibiotic treatment with tetracyclines, as achromycin, terramycin, aureomycin etc.
2. Eradication of weed hosts and diseased brinjal plants from the field.
3. Control of insect vectors by insecticides.

### Sandal Spike Disease

Sandal (*Santalum album* L.) is an economically important tree in India. It is grown extensively in the southern parts of the country, in states of Karnataka, Tamil Nadu and Kerala. Spike disease is the most destructive disease of the plant.

Shortening of the internodes: This results in crowding of leaves on the leaf bearing branches. The infected leaves give an appearance of spike-like structures. In advanced stages of the disease, the leaves become yellowish and finally reddish. The flowers show phyllody (virescence). The diseased parts rarely bear any fruit.

The other form of the disease, known as "pendulous spike", characterized by continuous apical growth of individual shoots, results in a drooping habit of the shoot. The dormant buds do not develop or grow and, therefore, no resetting effect is seen. In this form of the disease, roots and haustoria are not damaged.

**The Causal Organism:** The disease is caused by mycoplasma-like organism. It is reported to infect a large number of plant species such as *Eucalyptus grandis, Vinca rosea, Zizyphus oenoplea, Dodonia viscose,* etc. The disease can be transferred from *Vinca rosea* to sandal and *vice versa* through dodder. The spike disease is spread through root contact and insect vectors. Leaf hoppers like *Nephotettix virescence* and *Moonia albimaculata* seem to be involved in transmission of the disease.

**Control of the Disease:** Treatment with tetracycline offers control of the disease to some extent. The systemic fungicide Benlate was also reported to bring about temporary control of disease. Application of other antibiotics like terramycin alone or in combination with Benlate has been recommended. Ledermycin alone or in combination with Benlate also has been recommended as a control measure. However, these treatments provide only temporary relief.

## 20.8.2 Viral Diseases of Plants

### Transmission of Viral Diseases

Plant viruses rarely come out of the plant spontaneously. Therefore, viruses are not disseminated by wind or water. Even when they are carried in debris or plant sap, they would cause infections only when they come in contact with the contents of wounded live cell. Viruses are transmitted from plant to plant (i.e., they actually spread) in a number of ways; vegetative propagation, mechanical transmission through sap, seed, pollen, insects, mites, nematodes, fungi, dodder etc.

**1. Vegetative propagation**

Any part of plant used for vegetative propagation will transmit the viruses to the progeny. Thus viruses are transmitted by budding, grafting, cuttings, tubers, corms, bulbs, rhizomes etc. This mode of transmission is most important for ornamental trees and shrubs (propagated by budding, grafting or cuttings) and field crops like potato and most florist's crops that are usually propagated by tubers; corms or cuttings particularly in trees, viruses are transmitted through natural root grafts of adjacent plants. For several tree viruses, natural grafts are the only known means of tree-to-tree spread of the virus.

**2. Mechanical transmission**

**Through sap:** Under natural conditions, direct transfer of sap through contact of one plant with another is uncommon. Such mechanical transmission may take place between closely placed plants. Strong wind may cause the leaves of adjacent plants to rub together, and if wounded, some of their sap is exchanged, thus transmitting any virus present in the sap. Potato Virus X (PVX) is most easily transmitted in this way.

Plants are wounded by man during cultural practices in field or greenhouse and some of the virus infected sap adhering to tools, hands or clothes is accidentally transferred to wounded plants. TMV on tobacco and tomato spread rapidly in this way. Occasionally, virus infected sap is transferred from one plant to another on the mouthparts or body of animals feeding on and moving among the plants. Mechanical transmission is the only authentic means to prove the ability of infection by a virus.

**Seed transmission:** This is not as common as the above said two methods. More than 100 viruses are reported to be transmitted by seed. However, only a small portion of (1–30%) the seeds derived from virus infected plants transmit the virus. In some, such as tobacco ringspot virus in soybean, almost 100% of seeds or infected plants transmit the virus. Barley stripe mosaic virus also has high transmission of 50–100% seeds. In most cases, the virus comes primarily from the ovule of infected plants.

**Pollen transmission:** Virus transmitted by pollen not only infects the seed and the seedling growing from it, but more important, it can also spread through fertilized flower and down into the mother plant, which thus becomes infected with the virus. This method is important in stone fruit ringspot virus of some plants.

**Mite transmission:** Members of the family Eriophidae are shown to transmit nine viruses, including wheat streak mosaic, peach mosaic and fig mosaic viruses. These mites have piercing and sucking mouth parts. Mite transmission appears to be specific, since each of the species has a restricted host range. In such cases, a particular mite is the only known vector for the virus or viruses transmitted by it. Some of the mite transmitted viruses are stylet-borne, whereas others are circulatory. One mite (Tetranychidae, spider mites) is known to transmit Potato virus Y.

**Nematode transmission:** About 12 viruses are shown to be transmitted by one or more species of three genera of soil inhabiting, ectoparasitic nematodes. Members of the genera Longidorus and Xiphinema are vector of tobacco ringspot, tomato ringspot, raspberry ringspot, cherry leaf roll, tomato black ring, grape fanleaf viruses etc., whereas those of the genus Trichodorus transmit tobacco rattle and pea early browning viruses. Nematodes feed on roots of infected plant and then move on to roots of healthy plants. Larvae as well as adults can acquire the viruses.

Four genera of nematodes, all belonging to Dorylaimoidea, are known to be the vectors of plant viruses, which are either polyhedral (neproviruses or nematode transmitted polyhedral viruses) or tubular (tobraviruses, netuviruses or nematode transmitted tubular viruses). Viruses are transferred into the host by the stylet of the nematodes. The nematodes that transmit plant viruses are *Xiphinema, Longidorus, Trichodorus* and *Paratirchodorus*.

**Fungus transmission:** *Olpidium,* a root infecting chytrid fungus, is known to transmit at least four viruses, tobacco necrosis, cucumber necrosis, lettuce big vein and tobacco stunt viruses. Four other fungi, *Synchytrium, Polymyxa, Spongospora* and *Pythium* transmit potato virus X, wheat mosaic virus, potato mop top virus and beet necrotic yellow vein virus, respectively. The viruses are present in or on the zoospores and resting spores.

The fungi belonging to class Plasmodiophormycetes (family Plasmodiophoraeceae) and class Chytridiomycetes (family Olpidiaceae and Synchytriaceae) transmit plant viruses. The species are *Spongospora subterranean, Polymyxa graminis, Polymyxa batae, Synchytrium endobioticum, Olpidium brassicae* and *Olpidium cucurbitacearum*. These fungi infect their host roots by zoospores, which carry the virus externally or internally.

**Dodder transmission:** Several viruses can spread through the bridge formed between plants by twining stems of the parasitic phanerogam, dodder (*Cuscuta* sp.). Many viruses spread in this way between plants of widely different taxonomic species.

**Insect transmission:** The most common and economically the most important method of transmission of viruses in the field is by insect vectors. The order Homoptera which includes aphids and leafhoppers contains the largest number and the most important insect vectors of plant viruses. White flies, the mealy bugs, scale insects and the treehoppers, belonging to this order also transmit viruses but are not economically as important as aphids and leafhoppers. Other insect vectors are true bugs (Hemiptera), thrips (Thysanoptera) beetles (Coleoptera) and the grasshoppers (Orthoptera). The most important vectors, i.e., aphids, leafhoppers and other members of Homoptera as well as true buds have piercing and sucking mouthparts; all other vectors have chewing mouthparts and the transmission by the latter is much less common.

On the basis of virus-vector relationship (where virus is carried by vector; and behaviour of virus within the vector), the viruses are also categorized as follows:

(a) *Non-persistent or stylet-borne viruses:* These are the viruses carried by insects (with sucking mouthparts) on their stylets.
(b) *Persistent or circulative viruses:* These are the viruses accumulated by the insects (with sucking mouth parts) internally and after their passage through insect tissues, they are introduced again into plants through the mouthparts.
(c) *Propagative viruses:* These are some of the circulative viruses that may multiply in their vectors.

The viruses transmitted by insect with chewing mouthparts may also be circulated or they may be carried on the mouthparts (i.e., stylet borne).

Aphids are the most important insect vectors of plant viruses and transmit the great majority of all stylet-borne viruses. More than 200 species are known as vectors of plant viruses. As a rule, several aphid species can transmit the same stylet-borne virus and the same aphid species can transmit several viruses, but in many cases the vector-virus relationship is quite specific. Aphids generally acquire the stylet-borne virus after a brief feeding period on the disease plant for only a few seconds (30 seconds or less) and can transmit the virus after transfer to and feeding on a healthy plant for a similar brief period. After acquisition, the aphids remain viruliferous from a few minutes to several hours, after which they can no longer transmit the virus. In some cases of aphid transmission of circulative viruses, aphids cannot transmit the virus immediately but must wait several hours after the feeding. However, once they begin to transmit the virus, they continue to do so for many days. The viruses are borne on the tips of the stylets. Several species of leafhoppers are involved in transmission of a number of plant viruses. Leafhopper transmitted viruses cause disturbance primarily in the phloem region. All leafhopper transmitted viruses are circulative and several are known to multiply in the vector i.e., are propagative. Most of the leafhoppers require a feeding period (one to several days) before they become viruliferous, but once they acquire the virus, they remain viruliferous for the rest of their lives.

## Tobacco Mosaic Disease

This is the best known of all virus diseases of plants and is worldwide in distribution. This disease affects more than 150 genera of primarily herbaceous, dicotyledonous plants including many vegetables (potato, tomato, cucurbits), flowers and weeds. There are serious losses in yield as well as quality in tobacco, tomato and some other crop plants. TMV affects plants by damage of leaf, flower and fruit and causes stunting of the plant.

**Symptoms:** The symptoms include various degrees of chlorosis, curling, mottling, dwarfing, distortion and blistering of leaves, dwarfing of the entire plant, dwarfing, distortion and discolouration of flowers and in some plants even development of necrotic areas of leaf (Fig. 20.6).

The most common symptom of tobacco is the appearance of mottled dark-green and light green areas on leaves. The dark green areas are thicker and appear somewhat elevated in a blister-like manner over the thinner, chlorotic, light green areas. Stunting of young plants is common and is accompanied by a slight downward curling and distortion of leaves that may become narrow and elongated rather than normal oval shape. The petioles may become enlarged (puckered) with enlarged capitate hairs. Old leaves may not show symptoms, young ones develop typical symptoms.

**Causal Organism:** *Tobacco Mosaic Virus* (TMV) is rod-shaped, 300 nm long by 15 nm in diameter. Protein (P) consists of 158 amino acids. The protein subunits are arranged in a helix. The nucleic acid (NA) is single-stranded RNA and consists of about 6400 nucleotides. The RNA strand also forms a helix parallel with that of protein and is located on the protein subunits and approx. 20 A out from the inner end of the protein subunits. The mol. wt. of each virus particle is between 39 and 40 million mol. wt. units.

**Fig. 20.6.** Tobacco mosaic virus disease.

TMV is one of the most thermostable viruses known, the thermal inactivation point of the virus in undiluted plant juice being 93°C. However, in dried infected leaves, the virus retains infectivity even when heated at 120°C for 30 min. Infected plant may contain up to 4g of virus per litre of plant juice and the virus retains infectivity even at dilution of 1:1,000,000. In ordinary plant sap, the virus is inactivated in 4–6 weeks, whereas in sterile bacteria free sap, the virus may survive for 5 years and in TMV infected leaves kept dry in the laboratory, the virus remains infectious for more than 50 years. TMV is transmitted readily through mechanical sap, grafting and dodder. It is not transmitted by insects, except occasionally through their contaminated feet and jaw. The most common method of transmission of TMC in field and greenhouse is through hands of workers handling infected and healthy plants.

**Disease Development:** TMV survives in infected leaves and stalks in the soil, on surfaces of contaminated seeds, on contaminated seedbed cloth, and in natural leaf and manufactured tobacco including cigarettes, cigar etc. The virus initially infects wounded tissues of tobacco seedlings in seedbed or transplants in the field. Then it spreads in the field throughout the season. TMV in all plants produces systemic infections, invading all parenchyma cells of plant. The virus moves from cell to cell through phloem. In the cytoplasm of cell, TMV appears as crystalline aggregates and amorphous bodies (x-bodies).

**Control:**

1. Sanitation is the main method. Crop should not be grown at least for two years in seedbeds or fields where diseased crop was grown. Removal of diseased plants and some solanaceous weeds harbouring the virus early in the season helps in reduction and elimination of subsequent spread of the virus.
2. Chewing and smoking of tobacco during handling of tobacco and other susceptible plants should be avoided.

3. Workers in the field must wash hands with 3% trisodium phosphate or soap.
4. Equipment and instruments used in plantations must be sterilized.
5. TMV resistant varieties of tobacco must be grown, though these may be of low quality.

## Yellow Vein Mosaic of Bhindi

This disease of *bhindi* (Okra) has taken a severe form in our country. Infection of four- to five-week-old plants results in usual retarded growth and only few leaves and fruits are formed on such plants, resulting in about 94% loss of the crop yield. The damage is relatively less on old plants.

**Symptoms:** The chief symptoms develop on leaves. There occurs vein clearing and veinal chlorosis of leaves. Infected leaves exhibit a very distinct yellow network of veins alongwith the thickened veins and veinlets. In severe cases, the chlorosis may involve interveinal areas resulting into yellowing of entire leaf. Fruits are dwarfed, malformed, distorted and yellow green in colour (Fig. 20.7).

**Causal Organism:** The Yellow Vein Mosaic Virus (YVMV) of *bhindi* is not transmitted mechanically through sap. It may be transmitted artificially through grafting. In the field, YVMV is transmitted by white fly, *Bremisia tabaci* and perhaps also by *bhindi* leafhopper, *Empoasca devastans*. There are several weed hosts of this virus. These are *Croton sparsiflora, Malvastrum tricuspidotum* and *Ageratum* sp., growing along the roadside and in wasteland areas.

**Control:**

1. Protection of crop from white flies and other insects by spraying with Follidol (0.3%) or other insecticides. Spray must be done early, just after the seedlings come out, say within three after germination.

**Fig. 20.7.** Yellow vein mosaic disease of *bhindi*.

2. Four to six sprays of systemic insecticides, ekatox, metasystox, rogor, etc.
3. Eradication of weed hosts.
4. Disease resistant cultivars. Some wild species of *Hibiscus* and *Abelmoschus* have been found as good source of resistance.

## Leaf Mosaic of Cucurbits

The leaf mosaic virus disease affects a number of cucurbits in almost every area where they are grown. This disease is very common on several cultivated cucurbits including fruit and vegetable crops. It has a wider host range. One of the viruses on cucurbits for example is the cucumber mosaic, which attacks a greater variety of vegetables, ornamentals and other plants. Among the most important vegetables and ornamentals attacks by this virus are cucumbers, melons, squash, peppers, spinach, tomato, celery, beets, banana, gladiolus, petunias etc.

**Symptoms:** Young seedlings are never attacked in the field during the first few weeks. Generally 5–6-week-old plants are attacked. Young leaves become mottled, distorted and wrinkled and their edges begin to curl downward. Plants become dwarfed and they produce few flowers and fruits. The plants develop bushy appearance, with leaves forming a rosette-like clump near the ground. Fruits develop pale green or white areas intermingled with dark green, raised areas.

**Causal Organism:** The virus is polyhedral, 300 nm dia. It consists of 180 protein subunits, single-stranded RNA and a hollow core. The mol. wt. is 5.8–6.7 million of which 18% is RNA and the rest 82% protein. The thermal inactivation point is about 70°C, while dilution end point is about 1: 10,000. There are many strains of cucumber mosaic virus alone.

The virus is readily transmitted by sap and also by many aphids. The virus survives in many perennial weeds, flowers and crop plants. Insect vectors and cultivation and handling of plants, especially at picking time spread the virus in field of healthy plants. Sometimes the whole field of cucurbits begin to turn yellow with mosaic, immediately after the first pick. The virus produces systemic infection in cucurbits and most other host plants.

**Control:**

1. Elimination of weed hosts.
2. Control of insect vectors.
3. Resistant varieties.

## Potato Spindle Tuber

The disease is caused by a viroid and causes severe losses. In some regions it is one of the most destructive diseases of potatoes. It attacks all varieties and spreads rapidly. Though this disease also attacks tomato, it is of little economic value in this crop.

**Symptoms:** Infected potato plants become erect, spindle-shaped and dwarfed. The leaves are small and erect and the leaflets are dark green, occasionally being rolled and twisted. The tubers are elongated with cylindrical middle and tapering ends. The tubers are smoother with more numerous, conspicuous and shallower eyes. Yield is reduced to about 25% or more.

**Causal Organism:** The *Potato Spindle Tuber Viroid* (PSTV) is the first recognized viroid. PSTV is an infectious RNA of low mol. wt., approx. 80, 000 daltons. The RNA is single-stranded molecule with extensive regions of base pairing. Under electron microscope, purified denatured PSTV appears as short strands about 50 nm long with a thickness of a double-stranded DNA. Sap from infected plant is infective even at dilutions of 1:1000 to 1:10,000. After heating at 75 to 80°C for 10 min, PSTV is quickly inactivated in expressed sap of infected plant. However infectivity remains intact in phenol treated sap. PSTV is mechanically transmissible. It spreads mainly by knives used to cut healthy and infected potato "seed" tubers and during handling and planting of the crop. This viroid is also transmitted by pollen and seed and by several insects including some aphids, grasshoppers, flea beetles and bugs.

The viroids infects and then spreads systemically throughout the plant. Little or nothing is known about the mechanism of spread within the plant and the development of symptoms.

**Control:** Planting of only PSTV free potato tubers in the field.

### 20.8.3 Bacterial Diseases of Plants

#### Bacterial Leaf Blight of Rice

The bacterial leaf blight (BLB) of rice, known since 1881 in Japan, became quite severe in India, since last 15 years or so, though reported first in 1951. Bhapkar et al. (1960) appear to be the first to have described the typical BLB symptoms in India. Previously it remained confined to Maharashtra, but later appeared in Bihar and northern part of the country. The disease causes much damage in the Tarai region of U.P. where weather conditions are usually favourable for its appearances and development (Fig. 20.8).

Damage is due to partial or complete blighting of the leaves or due to complete wilting of the affected tillers leading to unfilled grains. In the wilt or "Kresek" phase, which results from systemic infection in the nursery or from seed inoculum, the crop may dry up completely before seed maturation. In India, the per cent loss in the yield has been estimated from 6 to 60 per cent.

**Fig. 20.8.** Bacterial leaf blight of rice.

**Symptoms:** The symptoms vary with the stage of infection and the prevailing weather conditions. The most apparent symptom is the leaf blight phase of the disease. During this phase, linear yellow to straw-coloured stripes with wavy margin develop usually on both the edges of leaf. The stripes usually develop from the tip towards the base. Sooner or later, the tips dry and become twisted and the stripes extend in length and breadth to cover major area of leaf. In severe state, the leaf sheath and culms also become blighted, causing the death of tiller or whole clump. In dry weather, turbid drops of bacterial mass ooze out. This mass on drying appears as yellowish beads on leaf surfaces. The drops are washed down by rains and glumes may also get infected. Blight phase usually appears about 4–6 weeks after transplanting. The cut ends of diseased leaves ooze out the bacterial mass.

The most destructive phase of BLB is "Kresek" or wilt due to early systemic infection. Leaves of infected tiller suddenly become dry, turn greyish green and begin to roll along the midrib. Severely affected tillers wither away and finally the entire stool (plant) is killed. This symptom is common in tropical areas. This phase may not be confused with stem borer attack, where affected tillers can be easily pulled out, which is not possible with "Kresek".

**Causal Organism:** BLB is caused by *Xanthomonas campestris* p.v. *oryzae*. The bacterium is a Gram-negative rod, 0.5–0.8 × 1.0–2.0 μm, no spores and no chains, with a single polar flagellum. The bacterium is aerobic. Nitrates are not reduced. Thermal death point is 53°C.

The source of primary inoculum is not completely known. The bacterium in India is reported to occur on some grasses, the collateral hosts. The pathogen is reported to perpetuate in the diseased straw left over in the field and also in stubbles. In the south, where two or three crops of rice are cultivated successively, the pathogen may survive on crop debris in soil. There is evidence that bacterium is carried over through infection seeds, which initiate disease in the nursery. Secondary infection perhaps occurs through wounds and stromata by bacterial cells disseminated by wind-borne rain splashes and contact. Fallen decomposing leaves in stagnate waters of rice field is good site for its multiplication. Insects also play role in transmission.

In India, the disease is confined only to monsoon crops. Cloudy weather with a shower or high humidity are most conducive for disease development in U.P. Heavy nitrogen doses and poorly drained and shaded conditions aggravate the severity of disease even in moderately resistant varieties. Rainy weather, strong winds and temperature of 22–26°C favour the rapid spread of disease in the field.

**Control:**

1. Seed treatment i.e., soaking of seeds for 12 hours in 0.025% aqueous solution of Agrimycin (antibiotic) plus 0.05% wettable Ceresan and then transferring seeds to hot water at 52°–54°C for 30 minutes eradicated 95 per cent of infection from seeds. Dipping the seeds for eight hours in 0.1% ceresin wet plus streptomycin (0.3g/2.5 gallons of water) considerable checked initial infection.
2. Five foliar sprays of Agrimycin plus copper oxychloride at 12 days interval gave good control in the field.
3. Resistant varieties such as Variety N-22 is highly resistant, IR-20, IR22 and Ratna mildly resistant while IR-8, Jaya and IR-24 are mildly susceptible.

## Bacterial Leaf Streak of Rice

This is also an important and much destructive disease of rice in India and other paddy growing countries. The disease is caused by a bacterium, *Xanthomonas campestris* p.v. *oryzicola* (Fig. 20.9).

**Fig. 20.9.** Bacterial leaf streak of rice.

## Tundu Disease of Wheat

This disease has several other names also such as yellow ear rot, bacterial rot and ear and yellow slime disease, and is common in India since quite a long time. It is often associated with a nematodal disease, the ear-cockle of wheat.

**Symptoms:** The early symptoms of tundu are wrinkling of lower and twisting of the middle leaves. This is accompanied or followed by the exudation of a bright yellow sticky slimy substance enveloping the entire ear. Stem, leaf sheaths and glumes become enveloped in this slime resulting in the retardation of plant growth and distortion of stem and ear. Under wet weather, the slime starts tickling down, while in dry conditions becomes a deeper yellow, hard and dry substance. The disease appears on crops reaching maturity.

**Causal Organism:** Tundu of wheat is caused by *Corynebacterium tritici*. The bacterium is a Gram-positive rod, 2.3 × 1.0 μm. Motility is by single polar flagellum.

There has been suggested an intimate etiological relationship between this pathogen and the nematode, *Anguina tritici* (the cause of ear cockle disease). It is presumed that the nematode mechanically transmits the bacterium. In some cases, the successful infection was possible when only galls (nematodal) as such form rotting ears were used as inoculum. This indicated that galls carried bacteria on their surfaces. The nematode is essential only for infection by the bacterium and not for later part of the life cycle.

**Control:**

1. Galls should be removed by floating the seeds in brine (40lb salt in 25 gallons of water).
2. Galls should be destroyed.

3. Affected plants should be destroyed and field be well drained.
4. Seeds should be obtained from disease-free areas.

### 20.8.4 Nematode Diseases of Plants

#### Root-Knot Nematode Disease

This disease of root system of several hundred plants is caused by a nematode, *Meloidogyne.* Root-knot nematodes occur throughout the world, but they occur more frequently and in greater number in areas with warm or hot climates and short or mild winters. These nematodes attack more than 200 species of plants including almost all cultivated plants. In our country, root knot nematodes attack a large number of vegetable crops causing considerable damage. The common species are *Meloidogyne javanica, M. incognita* and *M. arenaria.* They attack various cucurbits, potato, tomato, brinjal, bhindi, chilli, carrot, groundnut, radish, colocasia etc. They also attack a number of weed plants of different families.

Root knot nematodes damage the plant by devitalizing root tips and stopping their growth or causing excessive root formation, but primarily by causing swellings of roots, which not only deprive plant of nutrients but also reduce the market value due to disfiguration.

**Symptoms:** The above-ground symptoms are similar to those caused by many other root pathogens. The most characteristic symptoms are those appearing on the underground parts. Infected roots swell at the point of invasion and develop into the typical root knot galls, two or three times as large in diameter as the healthy roots (Fig. 20.10). Since same root is infected at several points, developing galls give the root a rough, clubbed appearance. In some cases, besides galls, several root branches may also arise from the upper part of the gall, which result into a dense, bushy root system.

**Fig. 20.10.** Root knot nematode disease of brinjal.

## Bacterial and Viral Diseases of Human and Animals

The pathogens are widely scattered amongst the viruses, the bacteria, the higher fungi, the protozoa and even somewhat higher forms of life such as worms. Infectious diseases are similar in all instances in that they represent a struggle between the cells of the pathogenic species and the host. An attempt is made to cover some of the bacterial and viral pathogens and infections in man and animals in this section.

Bacterial Pathogens to man are shown in Table 20.2.

In 1915, Twort and d'Herelle independently discovered that bacteria are susceptible to infection by ultramicroscopic, filter-passing agents, designated as bacteriophages (eaters of bacteria). The designation is often shortened as phages. They contain a single stranded (ss) DNA, double stranded (ds) DNA, ssRNA or dsRNA as genetic material.

**Table 20.2.** Diseases in humans and animals

| Disease | Causal organism |
|---|---|
| Pyemia, furanculosis, carnuncles | *Micrococcus pyogenes* var. *aureus* |
| Secondary infections | *Gaffkya tetragna* |
| Infection of teeth, tonsils, sinuses, wounds | *Streptococcus* |
| Bronchial pneumonia, septic sore throat and rheumatic fever | |
| Pneumonia (pneumococcoal) | *Diplococcus pneumoniae* |
| Gonorrhea | *Neisseria gonnorrhoeae* |
| Epidemic meningitis | *Neisseria meningitidis* |
| Infections of urinary tract | *Pseudomonas aeruginosa* and *Proteus* |
| Typhoid fever | *Salmonella typhimurium* |
| Food poisoning | *Salmonella ereaesuis* and *Clostridium botulinum* |
| Bacillary dysentery | *Shigella dysenteriae* |
| Plague | *Pasteurella pestis* |
| Wooping cough | *Hemophilus oertusis* |
| Anthrax | *Bacillus anthrasis* |
| Tetanus | *Clostridium tetani* |
| Gangrene | *Clostridium perfringens* |
| Diphtheria | *Corynebacterium diphtheriae* |
| Tuberculosis | *Mycobacterium tuberculosis* |
| Leprosy | *Mycobacterium leprae* |
| Cholera | *Vibrio cholerae* |
| Syphilis | *Treponema pallidum* |
| Relapsing fever | *Borrelia recurrentis* |
| Jaundice and meningitis | *Leptospira icterohemorrhagiae* |

The structural characteristics of bacteriophages have been determined through electron microscopy. All the phages contain a nucleic acid enclosed in a capsid which is made up of capsomers. The capsomers contain protomers i.e., protein subunits. Different families of bacteriophages are given in Table 20.3. Bradley (1967) described the following six morphological types of bacteriophages.

**Table 20.3.** Families of bacteriophages

| *Presence of nucleic acid* | *Families* | *Examples* |
|---|---|---|
| **Non-enveloped** | | |
| dsDNA | Myoviridae | P2*, T2** |
| | Styloviridae | λ |
| | Podoviridae | T7 |
| | Corticoviridae | PM2 |
| | Tectiviridae | PRDI |
| ssDNA | Inoviridae | MV-L1 type, fd type |
| | Microviridae | ϕX174 |
| ssRNA | Leviviridae | MS2 |
| **Enveloped** | | |
| dsDNA | Plasmaviridae | MV-L2 |
| dsRNA | Cystoviridae | ϕ6 |

* Isometric-head, ** Elongated head

When a small number of virulent bacteriophage particles is added to a growing liquid culture of susceptible bacteria, some of the bacterial cells become infected. The infected cells show no obvious changes for certain period, which commonly exists between 15 to 60 minutes depending on the nature of the bacterium and phage in question and the conditions of the cultivation. Then quite suddenly, the infected cells undergo lysis. The lysis of infected cell liberates a large number of new phage particles. These particles can in turn infect other cells in the population, with a repetition of the same cycle. Consequently, even if the number of infected phage particles originally introduced is small relative to the number of bacterial cells, practically the entire bacterial population may be destroyed in a matter of few hours.

The number of phage particles or infected bacterial cells in a suspension can be estimated by spreading an appropriate dilution of the suspension over the surface of an agar plate that is evenly inoculated with a thin suspension of susceptible bacteria. After appropriate incubation, the surface of the plate shows a confluent layer of bacterial growth, except at those points where phage particle or an infected cell has been formed as a result of the localized destruction of the film of the bacteria by successive cells of phage growth. By differential centrifugation, one can separate infected cells from phage particles in a mixture that contain both and enumerate each separately by this plating methods. The procedure plate counting thus provides a very simple method for studying quantitatively the growth of a bacteriophage.

# Utilization and Biotechnology of Microorganisms

## 21.1 INTRODUCTION

Microorganisms are omnipresent. They occur in soil, water, air, food etc. They are also present in arctic and antarctic zones, hot tropical areas, deserts, forest areas, on clothes etc. Microbes are present in upper horizon of atmosphere. Most of them are present on the surface as well as in the body tissues of man and animals. They are friendly as well as most deadly creatures.

## 21.2 MICROORGANISMS IN MEDICINE

The early recognition of the role of microbes in diseases led to the development of techniques for combating these microorganisms. One of the landmarks in the history of therapeutic microbiology is the discovery by Fleming in 1928 that the substance produced by the common fungus *Penicillium* is inhibitory for the growth of certain groups of pathogenic bacteria. The discovery of penicillin was a major breakthrough in medical microbiology and led to the search and development of a wide variety of antibiotics. The search for new antibiotics continues successfully. Even the agony due to viral infections and tumour causing agents can be conceived by the use of antibiotics. *Streptomyces* was known to produce Streptomycin.

Microorganisms have practical applications in medicine such as immunization against many diseases, the method of sterilization in surgery and the general sanitation followed in the hospitals etc.

## 21.3 MICROORGANISMS TO IMPROVE SOIL

Microbiological development in the later part of 19th century led to the realization that soil is a storehouse for various types of microorganisms. It was also realized at this time that the complex and vital processes occurring in cultivated soils are the result of a myriad of microorganisms present in the soil. The discovery of bacteria involved in nitrification and nitrogen fixation marks the beginning of soil microbiology. Soil contains an abundant amount of all the elements needed for plant growth, but these are present in unavailable forms. Certain specialized groups of microorganisms bring about the transformation of various elements required for plant growth. Thus, it was proved that the soil is not only a medium for microbial growth, but is also the result of microbial activity. This recognition led to the development of soil microbiology as well as attempts to improve soil fertility through a better understanding of the role of microbes in the mineral transformations.

Microorganisms are effective agents in the decomposition of organic matter and the return of minerals to the soil. The life of the green plants depend on these activities. Soil microorganisms are responsible for the change of the soil structure and texture and thereby affect the soil fertility. The growth and development of many plants are directly connected with the rhizosphere surrounding the root system of the plants. The atmosphere has about 80% nitrogen, which is not available to the plants. The plants cannot utilize this atmospheric nitrogen. Root nodule bacteria (rhizobia) present in soil can fix the atmospheric nitrogen symbiotically in root nodules of many leguminous plants and supply it to the plants. Similarly certain non-symbiotic bacteria like *Azotobacter, Azospirillum, Gluconacetobacter diazotrophicus, Beijerinckia* etc. have an ability to fix atmospheric nitrogen directly in the soil and they provide organic forms of nitrogen to the plants. These microorganisms play a key role in nitrogen balance in nature. Besides nitrification and nitrogen fixation, soil microorganisms are also responsible for various types of activities in soil such as denitrification, decomposition of complex polysaccharides like cellulose, starch, lignins etc., urea and proteins, fats etc. Certain bacteria like *Bacillus thuringiensis* play an important role in controlling certain insect pests and act as biocontrol agents.

Microorganisms are also involved in the study of geochemical reactions, degradation of petroleum products, waste water management, environmental quality improvement, retting of jute and coir fibres and several other branches of science and technology. They convert these products to either biomass or intermediary products that are secreted by the cells.

## 21.4 MICROORGANISMS IN INDUSTRY

Microorganisms have been associated with man for their mutual benefit. For example, yeasts have been known to be involved with man since times immemorial and have served a steady and useful purpose. They are involved in the fermentation processes of fruit juices, wine making, brewing and bread making etc. Pasteur showed that the process of grape juice fermentation was the result of microbial activity and the organism involved in bringing this desirable change in grape juice was identified as yeast. Pasteur's work showed that fermentations characterized by the formation of particular products such as alcohol, acetic acid, lactic acid, butyric acid, propionic acid, butanediol and other basic chemicals were dependent on the activities of different kinds of microorganisms, particularly by clostridia and bacilli. This recognition of the ability of microorganisms to bring about desirable chemical changes in appreciable amounts led to the development of industrial microbiology and the establishment of several industries.

Three important characteristics that have made microbes, particularly bacteria, useful in industry are: (1) their ability to grow rapidly using easily available and cheap raw materials, (2) their ability to maintain physiological constancy and (3) their ability to bring about biochemical transformations under simple cultural conditions. These properties have made the microorganisms highly preferable over synthetic processes. The wine and beer making industries and the baking industry are mainly dependent on the activities of the microorganisms. Acetone, butyl alcohol, lactic acid, acetic acid and many other industrial products such as enzymes, vitamins, hormones etc. are manufactured with the help of microorganisms. Antibiotics are always manufactured with the help of microorganisms.

Carotenoids and steroids can be obtained from fungi and algae. *Corynebacterium glutamicum* can produce large amounts of glutamic acid from sugar and ammonium salts. The development of microbial mutants has led to the possibility of large scale production of various amino acids, nucleotides and several other biochemical products. Enzymes like amylases for starch hydrolysis, proteinases for curing of leather, pectinases for producing fruit juices and several industrially important enzymes are obtained from microbial sources.

## 21.5 MICROORGANISMS IN DAIRY

Microorganisms have been associated with milk and milk products in dairy industry. Milk is a complete medium for the growth and reproduction of several types of bacteria, and in particular, the lactic acid bacteria. The lactic acid bacteria have the capacity to convert milk sugar into lactic acid and this lactic acid increases the acidity of milk leading to curdling of the milk casein. This phenomenon, although known and observed for several years, was explained on a scientific basis only after the studies on sour milk by Pasteur.

The art of cheese making developed about the same time as that of milk curdling. Cheese making also involves lactic acid bacteria and these are involved both in curdling as well as in bringing about the desirable flavours. In addition to bacteria, some moulds are also known to be important in cheese making.

Besides microorganisms that help in the conversion of milk into various other desirable products, there are many others, which cause the deterioration of milk and milk products. There are also many human diseases that are transmitted through milk to humans.

## 21.6 MICROORGANISMS IN FOOD

The association of microorganisms with food has been known for long, and methods of food preservation either by heating or by heating and sealing were developed as early as 19th century. Microbes have been known to play several different roles in food industry. The microbial cells may be used as food material or food supplement, or their cell components, particularly the enzymes, may be used as accessory agents in food industries. Microbes could also be agents of food spoilage. A number of other processes employed in food industries are dependent on microbial fermentations for preservation or characteristic flavour development.

## 21.7 MODERN GENETIC APPROACHES TO MODIFY MICROORGANISMS

Elucidation of gene transfer in bacteria by extrachromosomal elements has opened up the possibility of transferring foreign DNA into bacteria. Genetic manipulation allows the introduction of small pieces of genetic information, for example from man into bacteria, which can then synthesize the corresponding proteins. This provides the possibility of production of hormones, antigens, antibodies and other proteins like insulin in bacterial cultures, the transfer of nitrogen-fixing property to green plants and the cure of diseases due to biochemical defects. The recombinant DNA (rDNA) technology provides ample scope in the diagnosis and cure of hereditary defects and serious diseases. This technology is the most revolutionary technology in the 21st century.

## 21.8 UTILIZATION OF FUNGI AND FUNGAL BIOTECHNOLOGY

Communities of saprophytic, parasitic and endophytic fungi living in the forest ecosystem contribute significantly to the species diversity. Recent studies have revealed that tropical fungi colonizing diversified habitats harbour diverse microfungi in abundance (Hyde, 1997). It is becoming clear that both micro and macrofungi of the tropics are potential source of biotechnologically significant, industrially important and pharmaceutically valuable organic molecules (Cuomo et al., 1995). Realizing this, Rossman (1994) and Hawksworth (1997) have advocated complete and urgent recovery, inventory and investigation of mycobiota from forests and other localities of tropical belt, since this region of the world is presently under stress.

The microbiological conversion of carbohydrates of cellular material and other useful products has become increasingly important and interest in cell mass production is constantly growing. *Saccharomyces cerevisiae* plays a significant role in the formation of biomass from carbohydrates. Twenty amino acids were found to be the components of protein in yeast, and on an average yeast contains 10.8% nucleic acids. Lipid content of yeast is 8% and yeast lipids include sterols.

The filamentous fungi contain many species capable of hydrocarbon oxidation. However, because of their average growth form, they have seldom been extensively studied in the context of the production of microbial biomass with hydrocarbons as feedstock. One exception is the kerosene fungus, *Cladosporium (Amorphotheca) resinae*, which is important in the field of petroleum product contamination. Filamentous fungi have also been used to produce different types of metabolites from alkane media. *Cunninghamella, Mucor, Aspergillus, Botrytis, Fusarium, Paecilomyces, Sporotrichum, Verticillium* and others are reported to contain hydrocarbon assimilating species or strains. *Hansenula capsulata, Pichia lindnerii, Torulopsis glabrata* and *Candida boidinii* are selected as methylotrophic yeasts.

More than half the total production of plant residues remain unused, mainly straw and leaves and often wastes from agricultural, forest and industrial production processes. These different materials are partially burnt or disposed of on land for soil amelioration after being shredded or composted. By suitable treatment they can be converted into substrates for the cultivation of higher fungi. Fruit bodies serve as delicious food, spent mushroom compost can be utilized as humus fertilizer and widely grown substrates. This process has an advantage in which waste is profitable, removed and reintegrated by way of natural process into the ecosystem followed by the mobilization of cellulose or hemicellulose through mineralization. Hardly usable substrates are transformed into consumable protein rich biomass. Edible fungi represent a well characterized microbial biomass, which will be generally accepted by the consumer. The consumption of edible fungi as food and drug is closely related to the history of mankind. *Lentinus elodes* and *Volvariella volvacea* belong to those species which have been cultivated for 2000 years in Asiatic countries. *Agaricus bisporus* was first mentioned in the seventeenth century in France. More recently *Agaricus bitorquis* and *Pleurotus* species have also been produced commercially. Edible fungi are cultivated worldwide under various climatic conditions. In the last two decades, Taiwan has become the third largest producer of mushrooms in the world. China is an important producer of *Volvariella volvacea*. Mushrooms are considered more nutritious than many other vegetables. The nutritive value of some quality mushrooms is nearly equal to that of milk. These mushrooms contain high level of essential amino acids, less carbohydrate, more fibre, vitamins, unsaturated fatty acids with a mineral content exceeding that in fish and meat and twice that in most vegetables. Fruiting bodies, in general, on dry weight basis, contain about 32% protein, 2% fat, less carbohydrate and rest in ash constituting the minerals. More than 2000 species of mushrooms are reported to be edible throughout the world. Around 150 species belonging to 30 genera are regarded as prime edible mushrooms. Over 60 are cultivated commercially and have reached an industrial scale in many countries. Mushrooms have great demand in food industry because of flavourants associated with them. They have long been used for their medicinal and tonic properties in China, Japan and other countries. A number of mushrooms possess alkaloids or chemicals that are of pharmaceutical importance. Mushroom contains antitumour glucan and an immunomodulatory FIP-fve has recently been isolated from biotransformation of penicillin as biopolymer in bioabsorption and other uses. Mushrooms like *Ganoderma, Polyporus, Lentinus* and *Pleurotus* have been used as ornamentals. Species of *Ganoderma* and others have been employed in cosmetic industry. *Ganoderma lucidum* has been used as a biosorbent of copper and chromium. Spent mushroom compost consisting of degraded cellulose, hemicelluloses and lignin serves as an effective soil fertilizer and conditioner. Spent substrate has also been used as a source of saccharifying enzymes and production of single cell protein. Spent substrate contains more sugars, which are enzymatic hydrolysis of substrate by mushrooms. Spent substrate is a rich source for biogas production. Exploitation of many hidden activities of edible mushrooms is essential for mankind. Molecular tools like genetic engineering, protoplast fusion and others are useful for the development of highly productive newer strains of edible mushroom.

Soil fungi influence plant growth and health of the plant. Mycorrhiza, the plant root-fungus symbiotic association, plays a key role in the effective uptake of phosphorus by mycorrhizal roots besides enhancing plant growth.

Entomogenous fungi are found in all taxonomic groups, although the Deuteromycotina, Chytridiomycetes and Zygomycotina contain many of the useful species. *Verticillium lecanii* parasitic on aphids and white flies, *Hirsutella thompsonii* pathogen of eriophyid mites, *Beauvaria bassiana* and *Metarrhizium anisopliae* have received utmost consideration and were also commercialized. The key disadvantages of fungi as microbial pesticides involve the rather restricted range of environmental conditions suitable for their use.

The role of fungi is well established in fermentation technology, and using this industrial process, many metabolites are commercially exploited for their antibiotic properties. Cephalosporin, penicillin and griseofulvin are derived from fungi. Yeast forms an important agent in fermentation to produce ethanol from sugar or after hydrolysis from starch and cellulose. Fungi are involved in the production of foodstuffs, bread, fermented milk drink, yoghurt, cheese and various yeast preparations. There are many foods used in Asian and Oriental cuisines, for example tofu, tempeh and miso whose production is dependent on fungi. Many different industrial organic acids are produced by representatives of a wide range of fungal genera. These include citric acid (*Aspergillus, Mucor* and *Penicillium*), fumaric acid, and lactic acid (*Rhizopus*), gluconic and itaconic acids (*Aspergillus*). Rennin is produced by *Rhizomucor* and *Rhizopus,* riboflavin by *Ermothecium* and various fungi imperfecti are employed in steroid transformation. Production of mycelial paper involves the addition of mycelium of *Phytophthora* and other members of the mastigomycotina to wood pulp. Today three species of fungi, all ascomycetes, namely *Neurospora crassa, Aspergillus nidulans* and *Saccharomyces cerevisiae* are employed as model system in genetics. Transformation systems are available for all three species and it is easy to isolate their DNA. Fungi are good candidates for employing them in degrading refractory substrates, cellulose, lignin, chitin, keratin and other substrates. Fungi like *Aspergillus niger* and *A. oryzae* are regarded as safe by the food and drug administration. Moulds also secrete several hydrolytic enzymes that digest the substrate. Around 20 standard amino acids are produced commercially through fermentation using fungi. *Hansenula anomala* is used to produce tryptophan, *Saccharomyces cerevisiae* is used in the production of lysine, methionine is produced by *Candida tropicalis* and several marine fungi are found to accumulate alanine. Pullulan and Scleroglucan have greater significance as biopolymers. Several fungi are identified to produce lipids. *Saccharomyces, Torulopsis* and others have been found to be potential producers of glycerol. Several vitamins including vit. $B_{12}$, biotin, folic acid, pantothenic acid, pyridoxine, riboflavin and thiamine are the products of fungi. *Eremothecium* and *Ashbya* are important since these fungi can make up 2.5 and 7 g/litre of vit. $B_{12}$, respectively. *Blakeslea trispora,* a heterothallic fungus of mucorales group is extremely rich in mycelial β-carotene, as a potential source of carotenoids. Industrial enzymes are used in food processing and in particular amylase, glucoamylase and glucose isomerase are exploited for the production of glucose and fructose syrups.

The solubilization of insoluble phosphorus by the fungi is one of the vital microbial processes in soil. Several fungi bring about solubilization by producing various types of organic acids. Besides solubilizing phosphorus, these organisms release sufficient amounts of soluble phosphorus into its surroundings. The most efficient phosphate solubilizing fungi belong to *Aspergillus* and *Penicillium.* Fungal toxins play a significant role in the health of humans and animals. Aflatoxin elaborated by *Aspergillus flavus* is commonly found associated with animal feed and in some food stuffs also. Aflatoxin $B_1$, $B_2$, $B_{2a}$, $G_1$, $G_2$, $G_{2a}$, $M_1$ and $M_2$ have been isolated and characterized. Other aspergilli produce ochratoxin, sterigmatocystin which are commonly found associated with food stuffs and they seem to be carcinogenic. *Penicillium urticae, P. expansum* and others were found to produce patulin, a mycotoxin clinically found to be carcinogenic. Around 20 fungi have been found to cause mycotoxicoses in animals. Ergot fungus is known to elaborate at least seven alkaloids including LSD. Ergot itself has a respectable place in pharmacopia. Several fungi have been found to cause mycotic diseases and in particular in tropics. Coccidiodomycosis, Blastomycosis, Histoplasmosis, Aspergillosis, Conidiasis have also been recognized with increased frequency due to the loss of immunocompetence of individual host.

Cryptococcosis has a worldwide distribution and is found to be prevalent in India, and the organism *Cryptococcus neoformans* is being considered a diagnostic tool for AIDS. Fungal biomass produced from *Penicillium chrysogenum* is used as fertilizer (Bisol) or cattle feed. The same fungus is one component of biosorbant M, which enables the extraction of uranium and radium from atomic industrial waste water. Cyclosporin-A, the most often used powerful immunosuppressant is extracted from the members of hyphales. Genera such as *Arthrobotrys, Dactylella, Dactylaria* and *Monaorosporium* are predators used against pathogenic nematodes. *Trichoderma viride, T. harzianum* and *Verticillium biguttatum* are the common fungal antagonists used as biocontrol agents to control pathogenic types of *Fusarium, Rhizoctonia, Macrophomina* and *Sclerotium. Phellinus weirii,* which causes rot of telephone poles, is controlled by *Scytalidium. Fusarium oxysporum* var. *cannable* is used as bioherbicide for suppressing marijuana plant. *Colletotrichum gloeosporioides* and *C. truncatum* are also used as bioherbicides. Pulmonary Aspergillosis *(Aspergillus),* lymphatic sporotrichosis *(Sporothrix schenkii),* enquine lymphatic histoplasmosis *(Histoplasma),* onchomycosis *(Graphium)*, ringworm, Chromomycotina *(Phialophora)*, skin and respiratory allergies and such type of mycotic infection are of great importance in tropical health management. Such mycotic infectious need to be tackled with all seriousness.

Plant diseases such as blights, wilts, rust, smuts, cankers, leaf spots, blasts etc. on a number of crop plants and forest plants have become important as they dwindle country's economy. Biodeterioration is a biological processes that contributes to deterioration or destruction of material resulting in economic loss. Biodegradation refers to the biological breakdown of undesirable materials or compounds to harmless or tolerable products. A vast range of materials used by man from food, feed, fabric, wood to numerous agricultural materials and degradation of undesirable materials including sewage, coal mine, waste from paper pulp mills, sugar and oil refineries, industrial effluents, pesticides, detergents etc. is caused by fungi as fungi are second to none in their armoury of enzymes and ability to degrade substrates of all kinds.

Application of mycoherbicides to control weeds has been an age-old approach through classical methods, in which plant pathogens are released to control weeds through natural spread. Control of skeleton weed by *Puccinia condrillina* in Australia, *Puccinia abrupta* var. *parthenicola* for the control of *Parthenium hysterophorus, Alternaria alternata* and *Cercospra rodamanii* for the control of water hyacinth and others clearly indicate the biopotential of fungal pathogens in the biological control of weeds. A mycoherbicide is a formulation of a fungal pathogen that kills weed plants by causing disease at the lethal or threshold level. Thermophilic fungi have been playing key role in nature's economy ever since they were recognized. Thermophiles have the ability to degrade organic matter acting as biodeteriorators thus elaborating extracellular and intracellular enzymes, amino acids, antibiotics, phenolics, polysaccharides and sterols during process. They are also involved in the production of nutritionally enriched feeds and single cell protein besides their involvement in genetic manipulation. Fungi like *Thermomyces, Sporotrichum* and *Torula* are found to produce amylases. A number of thermophilics *viz: Acremonium cellulophilum, Aspergillus fumigatus, Chaetomium thermophile, Thermoascus aurantiacus* and others were reported to degrade cellulose and produce extracellular enzymes. Xylanases, lipases and proteases were also found to be elaborated by thermophilic fungi. Bioactive compounds such as antibiotics including myriocin are synthesized using fungi. Important cellulolytic and lignolytic enzyme producing thermophilic fungi like *Chaetomium cellulolyticum* and *Sporotrichum pulveulentun* are the most commonly used thermophiles for upgradation of animal feed and to produce single cell protein from lignocellulosic wastes.

The disposal of municipal wastes, refuse and industrial effluents has become a serious problem, and bad odour emitted is of serious concern. Fungi like *Phanerochaete, Mucor, Sporotrichum* and others have shown their abilities in treatment.

Plant disease epidemics have become important as they cause huge losses thus dwindling country's economy. In order to make agriculture sustainable, it is desirable that we should resort to integrated disease management (IDM), which would help not only in sustainability but also in preserving the ecological balance and protection of environment.

Keratinophilic fungi possess the specialized ability to degrade hard keratin and utilize it, as many untreated buildings of historic and artistic value with wall paintings such as old churches and caves provide favourable conditions for the growth of fungi. Deterioration of wall paintings by fungi is largely promoted by the dust that covers the painting. Species of *Penicillium, Aspergillus, Cladosporium, Aureobasidium, Alternaria* and *Phoma* are some of the fungal genera reported. Nearly 25 fungal species have been reported as responsible agents of deterioration of painting in Ajanta caves. Therefore, it is recommended to use a low molecular polymer biocide at a lower concentration to prevent fungal attack. Study of degradation of buildings and archaeological timber has revealed the association of *Serpula, Coniophora, Asterostroma* and other fungi.

## 21.9 ECONOMIC IMPORTANCE OF ALGAE

Seaweeds are used as food for domestic animals. Marine algae have immense value for their minerals and vitamins. The marine algae are of considerable importance in food. They are often mixed with rice and fish and serve as base for soups, condiments and salads. Red alga *Porphyra* forms a dietary item in Japan and Korea. Nori, the Japanese pasta, is rich in vitamins B and C. *Ulva* is used in salad and soups. Some products of algae such as gelatinous carbohydrates from *Gigartina* and *Chondrus* are used in puddings and as stabilizer in ice creams.

*Chlorella* is future hope of modern world as food supplement is cultivated in pilot plants. The alga is used as flour for making crackers and biscuits. *Chlorella* has proteins and carbohydrates. *Chlorella* is also used in aeronautical research. Blue-green alga, *Spirulina* tops the list of edible forms for single cell protein, which accounts for 40–50% protein besides being rich in vitamin $B_{12}$. Algae serves as primary food for fishes and other aquatic animals.

Alginic acid derivatives are extracted from brown algae. Alginate is used as thickness agent in food industry, for filling creams in cosmetics, textile industries etc. Carrageenan is a gelatinous carbohydrate used in puddings, ice cream and as clearing agent in beer preparation. Agar-Agar is the dried gel-like non-nitrogenous extract from red algae. It is used as solidifying agent.

Brown algae like *Laminaria, Fucus* are used in iodine extraction. Diatomite earth is used in industrial filtration processes and brewing industries. Because of their high iodine content, brown algae are used in various goitre medicine preparations. Chlorellin, an antibiotic, is extracted from *Chlorella.* Algae are used in water purification, sewage disposal and uptake of radio wastes. Blue-green algae are known for nitrogen fixation and are also used as biofertilizer for crop productivity, land reclamation etc. Algae produce growth promoting substances.

Harmful effects of the algae are death of fishes, animals and human beings due to toxins released by fish. Algae interfere with storage water reservoirs besides causing pollution.

## 21.10 UTILIZATION AND BIOTECHNOLOGY OF ACTINOMYCETES

True actinomycetes are bacteria that grow in the form of mycelia. Their natural occurrence is mostly confined to soil. They are gram-positive and are related to coryneform bacteria and mycobacteria by an almost continuous sequence of intermediate forms. They are aerobic with few exceptions. The name of this group is derived from the first described anaerobic species *Actinomyces bovis,* which causes actinomycosis, disease of cattle.

The genus *Streptomyces* occupies a major share among actinomycetes. The fragrance, which emanates from cultivated (ploughed) soil is due to the presence of genus *Streptomyces,* which produces geosmin. It is a 1,10–dimethyl 1–9–decalol. The knowledge of Streptomyces has advanced considerably because of their practical importance in the production of several effective antibiotics. Streptomycin, chloramphenicol, and aureomycin produced by *S. griseus, S. venezulae and S. aureofaciens*, respectively, are the most popular antibiotics among various antibiotics isolated from this genus.

Streptomycin consists of 3 groups, namely N-methyl-L-2 glucosamine, a methylpentose and an inositol derivative with 2 guanidyl residues. Streptomycin achieved its therapeutical success because it is effective against a number of acid-fast and gram-negative bacteria that were not sensitive to penicillin. It is also used in certain animal and plant diseases.

Chloromycetin (chloramphenicol) was first discovered in cultures of *S. venezuelae*, but it is also produced synthetically. It is very stable and effective against many gram-negative bacteria, spirochaetes, rickettsiae, actinomycetes and large viruses.

The tetracyclines are excreted by a number of streptomycetes including *S. aureofaciens.* They are closely related chemically, being derived from a naphthalene skeleton. The best known as chlortetracycline (aureomycin), oxytetracycline (terramycin) and tetracycline. They have a spectrum of effectiveness.

The macrolide antibiotics (erythromycin, carbomycin A, picromycin etc.) are of diverse origin and have a relatively high molecular weight and are characterised by the presence of macrocyclic lactone ring.

Actinomycin was isolated as first of the antibiotics produced by streptomycetes. It is actually a mixture of several compounds, all of them containing a phenoxazone chromophore, but all are substituted with different polypeptide chains.

One group of polypeptide antibiotics (gramicidin, S. polymyxins, bacitracin, ristocetin etc.), Polymyxin B, consists of a ring of seven amino acids with a side chain attached by a peptide bond. The polypeptide antibiotics have a high affinity for the cytoplasmic membrane, and are therefore toxic to bacteria and eukaryotes, which makes their effectiveness very much restricted. Some of the polypeptide antibiotics are used in research as ionophores because of their capacity for selective ion transport through biological membranes. Valinomycin promotes the transport of potassium through membranes.

Many streptomycetes degrade cellulose, chitin, and other recalcitrant natural substrates. Genus *Micromonospora* is a cellulose degrading organism and is widely distributed in soil and rotting aqueous sediments. Extracellular enzymes of microbial origin contributes to the attack on chitin. Those excreted by *Streptomyces griseus* could be separated into chitinase and chitobiase. Degradation apparently occurs by simultaneous attack of chitinase at many points along the polymer, yielding little N-acetyl-glucosamine and predominantly chitobiose and chitotriose as the products. The latter are then converted to monomers by chitobiase.

# Annexure

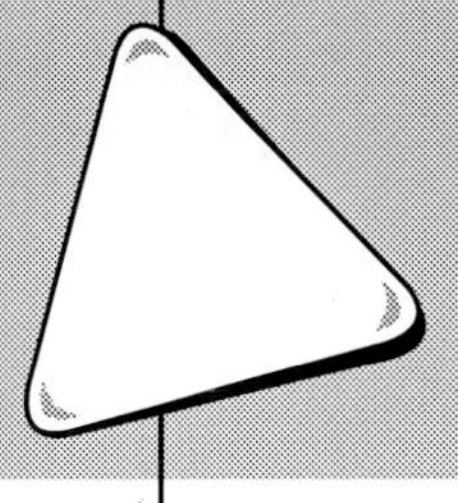

## COMPOSITION OF DIFFERENT MEDIA

### 1. Minimal Medium

This is a type of culture medium that lacks certain growth factors. This does not support the growth of auxotrophic strains i.e., a strain of microorganism that lacks the ability to synthesize one or more essential growth factors. It arises by mutations of a wild type strain (prototype).

### 2. Basal Medium

This is a type of medium that supports the growth of various nutritionally undemanding species. It includes peptone and nutrient agar, potato and nutrient agar, potato dextrose agar (PDA) and Sabouraud's agar media.

#### a) Peptone Water Agar Medium

*Composition:* 1.2% peptone, 0.5% sodium chloride (NaCl) and 1.5% of agar in 1 litre of water.

#### b) Nutrient Agar Medium

| *Composition* | : | |
|---|---|---|
| Beef extract | : | 3 g |
| Peptone | : | 5 g |
| Agar | : | 15 g |
| Distilled water | : | 1 litre |
| pH | : | 6.8–7.0 |

This is a simple medium and has been most generally used in bacteriological examination of water, sewage and food products and for isolating organisms in pure culture.

### c) Potato Dextrose Agar (PDA) Medium

| *Composition* | : | |
|---|---|---|
| Peeled potatoes | : | 200 g |
| Dextrose | : | 20 g |
| Agar | : | 20 g |
| Water | : | 1 litre |
| pH | : | 4.5 |

It is a general medium recommended for isolation of fungi.

To 200 g of washed, peeled and sliced potatoes, add 500 ml of water and boil gently till they are easily penetrated by a glass rod. Filter the material through muslin cloth. To 500 ml of boiled water, add 20 g agar slowly and mix it with a glass rod. Mix this agar solution with potato extract and make the volume to 1 litre and sterilize the medium at 15 lb pressure (121°C) for 15 minutes in an autoclave.

### d) Sabouraud's Agar Medium

| *Composition* | : | |
|---|---|---|
| Neopeptone | : | 10 g |
| Dextrose | : | 40 g |
| Agar | : | 15 g |
| Distilled water | : | 1 litre |
| pH | : | 5.5 |

This medium is recommended for the cultivation and growth of fungi, particularly those associated with skin infections.

## 3. Enriched Medium

Examples: Blood agar, serum agar, chocolate agar, Jensen's medium etc.

### a) Blood Agar

| *Composition* | : | |
|---|---|---|
| Infusion from beef heart | : | 500 g |
| Tryptose | : | 10 g |
| Sodium chloride | : | 5 g |
| Agar | : | 15 g |
| pH | : | 7.4 |

Blood agar base is recommended for use in isolation and cultivation of many fastidious pathogenic microorganisms.

### b) Chocolate Agar

| *Composition* | : | |
|---|---|---|
| Proteose peptone | : | 20 g |
| Dextrose | : | 0.5 g |
| Sodium chloride | : | 5 g |
| Disodium phosphate | : | 5 g |
| Agar | : | 15 g |
| pH | : | 7.0–7.2 |

This medium is proved to be satisfactory for the isolation of organisms from all types of gonococcal infections.

### c) Serum Agar

Bacto-beef blood serum is a dehydrated powdered form of fresh beef blood serum.

## 4. Selective Media

### a) Mannitol Salt Agar

| *Composition* | : | |
|---|---|---|
| Beef extract | : | 1 g |
| Proteose peptone | : | 10 g |
| Sodium chloride | : | 75 g |
| D-mannitol | : | 10 g |
| Agar | : | 15 g |
| Bacto-phenol | : | 0.025 g |
| Distilled water | : | 1 litre |
| pH | : | 7.2–7.4 |

This medium is used for isolation of pathogenic staphylococci.

### b) Mac Conkey's Agar Medium

| *Composition* | : | |
|---|---|---|
| Peptone | : | 17 g |
| Lactose | : | 10 g |
| Bile salts | : | 1.5 g |
| Sodium chloride | : | 5 g |
| Agar | : | 15 g |
| Neutral red | : | 0.03 g |
| Crystal violet | : | 0.001 g |

| | | |
|---|---|---|
| Distilled water | : | 1 litre |
| pH | : | 7.0–7.2 |

This medium is used for isolating members of Enterobacteriaceae.

### c) Jenson Medium

| | | |
|---|---|---|
| *Composition* | : | |
| Sucrose | : | 20 g |
| $K_2HPO_4$ | : | 1 g |
| $MgSO_4 . 7H_2O$ | : | 0.5 g |
| Nacl | : | 0.5 g |
| $FeSO_4$ | : | 0.1 g |
| $CaCO_3$ | : | 2.0 g |

### d) Crystal Violet Blood Agar

This medium is used for isolating *Azotobacter* sp.

## 5. Enrichment Medium

### a) Selenite F-Medium

This medium is used for isolating *Salmonella typhosa*

| | | |
|---|---|---|
| *Composition* | : | |
| Aqueous solution of peptone | : | (0.5%) |
| Lactose or mannitol | : | (0.4%) |
| Sodium selenite | : | (0.4%) |
| Disodium hydrogen phosphate or a combination of disodium hydrogen phosphate and sodium dihydrogen phosphate | : | (1.0%) |
| pH | : | 7.0 |

The medium should not be autoclaved, but may be steamed for 30 minutes. Incubation of the inoculated medium should be around 12 to 18 hours at 30°C.

### b) Alkaline Peptone Water

| | | |
|---|---|---|
| *Composition* | : | |
| Peptone | : | 1% |
| NaCl | : | 1% |
| Distilled water | | |
| pH | : | 8.6 to 9.0 |

## 6. Differential Medium

### a) Blood Agar

It is used for isolating and cultivating fastidious pathogenic microorganisms.

This medium contains oxalated blood. It is prepared by mixing 5–10 g blood drawn aseptically with a pre-autoclaved molten agar base such as trypticase-soy agar at 50°C.

### b) Chocolate Agar Medium

This medium is used for the isolation and culture of fastidious bacteria like *Haemophilus* sp., *Neisseria gonorrhoeae* etc. A medium like blood agar base or brain-heart infusion agar is autoclaved and cooled to 80°C. Bovine blood (1%) is added and the medium is shaken gently until chocolate-brown in colour is obtained. It may be dispensed into plates or tubes. Alternatively, a chocolate agar plate may be prepared by heating blood agar medium to around 80°C.

## 7. Biochemical Media

### a) Glucose Phosphate Peptone Water

Carbohydrates like glucose, galactose, arabinose, lactose, maltose, mannitol, mannose, rhamnose, sucrose, xylose etc are added to peptone water.

### b) Tryptone Water

| *Composition* | : | |
|---|---|---|
| Tryptone | : | 1–2% |
| NaCl in water | : | 0.5% |

### c) Oxidation-Fermentation Medium

Oxidation-fermentation test (Hugh and Leifson's test; O–F test) is used to indicate whether a given carbohydrate (usually glucose) is attacked oxidatively or fermentatively by a given bacterial strain.

### d) Simon's or Koser's Citrate Medium

| | | |
|---|---|---|
| NaCl | : | 0.5 |
| $MgSO_4$, $7H_2O$ | : | 0.02 g |
| $(NH_4)$ $H_2PO_4$ | : | 0.1 g |
| $K_2HPO_4$ | : | 0.1 g |
| Citric acid | : | 0.2 g |

| | | |
|---|---|---|
| Water | : | 100 ml |
| pH | : | 6.8 |
| Bromothymol blue | : | 8 mg |

This is a medium differentiating faecal coli and members of the aerogens group.

## 8. Transport Media

Any medium used specially for the transportation and/or temporary storage of material (swabs) from which the isolation of particular organism(s) is attempted is known as transport medium. e.g., Monsur transport medium.

### a) Yeast Extract Mannitol (YEM) Medium

This medium is used for isolating legume bacteria (rhizobia)

| | | |
|---|---|---|
| $K_2HPO_4$ | : | 0.5 g |
| $MgSO_4,7H_2O$ | : | 0.2 g |
| NaCl | : | 0.1 g |
| Mannitol | : | 10.0 g |
| Yeast extract | : | 1.0 g |
| Distilled water | : | 1000 ml |
| Congo red (1%) | : | 2.5 ml (only for solid medium during isolation) |

### b) Peptone-Glucose Agar Medium

| | | |
|---|---|---|
| Peptone | : | 10.0 g |
| Glucose | : | 5.0 g |
| Agar | : | 15.0 g |
| Distilled water | : | 1000 ml |

### c) Jensen's Medium

Nitrogen-free media for growing seedlings

| | | |
|---|---|---|
| $CaHPO_4$ | : | 1.0 g |
| $K_2HPO_4$ | : | 0.2 g |
| $MgSO_4$, $7H_2O$ | : | 0.2 g |
| NaCl | : | 0.2 g |
| $FeCl_3$ | : | 0.1 g |
| Agar | : | 15.0 g |
| Distilled water | : | 1000 ml |

*Note:*

1. Make a stock solution of minor elements containing B: 0.05 g ($H_3BO_3$), Mn: 0.005% ($MnSO_4$), Mo: 0.005% ($Na_2MoO_3$), Cu: 0.002% ($CuSO_4$), (Gibson, 1963). Add 1 ml of stock solution to 1 litre of the medium.
2. Adjust pH to 6.5–7.0.

## STAINING

### Simple Staining

- Staining solution of methylene blue
- The bacteria stain as deep blue

### Negative Staining

- Nigrosine solution
- Cells appear as colourless on a blue background

### Gram Stain

***Reagents***

| | | |
|---|---|---|
| 1. Crystal violet | : | 10 g |
| Ammonium oxalate | : | 4 g |
| Ethyl alcohol | : | 100 ml |
| Distilled water | : | 400 ml |
| 2. Iodine solution | | |
| Iodine | : | 1 g |
| Potassium iodide | : | 2 g |
| Ethyl alcohol | : | 25 ml |
| Distilled water | : | 200 ml |
| 3. Ethyl alcohol | : | 95% |

### Counter Stain

| | | |
|---|---|---|
| 2.5% safranin in ethyl alcohol | : | 10 ml |
| Distilled water | : | 100 ml. |

The bacteria that appear blue is colour are referred as gram-positive and those that appear pink in colour are referred as gram-negative bacteria.

## Acid Fast Staining

Carbol fuchsin
Acid-alcohol (3% HCl + 95% alcohol)
Methylene blue

Acid fast cells appear as red coloured and non-acid fast cells appear as blue coloured structures.

## Bacterial Cell Wall Staining

Aqueous congo red solution – 1 drop
Cetyl pyridinium chloride (0.34%)
Loeffler's methylene blue solution

The cell wall stains red and cytoplasm stains blue.

## Cytoplasmic Membrane Staining

Potassium nitrate solution
Bouin's fixative solution
Victoria blue solution

The cytoplasmic membrane appears as a deep blue outer layer.

## Bacterial Spore Staining

Malachite green (5% aqueos)
Safranin (0.5% aqueous)

Bacterial endospores stain green and the vegetative cells stain red.

## Capsule Staining

Congo red
Acid-alcohol
Acid fuchsin

The capsules appear as colourless surrounding red cells against dark blue background.

## Flagella Staining

Zeihl's carbol fuchsin
Alcohol (95%)

Flagella appear as pink-stained structures.

## Bacterial Nuclear Material Staining

1% HCl

Stains purple surrounded by a colourless zone of cytoplasm. The cell membrane appears as faintly purple layer.

## Fungal Staining

Lactophenol cotton blue
Alcohol (70%)

The fungal cytoplasm appears as lightly stained blue region surrounded by light blue background on the slide.

## Staining of Actinomycetes

Bismark brown solution (0.1% W/V)
Toluidine blue stain (0.1% W/V)
Ammonium sulphate (saturated)

Mycelium stains light yellow and spores blue with red granules.

## Staining of Spirochetes

Fontana's fixative
Fountana's ammoniacal silver nitrate
Fontana's mordant
Canada balsam

Spirochetes appear dark brown stained spiral-shaped cells against a light yellow background.

## Staining of Mycoplasmas

| | | |
|---|---|---|
| Methylene blue | : | 2.5 g |
| Azurell | : | 1.25 g |
| Maltose | : | 10.0 g |
| Sodium carbonate | : | 0.25 g |
| Distilled water | : | 1000 ml |

Mycoplasma colonies appear with distinct blue stained centres and light blue peripheries.

## Staining of Arbuscular Mycorrhizal Fungal (AMF) Infected Roots

10% KOH
80% Lactic acid

Glycerin
95% ethyl alcohol
Chlorazol black E
Chloral hydrate
Basic fuchsin

Mounting fluid (chloral hydrate 20g; gum arabic 20g; glycerin 20g; glucose syrup 3ml; basic fuchsin 10 drops (0.3 g in 10 ml 95% ethyl alcohol); 35 ml water)

# Glossary

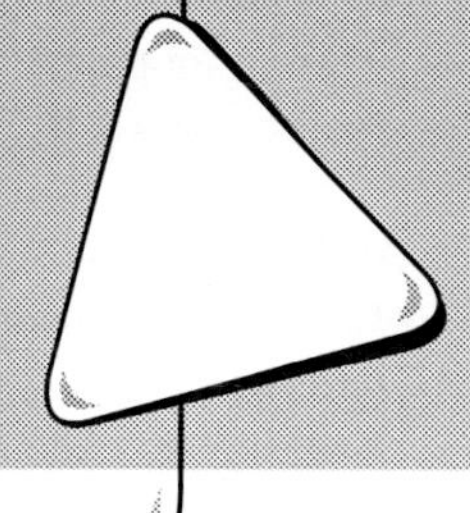

**Acetyl coenzyme A (acetyl CoA)** A substance composed of an acetyl group and a carrier molecule called coenzyme A (CoA); provides the means for pyruvic acid to enter the Krebs cycle.

**Acid-fast stain** A differential stain used to identify bacteria that are not decolourized by acid-alcohol.

**Acquired immunity** The ability, obtained during the life of the individual, to produce specific antibodies.

**Actinomycetes** Gram-positive bacteria that tend to form branching filaments; may form true mycelia; may produce conidiospores.

**Activated sludge** Aerobic digestion used in secondary sewage treatment.

**Adenine** A purine nucleic acid base that pairs with thymine in DNA and uracil in RNA.

**Adenosine diphosphate (ADP)** The substance formed when ATP is split and energy is released.

**Adenosine triphosphatase (ATPase)** The enzyme that catalyzes the following reactions: $ADP + P \rightarrow ATP$ and $ATP \rightarrow ADP + P$.

**Aerobe** An organism requiring $O_2$ to grow.

**Aerobic respiration** Respiration in which the final electron acceptor in the electron transport chain is oxygen.

**Aflatoxin** $C_{17} H_{10} O_6$, a carcinogenic toxin produced by *Aspergillus flavus*.

**Agar** A complex polysaccharide derived from a marine alga and used as a solidifying agent in culture media.

**Agglutination** A joining together or clumping of cells.

**Alcohol** An organic molecule with the functional group—OH.

**Aldehyde** An organic molecule with the functional group—C-H-O.

**Algae** A group of photosynthetic eukaryotes; some are included in the Kingdom Protista, and some in the Kingdom Plantae.

**Alkaline** having more $OH-$ ions than $H+$ ions; pH is greater than 7.

**Allergen** An antigen that evokes a hypersensitivity response.

**Amino group**—$NH_2$

**Ammonification** Removal of amino groups from amino acids to form ammonia.

**Amoeba** An organism belonging to the Kingdom Protista that moves by means of pseudopods.

**Amphitrichous** Having tufts of flagella at both ends of a cell.

**Anabolism** All synthesis reactions in a living organism.

**Anaerobe** An organism that does not require oxygen for growth.

**Anaerobic respiration** Respiration in which the final electron acceptor in the electron transport chain is an inorganic molecule other than oxygen; for example a nitrate or sulphate.

**Angstrom (A°)** A unit of measurement equal to $10^{-10}$ m, $10^{-4}$ μm and $10^{-1}$nm; no longer an official unit.

**Animal virus** A virus that multiplies in animal tissues.

**Anoxygenic** Not producing molecular oxygen; typical of bacterial photosynthesis.

**Antagonism** Active opposition; for example, between two drugs or two microbes.

**Antibiotic** An antimicrobial agent produced naturally by a bacterium or fungus.

**Antibody** A protein produced by the body in response to an antigen and capable of combining specifically with that antigen.

**Anticodon** The three nucleotides by which a transfer RNA recognizes an RNA codon.

**Antigen** Any substance that, when introduced into the body causes antibody formation and reacts only with its specific antibody.

**Antiseptic** A chemical disinfectant for disinfection of the skin, mucous membranes, or other living tissues.

**Antiserum** A solution containing antibodies.

**Apoenzyme** The protein portion of an enzyme, which requires activation by acoenzyme.

**Arbuscule** Fungal mycelia in plant root cells.

**Archaebacteria** Prokaryotic organisms lacking peptidoglycan.

**Arthrospore** An asexual fungal spore formed by fragmentation of a septate hypha.

**Artificially acquired active immunity** The production of antibodies by the body in response to a vaccination.

**Artificially acquired passive immunity** The transfer of humoral antibodies formed by one individual to a susceptible individual, accomplished by injection of antiserum.

**Ascospore** A sexual fungal spore produced in an ascus, formed by the Ascomycetes class of fungi.

**Asexual reproduction** Reproduction without opposite mating strains.

**Attenuation** Lessening of virulence of a microorganism. Also, a regulatory mechanism for protein synthesis.

**Autoclave** Equipment for sterilization by steam under pressure usually operated 15 psi and 121°C.

**Autotroph** An organism that uses carbon dioxide as its principal carbon source.

**Auxotroph** A mutant microorganism with a nutritional requirement not possessed by the parent.

**Axial filament** The structure for motility found in spirochetes.

**Bacillus** Any rod-shaped bacterium; when written as a genus refers to rod-shaped, endospore-forming, facultatively anaerobic, gram-positive bacteria.

**Bacteria** All living organisms with prokaryotic cells.

**Bacterial growth curve** A graph indicating the growth of a bacterial population over time.

**Bacterial species:** A population of cells with similar characteristics. The second name in scientific binomial nomenclature.

**Bactericidal** Capable of killing bacteria.

**Bacteriocin** Toxin protein produced by bacteria that kill other bacteria.

**Bacteriophage (phage)** A virus that multiplies in bacterial cells.

**Bacteriostatic** Capable of inhibiting bacterial growth.

**Bacteroid** Enlarged Rhizobium cells found in root nodules.

**Basal body** A structure that anchors flagella to the cell wall and plasma membrane.

**Basidiospore** A sexual fungal spore produced in a basidium, characteristic of the Basidiomycetes.
**Basidium** A structure that produces basidiospores; found in Basidiomycetes.
**Bergey's manual** The standard taxonomic reference on bacteria.
**Binary fission** Bacterial reproduction by division into two daughter cells.
**Binomial nomenclature** The system of having two names (genus and specific epithet) for each organism.
**Biochemistry** The science of chemical processes in living organisms.
**Bioconversion** Changes in organic matter brought about by the growth of microorganisms.
**Biogeochemical cycles** The recycling of chemical elements by microorganisms for use by other organisms.
**Biomass** Organic matter produced by living organisms and measured by weight.
**Biotechnology** The use of microbes in industry.
**Blastospore** An asexual fungal spore produced by budding from the parent cell.
**Bloom (algal)** Abundant growth of microscopic algae, producing visible colonies in nature.
**Budding** Asexual reproduction beginning as a protuberance from the parent cell that grows to become a daughter cell; also, release of an enveloped virus through the plasma membrane of an animal cell.
**Cancer** A malignant, invasive cellular tumour that has the capability of spreading throughout the body or body parts.
**Capsid** The protein coat of a virus that surrounds the nucleic acid.
**Capsomere** A protein subunit of a capsid.
**Capsule** An outer, viscous covering on some bacteria composed of a polysaccharide or polypeptide.
**Carbohydrates** Organic compounds composed of carbon, hydrogen, and oxygen, with the hydrogen and oxygen present in a 2:1 ratio; includes starches, sugars and cellulose.
**Carbon cycle** The series of processes that converts carbon dioxide to organic substances and back to carbon dioxide in nature.
**Carboxyl group**—COOH.
**Carboxysome** Prokaryotic inclusion containing ribulose 1, 5-diphosphatecarboxylase.
**Carcinogen** Any cancer-producing substance.
**Casein** Milk protein.
**Catabolism** All decomposition reactions in a living organism.
**Catalase** An enzyme that catalyzes the breakdown of hydrogen peroxide to water and oxygen.
**Catalyst** A substance that affects the rate of a chemical reaction, usually increasing the rate, but isn't changed in the reaction.
**Cellulose** A glucose polysaccharide that is the main component of plant cell walls.
**Cell wall** The outer covering of most bacterial, fungal, algal, and plant cells: in eubacteria, it consists of peptidoglycan.
**Cephalosporin** An antibiotic produced by the fungus *Cephalosporium* that inhibits the synthesis of gram-positive bacterial cell walls.
**Chemical energy** The energy of a chemical reaction.
**Chemoautotroph** An organism that uses an inorganic chemical as an energy source and carbon dioxide as a carbon source.
**Chemotherapy** Treatment of a disease with chemical substances.
**Chemotroph** An organism that uses oxidation-reduction reactions as primary energy source.
**Chitin** A glucosamine polysaccharide that is the main component of fungal cell walls and arthropod skeletons.
**Chloramphenicol** A broad spectrum bacteriostatic chemical.

**Chlorophyll a** The light-absorbing pigment in cyanobacteria, algae, and plants.
**Chromosome** The structure that carries hereditary information.
**Cilia** Relatively short cellular projections that move in a wave-like manner.
**Cisternae** stacked elements of the Golgi complex.
**Clone** A population of cells which are identical to the parent cell.
**Coccus** A spherical or ovoid bacterium.
**Codon** A group of three nucleotides in mRNA that specifies the insertion of an amino acid into a protein.
**Coenocytic hyphae** Fungal filaments that are not divided into uninucleate cell-like units because they lack septa.
**Coenzyme** A non-protein substance that is associated with and that activates an enzyme.
**Coliforms** Aerobic or facultatively anaerobic, gram-negative, nonspore-forming, rod-shaped bacteria that ferment lactose with acid and gas formation within 48 hours at 35ºC.
**Colony** A clone of bacterial cells on a solid medium that is visible to the naked eye.
**Communicable disease** Any disease that can be spread from one host to another.
**Compound light microscope** An instrument with two sets of lenses that uses visible light at the source of illumination.
**Condenser** A lens system located below the microscope stage that directs light rays through the specimen.
**Conidiophore** An aerial hypha bearing conidiospores.
**Conjugation** The transfer of genetic material from one cell to another involving cell-to-cell contact.
**Conjugative plasmid** A plasmid with genes for carrying out conjugation.
**Constitutive enzyme** An enzyme that is produced regardless of how much substrate is present.
**Contagious disease** a disease that is easily spread from one person to another.
**Cristae** Folding of the inner membrane of the mitochondrion.
**Crossing over** Process by which a portion of a chromosome is exchanged with a portion of another chromosome.
**Culture** Microorganisms that grow and multiply in a container of culture medium.
**Culture medium** The nutrient material prepared for growth of microorganisms in a laboratory.
**Curd** The solid part of milk that separates from the liquid in the making of cheese, for example.
**Cyanobacteria** Oxygen-producing photoautotrophic prokaryotes; formerly called blue-green algae.
**Cyclosporine** A drug that suppresses the T-cell response.
**Cytochrome oxidase** An enzyme that oxidizes cytochrome c.
**Cytoplasm** In a prokaryote, everything inside the plasma membrane; in a eukaryote, everything inside the plasma membrane and external to the nucleus.
**Cytosine** A pyrimidine nuclei acid base that pairs with guanine.
**Darkfield microscopy** A microscope that has a device to scatter light from the illuminator so that the specimen appears white against a black background.
**Decolourization** The process of removing stain.
**Dehydration** The removal of water.
**Denitrification** The reduction of nitrates to nitrites or nitrogen gas.
**Deoxyribose** A five-carbon sugar contained in DNA nucleotides.
**Dermatophyte** A fungus that causes a cutaneous mycosis.
**Detergent** Any substance that reduces the surface tension of water.
**Diagnosis** Identification of a disease.
**Dioecious** Referring to organisms in which organs of different sexes are located in different individuals.
**Diplococci** Cocci that divide and remain attached in pairs.

**Dipicolinic acid** Chemical substances found in bacterial endospores and not in vegetative cells.

**Disaccharide** A sugar consisting of two monosaccharides.

**Disease** Any change from a state of health.

**Disinfectant** Any substance used on inanimate objects to kill or inhibit the growth of microorganisms.

**Dysentery** A disease characterized by frequent, watery stools with blood and mucus.

**Electron** A negatively charged particle in motion around the nucleus of an atom.

**Electron acceptor** An ion that picks up an electron that has been lost from another atom.

**Electron microscope** A microscope that uses a flow of electrons instead of light to produce an image.

**Electron transport chain** A series of compounds that transfer electrons from one compound to another, generating ATP by oxidative phosporylation.

**Electrophoresis** The separation of substances by their rate of movement through an electric field.

**ELISA** A group of serologic tests that use enzyme reactions as indicators.

**Encephalitis** Inflammation of the brain.

**Endemic disease** A disease that is constantly present in a certain population.

**Endospore** A resting structure formed inside some bacteria.

**Enrichment culture** A culture medium used for preliminary isolation that favours the growth of a particular microorganism.

**Enterotoxin** An exotoxin that causes gastroenteritis; produced by *Vibrio, E. coli* etc.

**Entner-Doudoroff pathway** An alternative pathway for the oxidation of glucose to pyruvic acid.

**Enzyme** A protein that catalyzes chemical reactions in a living organism.

**Epidemic disease** A disease acquired by many people in a given area in a short time.

**Epidemiology** The science dealing with when and where diseases occur and how they are transmitted.

**Eubacteria** Prokaryotic organisms; possess petidoglycan cell walls.

**Eukaryote** A cell with DNA enclosed within a distinct membrane-bounded nucleus.

**Fermentation** The enzymatic degradation of carbohydrates in which the final electron acceptor is an organic molecule, ATP is synthesized by substrate-level phosphorylation and oxygen is not required.

**F factor** Fertility factor; a plasmid found in the donor cell in bacterial conjugation.

**Flagella** Thin appendages that arise from one or more locations on the surface of cell and are used for cellular locomotion.

**Flavoprotein** Protein with flavin coenzyme that functions as an electron carrier in respiration.

**Fluorescence** The ability to give off light of one colour when exposed to light of another colour.

**Fluorescent microscope** A microscope that uses an ultraviolet light source to illuminate specimens that will fluoresce.

**Fungi** Organisms that belong to Kingdom Fungi; eukaryotic chemheterotrophs.

**Gamete** A male or female reproductive cell.

**Gametocyte** A male or female protozoan cell.

**Gene** A segment of DNA or a sequence of nucleotides in DNA that codes for a functional product.

**Generalized transduction** Transfer of bacterial chromosome fragments from one cell to another by bacteriophage.

**Genetic engineering** Manufacturing and manipulating genetic material *in vitro.*

**Genetics** The science of heredity.

**Genotype** The genetic makeup of an organism.

**Genus** The first name of the scientific name (bionomial); the taxon between family and species.

**Germ** Part of an organism capable of developing.

**Germ theory** The principle that microorganisms cause disease.

**Globulin** The protein type to which antibodies belong.

**Glycerol** An alcohol; $C_3H_5(OH)_3$.
**Glycolysis** The main pathway for the oxidation of glucose to pyruvic acid.
**Golgi complex** An organelle involved in the secretion of certain proteins.
**Gram-stain** A differential stain that divides bacteria into two groups, gram-positive and gram-negative.
**Guanine** A purine nucleic acid that pairs with cytosine.
**Halophile** An organism that grows in high concentration of salt.
**Haploid** A eukaryotic cell or organism with one of each type of chromosome.
**Heterocyst** A large cell in certain cyanobacteria; site of nitrogen fixation.
**Histones** Proteins associated with DNA in eukaryotic chromosomes.
**Host** An organism infected by a pathogen.
**Humus** Organic matter that remains in soil following partial decomposition.
**Hyphae** Long filaments of cells in fungi or actinomycetes.
**IgA** The class of antibodies in secretions.
**IgD** Antibodies found on B cells.
**IgE** The class of antibodies involved in hypersensitivities.
**IgG** The most abundant antibodies found in serum.
**Immunity** The body's defence against a particular microorganism.
**Immunization** A process that produces immunity.
**Incubation period** The time interval between the actual infection and first appearance of any signs or symptoms of disease.
**Infection** Growth of microorganisms in the body.
**Inoculate** To introduce microorganisms into a culture medium or host.
**Intoxication** Poisoning.
**Ionization** Separation of a molecule into groups of atoms with electrical charges.
**Kingdom** The highest category in the taxonomic hierarchy of classification.
**Koch's postulates** Criteria used to determine the causative agent of infectious diseases.
**Kreb's cycle** A pathway that converts two carbon compounds to carbon dioxide, transferring electrons to NAD + and other carriers.
**Lag phase** The time interval in bacterial growth curve with no growth.
**L form** A natural bacterium with defective cell wall.
**Ligase** An enzyme that joins together pieces of DNA.
**Lipase** An exoenzyme that breaks down fats into their component fatty acids and glycerol.
**Lyophilization** Free drying; freezing a substance and subliming the ice in a vacuum.
**Lysogeny** A state in which phage DNA is incorporated into the host cell without lysis.
**Lysozyme** An enzyme capable of lysing bacterial cell walls.
**Lytic cycle** A sequence for replication of phages that results in host cell lysis.
**Meiosis** The process that leads to the formation of haploid gametes in a diploid organism.
**Mesophile** An organism that grows between 25°C and 40°C.
**Microorganism** A living organism too small to be seen with naked eye; includes bacteria, fungi, protozoan and microscopic algae; also includes viruses.
**Microtubule** The structure of the proteins comprising eukaryotic flagella and cilia.
**Mitochondria** Organelles containing the respiratory ATP synthesizing enzymes.
**Mitosis** The division of the cell nucleus, often followed by division of the cytoplasm of the cell.
**Mixed culture** A culture containing more than one kind of microorganism.
**Molecular biology** The science dealing with DNA and protein synthesis of living organisms.
**Molecule** A combination of atoms forming a specific chemical compound.

**Monomers** The units that combine to form polymers.
**Morphology** The external appearance.
**Mortality** Deaths from a specific notifiable disease.
**Most probable number (MPN)** A statistical determination of the number of coliforms per 100 ml of water or food.
**Motility** The ability of an organism to move itself.
**Mutagen** An agent in the environment that brings about mutations.
**Mutation** Any change in the base sequence of DNA.
**Mutation rate** The probability of a gene mutating each time a cell divides.
**Mutualism** A symbiosis in which both organisms are benefited.
**Mycelium** A mass of long filaments of cells that branch and intertwine, typically found in moulds.
**Mycetoma** A chronic infection caused by certain fungi and *Nocardia* characterized by a tumour-like appearance.
**Mycology** The science dealing with fungi.
**Mycorrhiza** A fungus growing in symbiosis with plant root hairs.
**Mycosis** A fungal infection.
**Mycotoxin** A toxin produced by a fungus.
**Nitrogen cycle** The series of processes that converts nitrogen ($N_2$) to organic substances and back to nitrogen in nature.
**Nitrogen fixation** The conversion of nitrogen ($N_2$) into ammonia.
**Nomenclature** The system of naming things.
**Nonsense codon** A special terminator codon that does not code for any amino acid.
**Normal flora** Microorganisms that colonize an animal without causing disease.
**Nuclear envelope** The double membrane that separates the nucleus from the cytoplasm in the eukaryotic cell.
**Nucleic acid** A macromolecule consisting of nucleotides; for example, RNA and DNA.
**Nucleiod** The region in a bacterial cell containing the chromosome.
**Nucleoli** Areas in a eukaryotic nucleus where rRNA is synthesized.
**Nucleoplasm** The gel-like fluid within the nuclear envelope.
**Nucleoprotein** A macromolecule consisting of a protein and nucleic acid.
**Nucleoside** A compound consisting of a purine or pyrimidine and pentose sugar.
**Nucleotide** A compound consisting of a purine or pyrimidine base, a five carbon sugar, and a phosphate.
**Nucleus** The part of a eukaryotic cell that contains the genetic material; also, the part of an atom consisting of the protons and neutrons.
**Numerical taxonomy** A method of comparing organisms on the basis of many characteristics.
**Nutrient broth (agar)** A complex medium made of beef extracts.
**Objective lens** In a compound light microscope, the lens closest to the specimen.
**Obligate anaerobe** An organism that is unable to use oxygen.
**Operon** The operator site and structural genes it controls.
**Organelles** Membrane-bounded structures within eukaryotic cells.
**Oxidation** The removal of electrons from a molecule or the addition of oxygen to a molecule.
**Oxidative phosphorylation** The synthesis of ATP coupled with electron transport.
**PAGE** Polyacrylamide gel electrophoresis, *see* electrophoresis.
**Parasite** An organism that derives nutrients from a living host.
**Parasitism** A symbiosis in which one organism exploits another without providing any benefit in return.
**Pasteurization** The process of mild heating to kill particular spoilage organisms or pathogens.

**Pathogen** A disease causing organism.

**Pathology** Study of the causes and development of the disease.

**Penicillins** A group of antibiotics produced either by *Penicillium* or by adding side chains to the β-lactam ring.

**Pentose phosphate pathway** A metabolic pathway that can occur simultaneously with glycolysis to produce pentoses and NADH without ATP production.

**Peptide** A chain of two, three or more amino acids.

**Peptide bond** A bond joining the amino group of one amino acid to the carboxyl group of a second amino acid with the loss of a water molecule.

**Peptidoglycan** The structural molecule of bacterial cell walls consisting of the molecules N-acetylglucosamine, N-acetylmuramic acid, tetra peptide side chain, and peptide side chain.

**Peptone** Short chains of amino acids produced by the action of acids or enzymes on proteins.

**Peritrichous** Having flagella distributed over the entire cell.

**Peroxide** An oxygen oxide consisting of two atoms of oxygen.

**pH** The symbol for hydrogen ion concentration; a measure of the relative acidity of a solution.

**Phage** *See* bacteriophage.

**Phase-contrast microscope** A compound light microscope that allows examination of structures inside cells through the use of a special condenser.

**Phenol** $C_6H_5OH$; carbolic acid.

**Phosphate group** A portion of a phosphoric acid molecule attached to some other molecule.

**Phospholipid** A complex lipid composed of glycerol, two fatty acids, and a phosphate group.

**Phosphorylation** The addition of phosphate group to an organic molecule.

**Photoautotroph** An organism that uses light for its energy source and carbon dioxide as its carbon source.

**Photohetrotroph** An organism that uses light for its energy source and an organic carbon source.

**Photophosphorylation** The production of ATP by photosynthesis.

**Photosynthesis** The light-driven synthesis of carbohydrate from carbon dioxide.

**Phototaxis** The response to the presence of light.

**Phototroph** An organism that uses light as its primary energy source.

**Phylogeny** The evolutionary history of a group of organisms.

**Phylum** A taxonomic classification between kingdom and class.

**Pinocytosis** The engulfing of small molecules by infolding the plasma membrane.

**Plantae** A kingdom composed of multicellular eukaryotes with cellulose cell walls.

**Plant virus** A virus that multiplies in plant tissues.

**Plaque** A clearing in a confluent growth of bacteria due.to lysis by phages.

**Plaque forming units** Visible plaques counted, perhaps due to more than one phage.

**Plasma membrane** The selectively permeable membrane enclosing the cytoplasm of a cell; the outer layer in animal cells, internal to the cell wall in other organisms.

**Plasmid** A small cyclic DNA molecule in bacteria replicating independent of the chromosome.

**Plasmodium** A multinucleated mass of protoplasm; when written as a genus refers to the aetiology of malaria.

**Plasmolysis** Loss of water from a cell in a hypertonic environment.

**Plate count** A method of determining the number of bacteria in a sample by counting the number of colony-forming units on a solid culture medium.

**Pleomorphic** Having many shapes.

**Point mutation** *See* base substitution.

**Poly-β-hydroxybutyric acid** A fatty acid storage material unique to bacteria.

**Polymer** A molecule consisting of a sequence of similar units or monomers.

**Polymerase** An enzyme that synthesizes specific polymers.

**Polypeptide** A chain of amino acids.

**Pour plate method** A method of inoculating a solid nutrient medium by mixing bacteria in the melted medium and pouring the medium into a Petri plate to solidify.

**Precipitation reaction** A reaction between soluble antigens and multivalent antibodies to form aggregates.

**Primary infection** An acute infection that causes the initial illness.

**Prokaryote** A cell whose genetic material is not enclosed in a nuclear envelope.

**Promoter site** The starting point on DNA for transcription of RNA by RNA polymerase.

**Prophage** Phage DNA inserted into the host cell's DNA.

**Prosthetic group** A coenzyme that is bound tightly to its attached apoenzyme.

**Protein** A large molecule containing carbon, hydrogen, oxygen, and nitrogen (and sulphur); some have a globular structure and others are pleated sheets.

**Proteolytic** An enzyme capable of hydrolyzing a protein.

**Protista** The kingdom to which protozoans belong; unicellular, eukaryotic organisms.

**Proton** A positively charged particle in the nucleus of an atom.

**Protoplast** A gram-positive bacterium without a cell wall; a plant cell lacking a cell wall.

**Protozoa** Unicellular eukaryotic organisms belonging to the Kingdom Protista.

**Psychrophile** An organism that grows best at 15°C and does not grow above 20°C.

**Purines** The class of nucleic acid bases that includes adenine and guanine.

**Pyocyanin** Blue-green pigment produced by *Pseudomonas aeruginosa.*

**Pyrimidines** The class of nucleic acid bases that includes uracil, thymine and cytosine.

**Quaternary ammonium compound (quat)** A cationic detergent with four organic groups attached to a central nitrogen atom; used as a disinfectant.

**Reactants** Substances that are combined in a chemical reaction.

**Recalcitrant** Being resistance to degradation.

**Receptor** An attachment for a pathogen on a host cell.

**Recipient cell** A cell that receives DNA from a donor cell in recombination.

**Recombinant DNA** A DNA molecule produced by recombination.

**Recombinant DNA techniques** *See* Genetic engineering.

**Recombinant RNA technology** Techniques used to make RNA molecules.

**Recombination** The process of joining pieces of DNA from different sources.

**Reduction** The addition of electrons to a molecule; the gain of hydrogen atoms.

**Replica plating** A method of inoculating a number of solid minimal culture from an original plate of complete medium; mutant colonies that don't grow on the minimal media can be selected from the original plate.

**Repressor** A protein that binds to the operator site to prevent transcription.

**Resistance** The ability to ward off diseases through non-specific and specific defences.

**Resistance (R) factor** A bacterial plasmid carrying genes that determine resistance to antibiotics.

**Resolution** The ability to distinguish fine detail with a magnifying instrument.

**Respiration** An ATP-generation process in which chemical compounds are oxidized and the final electron acceptor is usually an inorganic molecule; also, the process by which living organisms produce carbon dioxide.

**Restriction enzymes** Enzymes that cut double-stranded DNA leaving staggered ends.

**R genes** Genes carried on the R factor that code for enzymes that inactivate certain drugs.
**Rhizosphere** The region in soil where the soil and roots make contact.
**Riboflavin** B vitamin that functions as a flavoprotein.
**Ribonucleic acid (RNA)** The class of nucleic acids that comprises messenger RNA, ribosomal RNA and transfer RNA.
**Ribose** A five-carbon sugar that is part of ribonucleotide molecules and RNA.
**Ribosomal RNA (rRNA)** The RNA molecules that form the ribosomes.
**Ribosomes** The sites of protein synthesis in cell composed of RNA and protein.
**RNA-dependent RNA polymerase** An enzyme that synthesizes a complementary RNA from an RNA template.
**RNA primer** A short strand of RNA used to start synthesis of the lagging strand of DNA.
**Saccharide** Sugar general formula $(CH_2O)$ n.
**Salt** A substance that dissolves in water to cations and anions, neither of which is H+ or OH−
**Sanitation** Removal of microbes from eating utensils and food preparation areas.
**Saprophyte** An organism that obtains its nutrients from the organic matter.
**Sarcina** A group of eight bacteria that remain in a packet after dividing.
**Saturation** The condition in which the active site on an enzyme is occupied by the substrate or product at all times.
**Scanning electron microscope** An electron microscope that provides three-dimensional views of the specimen magnified about 10,000 times.
**Schaeffer-Fulton stain** An endospore stain that uses malachite green to stain the endospores and safranin as a counterstain.
**Sclerotia** The reddish hardened ovaries of a grain that are filled with mycelia of the fungus *Claviceps purpurea.*
**Serial dilution** The process of diluting a sample several times.
**Serum** The liquid remaining after blood plasma which contains immunoglobulins.
**Sewage** Domestic waste water.
**Sexual reproduction** Reproduction that requires two opposite mating strains, usually designated as male and female.
**Simple stain** A method of staining microorganisms with a single basic-dye.
**Single cell protein (SCP)** A food substitute consisting of microbial cells.
**Slide agglutination test** A method of identifying an antigen by combining it with a specific antibody in a slide.
**Soil** A mixture of solid inorganic matter, water, organic matter, and living organisms.
**Solvent** A dissolving medium.
**Specialized transduction** The process of transferring piece of cell DNA adjacent to a prophage to another cell.
**Species** The most specific level in the taxonomic hierarchy.
**Spheroplast** A gram-negative bacterium lacking a complete cell wall.
**Spirillum** A spiral or corkscrew-shaped bacterium.
**Spirochete** A corkscrew-shaped bacterium with an axial filament.
**Spontaneous generation** The idea that life could arise spontaneously from nonliving matter.
**Spontaneous mutation** A mutation that occurs without mutagen.
**Sporangiophore** Aerial hypha supporting a sporangium.
**Sporangiospore** An asexual fungal spore formed within sporangium.
**Sporangium** A sac containing one or more spores.

**Spore** A reproductive structure formed by fungi and actinomycetes.

**Staphylococci** A broad sheet of spherical cells.

**Stationary phase** The period in a bacterial growth curve when the number of cells dividing equals the number dying.

**STD** Sexually transmitted disease.

**Stem cells** Foetal cells that give rise to bone marrow, blood cells, and B and T cells.

**Sterile** Free of microorganisms.

**Steroids** A specific group of chemical substances, including cholesterol and hormones.

**Strain** A group of cells all derived from a single cell.

**Streak plate method** A method of isolating a culture by spreading microorganisms over the surface of a solid culture.

**Streptococci** Cocci that remain attached in chains after cell division.

**Structural gene** A gene that codes for an enzyme.

**Substrate** Any compound with which an enzyme reacts.

**Substrate-level phosphorylation** The synthesis of ATP by direct transfer of a high-energy phosphate group from an intermediate metabolic compound to ADP.

**Sulphahydryl group** –SH.

**Sulphonamides** Bacteriostatic compounds that interfere with folic acid synthesis by competitive inhibition.

**Superoxide dismutase** Enzyme that destroys $O_2^-$; $O_2^- + O_2^- + 2H^+ \rightarrow H_2O_2 + O_2$.

**Susceptibility** The lack of resistance to a disease.

**Symbiosis** The living together of two different organisms.

**Symptom** A change in body function that is felt by a patient due to a disease.

**Synergistic effect** The principle whereby the effectiveness of two drugs used simultaneously is greater than either drug used alone.

**Synthetic drug** A chemotherapeutic agent that is prepared from chemicals in a laboratory.

**Systemic (generalized) infection** An infection throughout the body.

**Taxon** A taxonomic category.

**Taxonomy** The science of classification.

**Teichoic acid** A polysaccharide found in gram-positive cell walls.

**Temperate phage** A bacteriophage existing in lysogeny with a host cell.

**Tetracyclines** Broad spectrum antibiotics that interfere with protein synthesis.

**Tetrahedron** A four-sided solid structure.

**Thallus** The entire vegetative structure or body of a fungus, lichen, or alga.

**Thermal death point (TDP)** The temperature required to kill all the bacteria in a liquid culture in 10 minutes at pH 7.

**Thermophile** An organism whose optimum growth temperature is between 50°C and 60°C.

**Thylakoid** A chlorophyll-containing membrane in a chloroplast.

**Tolerance** A state of immunological non-responsiveness to self antigens.

**Toxin** Any poisonous substance produced by a microorganism.

**Trace element** A chemical element required in small amounts for growth.

**Transamination** The transfer of an amino group from an o acid to an organic acid.

**Transcription** The process of synthesizing RNA from a *r* DNA template.

**Transduction** Transfer of DNA from one cell to another by a bacteriophage. *See also* generalized transduction, specialized transduction.

**Transfer RNA (tRNA)** The class of molecules that brings amino acids to the site where they are incorporated into proteins.

**Transformation** The process in which genes are transformed from one bacterium to another as "naked" DNA in solution; also, the changing of a normal cell into a cancerous cell.
**Translation** The use of RNA as a template in the synthesis of protein.
**Transposon** A small piece of DNA that can move from one region of the DNA molecule to another.
**Tube-agglutination test** A method for determining antibody titer using serial dilutions of serum mixed with antigen.
**Tumour** Excessive tissue caused by uncontrolled cell growth.
**Turbidity** The cloudiness of a suspension.
**Ubiquinones** Low molecular weight, non-protein carriers in the electron transport chain.
**Ultrastructure** Fine detail not seen with a compound light microscope.
**Ultraviolet (UV) radiation** Radiation from 10 to 390 nm.
**Uracil** A pyrimidine nucleic acid base in RNA that pairs with adenine.
**Vaccination** The process of conferring immunity using a vaccine.
**Vaccine** A preparation of killed, inactivated, or attenuated microorganisms or toxoids to induce artificially acquired active immunity.
**Vacuole** An intracellular inclusion, in eukaryotic cells, surrounded by a plasma membrane containing raw food; in prokaryotic cells, surrounded by a proteinous membrane containing gas.
**Vector** An arthropod that carries disease-causing microorganisms from one host to another.
**Vegetative cells** Cells involved with obtaining nutrients, as opposed to reproduction or resting.
**Vesicle** The expanded, terminal area of the Golgi complex; also, a fluid-filled blister. In prokaryote, a protein covered hollow cylinder in a gas vacuole.
**Vibrio** A curved or comma-shaped bacterium.
**Virion** A fully developed complex viral particle.
**Viriod** An infectious piece of "naked" RNA.
**Virus** A submicroscopic, parasitic, filtrable agent consisting of a nucleic acid surrounded by a protein coat.
**Visible light** Radiation from 400 to 700 nm, which the human eye can see.
**Volutin** Stored phosphate in a prokaryotic cell.
**Yeast** A unicellular fungus belonging to the Phylum Ascomycetes.
**Zygospore** A sexual fungal spore characteristic of the Zygomycetes.
**Zygote** A diploid cell produced by the fusion of two haploid gametes.

# Selected References

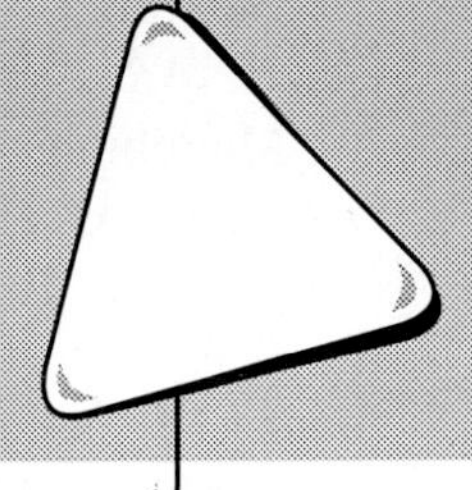


Albert G. Moat, John W. Foster and Michael P. Spector. (2002). Microbial Physiology. Wiley and Sons Inc. Publ. New York, Singapore.

Alexander, M. (1977). Introduction to Soil Microbiology. Wiley, New York.

Alexopoulos, C.J., Minus, C. W., Blackwell, M. (1996). Introductory Mycology. John Wiley & Sons Inc.

Banwart, G.J. (1999). Basic Food Microbiology, 2nd Edition, Van Nostrand Reinhold Co. New York.

Berdell R. Funke and Christine L. Case. (1989). Microbiology: An Introduction. 3rd Edition. The Benjamin/Cummings Publ. Co. Inc. New York.

Biswas, S.B. and Biswas, A. (2000). An Introduction to Viruses. Vikas Publ. House Pvt. Ltd. New Delhi.

Brock, T.D. (1979). Biology of Microorganisms, 3rd Edition. Englewood Cliffs, Prentice Hall.

Casida, L.E. Jr. (1993). Industrial Microbiology. Wiley, Eastern Ltd., New Delhi.

Clifton, C.E. (1958). Introduction to the Bacteria. New York, USA, McGraw-Hill Book Co. Inc

Cochran, W.G. 1950. "Estimation of bacterial densities by means of the most probable number". *Biometrics*, **6**: 105–116.

David L. Nelson and Michael M. Cox. (1999). Principles of Biochemistry. 4th Edition. W.H. Freeman & Co., New York,

Davis, B.D., Dulbecco, R., Eisen, H.N. and Ginsberg, H.S. (1980). Microbiology, 3rd Edition, Harper and Row, New York.

Demain, A.L. and Soloman, N.A. (1986). "Industrial Microbiology". *Sci. Am.*

Difco Manual (1966). Detroit, Michigan USA, Difco Laboratories

Frobischer, M. (1968). Fundamentals of Microbiology. W.B. Saunders Co., London, U.K.

George M. Garrity, David R. Boone and Richard W. Castenholz. (2001). Bergey's Manual of Systematic Bacteriology, Volume 1: The *Archaea* and the deeply branching and phototrophic *Bacteria,* 2nd Edition. Springer, New York.

George M. Garrity, Don J. Brenner, Noel R. Krieg and James T. Staley. (2005). Bergey's Manual of Systematic Bacteriology, Volume 2: The *Proteobacteria,* 2nd Edition. Springer, New York.

Glazer, A.N. and Nikaido, H. (1995). Microbial Biotechnology: Fundamentals of Applied Microbiolgy. W.H. Freeman, New York.

Gottschalk, G. (1979). Bacterial Metabolism. Springer; Heidelberg.

Gunasekaran, P. (1996). Laboratory Manual in Microbiology. New Age International (P) Ltd. Publisher.

Joanne M. Willey, Lansing M. Prescott, Linda Sherwood, Christopher J. Woolverton. (2009). Prescott's Principles of Microbiology. McGraw-Hill Higher Education, New York.

Kathleen P. Talaro. (2009). Foundations in Microbiology. McGraw-Hill, New York.

Krieg, N.R. and Holt, J.G. (1984). Bergey's Manual of Systematic Bacteriology, Vol. 1, Baltimore, Williams and Wilkins.

Kuby, J. (1994). Immunology. 2nd Edition. W.H. Freeman, New York.

Lehringer, A.L. (1982). Principles of Biochemistry. New York, Worth.

Luria, S.E., Darnell, J.E. Jr., Baltimore, D. and Campbell, A. (1978). General Virology, 3rd Edn., Wiley, New York.

McElroy, W.D. (1969). Cell Physiology and Biochemistry. Prentice-Hall of India Pvt. Ltd., New Delhi, India.

Pelczar, M.J., Jr, Reid, R.D. and Chan, E.C.S. (1977). Microbiology, 4th Edn., McGraw-Hill, New York.

Postgate, J.R. (1982). The Fundamentals of Nitrogen Fixation. Cambridge University Press, London, U.K.

Rangaswami, G. and Bagyaraj, D.J. (2002). Agricultural Microbiology. Prentice-Hall of India Pvt. Ltd.

Ratledge, C. and Kristiansen, B. (2001). Basic Biotechnology. Cambridge University Press, Cambridge, U.K.

Richard, M. and Twyman. (1999). Advanced Molecular Biology. Viva Books Pvt. Ltd.

Ronald M. Atlas and Richard Bartha. (1998). Microbial Ecology. 4th Edition. Pearson Education Inc. Singapore.

Samuel C. Prescott and Cecil G. Dunn. (2002). Industrial Microbiology. Agribios Jodhpur, (India).

Schlegel, H.G. (1988). General Microbiology. Cambridge, U.K., Cambridge University Press.

Singleton, P. and Sinsbury, D. (1987). Dictionary of Microbiology and Molecular Biology. 2nd Edn., John Wiley and Sons (SEA) Pvt. Ltd., Singapore.

Stanier, R.Y., Adelberg, E.A. and Ingraham, J.L. (1978). Gerneral Microbiology, 4th Edition, MacMillan, London.

Stanley R. Major, John E. Cronan and David Friefielder. (1994). Microbial genetics. Jones and Barlett Publ. London.

Starr, M.P., Stolp, H., Truper, H.G., Balows, A. and Schlegel, H.G. (1981). The Prokaryotes: A Hand Book on Habitats, Isolation and Identification of Bacteria. Vols I and II. Springer, Heidelberg.

Subba Rao, N.S. (1999). Soil Microbiology. Oxford and IBH Publ. Co. Pvt. Ltd. New Delhi.

Thimann, K.V. (1968). The Life of Bacteria. MacMillan Co.Ltd., New York, U.S.A.

Tilak, K.V.B.R., Pal, K.K. and Dey, R. (2010). Microbes for Sustainable Agriculture. I.K. International Publishing House Pvt. Ltd., New Delhi, India.

Van Den Hoek, Mann, D.G., and Jahns, H.M. (1995). Algae—an Introduction to Physiology. Cambridge Univ. Press, Cambridge.

Waksman, S.A. (1952). Soil Microbiology. John Wiley and Sons, New York, USA.

Watson, J.D. (1976). Molecular Biology of the Gene. 3rd Edition, Benjamin, New York.

Webster, J. (1980). Introduction to Fungi, 2nd Edition. Cambridge University Press, London.

Yves V. Brun and Lawerence J. Shimkets. (2002). Prokaryotic Development. ASM Press, Washington DC.

# Index